The Biology of Crustacea

VOLUME 1

SYSTEMATICS, THE FOSSIL RECORD, AND BIOGEOGRAPHY

The Biology of Crustacea

Editor-in-Chief

Dorothy E. Bliss

Department of Invertebrates
The American Museum of Natural History
New York, New York*

*Present address: Brook Farm Road, RR5, Wakefield, Rhode Island 02879

The Biology of Crustacea

VOLUME 1

Systematics, the Fossil Record, and Biogeography

Edited by

LAWRENCE G. ABELE
Department of Biological Science
Florida State University
Tallahassee, Florida

ACADEMIC PRESS 1982
A Subsidiary of Harcourt Brace Jovanovich, Publishers
New York London
Paris San Diego San Francisco São Paulo Sydney Tokyo Toronto

ACADEMIC PRESS, INC.
111 Fifth Avenue, New York, New York 10003

United Kingdom Edition published by
ACADEMIC PRESS, INC. (LONDON) LTD.
24/28 Oval Road, London NW1 7DX

Library of Congress Cataloging in Publication Data
Main entry under title:

The Biology of Crustacea.

Includes bibliographies and indexes.
Contents: v. 1. Systematics, the Fossil Record, and Biogeography/Dorothy E. Bliss/editor-in-chief, Lawrence G. Abele
1. Crustacea. I. Bliss, Dorothy E. II. Abele, Lawrence G.
QL435.B48 595.3 82-4058
ISBN 0-12-106401-8 (v. 1) AACR2

PRINTED IN THE UNITED STATES OF AMERICA

82 83 84 85 9 8 7 6 5 4 3 2 1

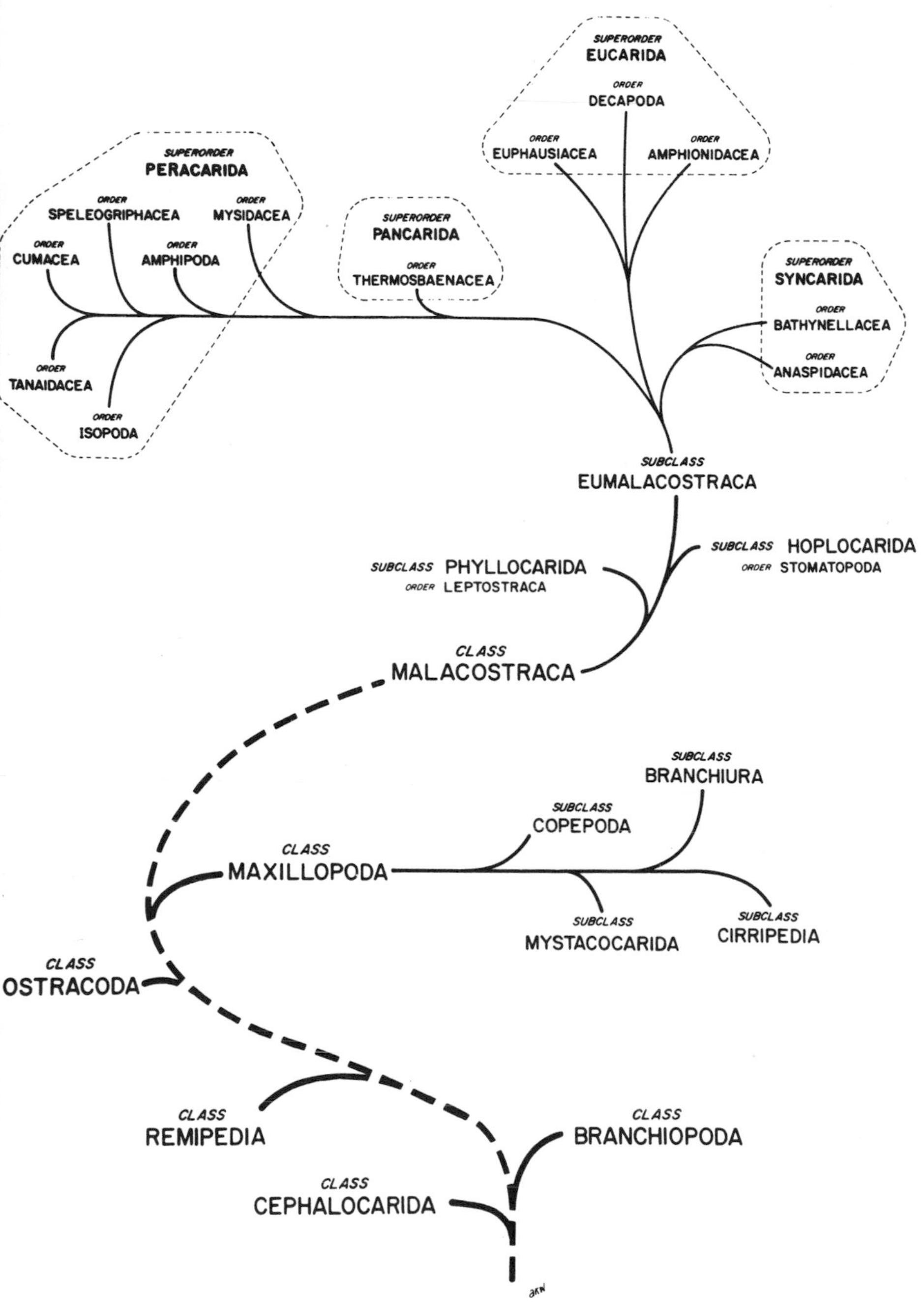

A visual representation of the Bowman and Abele classification. This is not intended to indicate phylogenetic relationships and should not be so interpreted. The dashed line at the base emphasizes uncertainties concerning the origins of the five classes and their relationships to each other.

Contents

3 Origin of the Crustacea

JOHN L. CISNE

4 The Fossil Record and Evolution of Crustacea

FREDERICK R. SCHRAM

5 Evolution within the Crustacea

ROBERT R. HESSLER, BRIAN M. MARCOTTE,
WILLIAM A. NEWMAN, and ROSALIE F. MADDOCKS

Part 1: General: Remipedia, Branchiopoda, and Malacostraca (Hessler)

List of Contributors

Numbers in parentheses indicate the pages on which the authors' contributions begin.

Lawrence G. Abele (1, 241), Department of Biological Science, Florida State University, Tallahassee, Florida 32306

Thomas E. Bowman (1), Division of Crustacea, National Museum of Natural History, Washington, D.C. 20560

John L. Cisne (65), Department of Geological Sciences and Division of Biological Sciences, Cornell University, Ithaca, New York 14853

Robert R. Hessler (149), Scripps Institution of Oceanography, La Jolla, California 92093

Patsy A. McLaughlin (29), Department of Biological Sciences, Florida International University, Miami, Florida 33199

Rosalie F. Maddocks (221), Department of Geology, University of Houston, Houston, Texas 77004

Brian M. Marcotte (185), Marine Sciences Centre, McGill University, Montreal, Quebec, H3A 2B2, Canada

William A. Newman (197), Scripps Institution of Oceanography, La Jolla, California 92093

Frederick R. Schram (93), Department of Geology, Natural History Museum, San Diego, California 92112

George T. Taylor (29), Department of Biological Sciences, Florida International University, Miami, Florida 33199

Martin L. Tracey (29), Department of Biological Sciences, Florida International University, Miami, Florida 33199

General Preface

In 1960 and 1961, a two-volume work, "The Physiology of Crustacea," edited by Talbot H. Waterman, was published by Academic Press. Thirty-two biologists contributed to it. The appearance of these volumes constituted a milestone in the history of crustacean biology. It marked the first time that editor, contributors, and publisher had collaborated to bring forth in English a treatise on crustacean physiology. Today, research workers still regard this work as an important resource in comparative physiology.

By the latter part of the 1970s, a need clearly existed for an up-to-date work on the whole range of crustacean studies. Major advances had occurred in crustacean systematics, phylogeny, biogeography, embryology, and genetics. Recent research in these fields and in those of ecology, behavior, physiology, pathobiology, comparative morphology, growth, and sex determination of crustaceans required critical evaluation and integration with earlier research. The same was true in areas of crustacean fisheries and culture.

Once more, a cooperative effort was initiated to meet the current need. This time its fulfillment required eight editors and almost 100 contributors. This new treatise, "The Biology of Crustacea," is intended for scientists doing basic or applied research on various aspects of crustacean biology. Containing vast background information and perspective, this treatise will be a valuable source for zoologists, paleontologists, ecologists, physiologists, endocrinologists, morphologists, pathologists, and fisheries biologists, and an essential reference work for institutional libraries.

In the preface to Volume 1, editor Lawrence G. Abele has commented on the excitement that currently pervades many areas of crustacean biology. One such area is that of systematics. The ferment in this field made it difficult for Bowman and Abele to prepare an arrangement of families of Recent Crustacea. Their compilation (Chapter 1, Volume 1) is, as they have stated, "a compromise and should be until more evidence is in." Their arrangement is likely to satisfy some crustacean biologists, undoubtedly not all. Indeed, Schram (Chapter 4, Volume 1) has offered a somewhat different arrangement. As generally used in this treatise, the classification of Crustacea follows that outlined by Bowman and Abele.

Selection and usage of terms have been somewhat of a problem. Ideally, in a treatise, the same terms should be used throughout. Yet biologists do not agree on certain terms. For example, the term *ostracode* is favored by systematists and paleontologists, *ostracod* by many experimentalists. A different situation exists with regard to the term *midgut gland,* which is more acceptable to many crustacean biologists than are the terms *hepatopancreas* and *digestive gland.* Accordingly authors were encouraged to use *midgut gland.* In general, however, the choice of terms was left to the editors and authors of each volume.

In nomenclature, consistency is necessary if confusion as to the identity of an animal is to be avoided. In this treatise, we have sought to use only valid scientific names. Wherever possible, synonyms of valid names appear in the systematic indexes. Thomas E. Bowman and Lawrence G. Abele were referees for all taxonomic citations.

Every manuscript was reviewed by at least one person before being accepted for publication. All authors were encouraged to submit new or revised material up to a short time prior to typesetting. Thus, very few months elapse between receipt of final changes and appearance of a volume in print. By these measures, we ensure that the treatise is accurate, readable, and up-to-date.

Dorothy E. Bliss

General Acknowledgments

In the preparation of this treatise, my indebtedness extends to many persons and has grown with each succeeding volume. First and foremost is the great debt owed to the authors. Due to their efforts to produce superior manuscripts, unique and exciting contributions lie within the covers of these volumes.

Deserving of special commendation are authors who also served as editors of individual volumes. These persons have conscientiously performed the demanding tasks associated with inviting and editing manuscripts and ensuring that the manuscripts were thoroughly reviewed. In addition, Dr. Linda H. Mantel has on innumerable occasions extended to me her advice and professional assistance well beyond the call of duty as volume editor. In large part because of the expertise and willing services of these persons, this treatise has become a reality.

Several biologists have provided valuable help of one sort or another during the preparation of these volumes. Worthy of special mention are Raymond B. Manning and John H. Welsh. Also deserving of thanks and praise are scientists who gave freely of their time and professional experience to review manuscripts. In the separate volumes, many of these persons are mentioned by name.

Thanks are due to all members of the staff of Academic Press involved in the preparation of this treatise. Their professionalism and encouragement have been indispensable.

Finally, no acknowledgments by me would be complete without mention of the help provided by employees of the American Museum of Natural History, especially those in the Department of Invertebrates and in the Museum's incomparable Library.

Dorothy E. Bliss

Preface to Volume 1

This is an exciting time for the study of Crustacea. This past decade has witnessed new discoveries and an incredible growth of knowledge in many areas. A new class of crustaceans, the Remipedia, was discovered in 1980. The application of X-ray techniques to the study of fossils has yielded new information on the head and limb structures of trilobites, thereby clarifying their relationship to the crustaceans. An extensive fossil find, the Middle Pennsylvanian Mazon Creek Essex fauna, combined with extensive revision of the diverse European Paleozoic faunas, has extended our knowledge of late Paleozoic Crustacea. New genetic techniques have been applied to the systematics of crustaceans, and the discovery of taxa in new localities has changed our view of crustacean biogeography. These data and many new ideas are presented by the authors of this volume.

In Chapter 1, Bowman and Abele accepted the unenviable task of compiling the first modern complete list of the families of Recent Crustacea. Their approach to higher levels of classification is a compromise, and should be, until more evidence is in. The reader will find much of the controversy on classification and phylogeny summarized by the contributors to Chapters 3, 4, and 5. Finally, Bowman and Abele's Table I summarizing numbers of species and general habits of the orders of crustaceans will surely be widely used.

The increasingly diverse methods used by crustacean systematists are reviewed by McLaughlin, Taylor, and Tracey in Chapter 2. These authors summarize collection, preservation, and staining techniques and review descriptive taxonomy. They then describe statistical techniques, electron microscopy, and electrophoresis as these techniques have been applied by systematists. This chapter should serve as an important reference for the future.

Cisne in Chapter 3 presents information relating to the origin of the Crustacea and the relationships that may exist among the major arthropod groups. He reviews the morphology of trilobitomorphs comparing them to early Cambrian pseudocrustaceans and other enigmatic groups. He concludes that crustaceans arose from a trilobite-like ancestor near the beginning of the Cambrian and that Trilobita, Chelicerata, and Crustacea shared a common ancestor.

Schram, in Chapter 4, comes to a different conclusion in his introduction to fossil crustaceans. He argues that the Crustacea form an independent phylum. Crustaceans have an extensive fossil record and these data, especially the Mazon Creek and European Paleozoic fossils, are reviewed in terms of their evolutionary relationships.

Chapter 5 represents independent contributions by Hessler (an overview), Newman (Cirripedia), Marcotte (Copepoda), and Maddocks (Ostracoda). Hessler provides the third view (see Chapters 3 and 4) of an ancestral crustacean and discusses the major schemes that have been proposed for evolution within the Crustacea. Newman, in his discussion of the Cirripedia, disagrees with Hessler's assessment of the Maxillopoda and presents an overview of evolution within the Cirripedia. The relative isolation of the Ostracoda from other crustaceans is emphasized by Maddocks. She nicely reviews the controversy on homologies of ostracode limbs and summarizes phylogeny and the fossil record in her Fig. 1. Marcotte concentrates on relationships within the diverse Copepoda, suggesting features that an ancestral copepod may have displayed.

Abele reviews the general distribution of crustaceans in Chapter 6. He begins by summarizing various patterns of numbers of species shown by crustaceans and finds that crustaceans with direct development (Peracarida) have the greatest number of species in the temperate zone, whereas crustaceans with planktonic larval development have their greatest diversity in the tropics. He then reviews the distribution of various taxonomic groups.

I thank the contributors for their patience and assistance in the preparation of this volume and especially the following individuals who served as outside reviewers: T. E. Bowman, F. A. Chace, Jr., B. Felgenhauer, D. Fry, G. Fryer, S. Gilchrist, R. Gore, R. Hessler, H. Hobbs, Jr., B. Kensley, L. Kornicker, R. Manning, P. McLaughlin, F. Schram, H. Spivey, and D. Thistle. Finally, a special thanks to Sandra Gilchrist for invaluable assistance.

Lawrence G. Abele

Contents of Volumes 2–4

Volume 4: Neural Integration and Behavior

Edited by David C. Sandeman and Harold L. Atwood

The Biology of Crustacea

VOLUME 1

SYSTEMATICS, THE FOSSIL RECORD, AND BIOGEOGRAPHY

1

Classification of the Recent Crustacea

THOMAS E. BOWMAN AND
LAWRENCE G. ABELE

The classification of Crustacea that follows is offered with the full knowledge that it will not and is not intended to reconcile the strong differences of opinion currently held on the classification of the Arthropoda, the rank to be assigned to the Crustacea, and some details of the classification of the Crustacea.

One school divides the old phylum Arthropoda into three (Manton, 1977) or four (Schram, 1978) phyla. Another school (Hessler and Newman, 1975) retains the phylum Arthropoda, assigning subphylum status to the "tip-biters" (Uniramia) and "base-biters" (Trilobitomorpha + Chelicerata + Crustacea). During the preparation of this chapter we have been buffeted by both sides, but we do not feel obliged to take sides in a dispute in which the evidence is only partly in. We avoid taking a position by offering the reader the option of selecting the hierarchal category he or she prefers for the Crustacea. The category selected need not affect the internal classification of the Crustacea. In defense of our irresolution, we point out that the subject of this chapter is the classification of the Crustacea, not the classification of the Arthropoda.

The reality of the lack of agreement on whether the Arthropoda are monophyletic or polyphyletic is evident from the varying opinions of contributors to the recent book, "Arthropod Phylogeny" (Gupta, 1979). Resolution of the issue, if it is to come at all, awaits the future.

Our classification is basically similar to that of Moore and McCormick

THE BIOLOGY OF CRUSTACEA, VOL. I

ISBN 0-12-106401-8

(1969), but it incorporates recent changes in the classification of certain taxa.

The arrangement of the Ostracoda, except Myodocopa, follows the views of Hartman and Puri (1974) and McKenzie (1977). The classification of the Myodocopa is essentially that of Kornicker (1975) and Kornicker and Sohn (1976). We include the Palaeocopa in our classification of Recent Ostracoda because of the discovery by Hornibrook (1949, 1963) of a few specimens that he considered to be living relicts of an essentially Paleozoic group. Assessment of Hornibrook's conclusions must await the availability of adequate material. Ostracode superfamily names in current use do not comply with Recommendation 29A of the ICZN for the termination "-oidea" but instead end in "-acea." Our classification conforms with the Recommendation, and endings of ostracode superfamily names are changed accordingly.

The classification of the Copepoda follows the revision recently proposed by Kabata (1979), which makes major changes in the Cyclopoida and Caligoida, largely on the basis of mouthpart structure. To Kabata's seven orders we add Boxshall's (1979) recently proposed order Mormonilloida. We follow Andronov's (1974) division of the Calanoida into nine superfamilies, but the family Pseudocalanidae Sars 1902 is replaced by the older name Clausocalanidae Giesbrecht, 1892. In the Harpacticoida the classification is that of Lang (1948a), updated by Bodin (1979); however, Lang's "Suprafam." endings "-idimorpha" are changed to "oidea." Following Lang (1948b), the order Notodelphyoida Sars (1921) is not recognized here; its families are assigned to the Cyclopoida. In the Monstrilloida the currently used family name Thespesiopsyllidae Sewell 1949 is replaced by the earlier Thaumatopsyllidae Sars 1913. *Thaumatopsyllus* was rejected by Wilson (1924) and replaced by *Thespesiopsyllus* because the former had been published by T. Scott (1894) as a synonym of the harpacticoid genus *Aegisthus*. However, publication as a synonym does not make a name available [ICZN 11 (d)].

In the Cirripedia the order Apoda is omitted, since its only genus *Proteolepas,* has been shown by Bocquet-Védrine (1972) to be an epicaridean isopod of the family Crinoniscidae.

We follow Schram (1969a,b) in removing the Hoplocarida from the Eumalacostraca and recognizing it as a distinct subclass of Malacostraca, knowing that some carcinologists will disagree. This removal is supported by studies on the functional morphology of the mouthparts and the proventriculus (Kunze, 1981). In the Syncarida, the order Stygiocaridacea is reduced to a family of the Anaspidacea, following Schminke (1978), and we do not think there is convincing evidence for recognizing the superorder Podophallocarida for the Bathynellacea as proposed by Serban (1970).

In the Amphipoda, the characters claimed to separate the Ingolfiellidea from the Gammaridea are either found in some of the Gammaridea or are adaptations to the interstitial habitat. Hence we follow the course suggested by Dahl (1977) and incorporate the Ingolfiellidea within the Gammaridea as the families Ingolfiellidae and Metingolfiellidae.

The arrangement of the Hyperiidea follows Bowman and Gruner (1973), but classification of the Gammaridea presents a dilemma. Much rearrangement of families and superfamilies is now being advocated by specialists, but the various classifications being proposed have major differences that appear to be irreconcilable at present. Examples of conflicting views are found in the different approaches to subdividing the old, unwieldy family Gammaridae by Bousfield (1977, 1979) and by Barnard and Karaman (1980). A modern consensus not being available, we follow herein the slightly outdated but still comprehensive and valuable synopsis of Barnard (1969), currently being revised by Barnard and Karaman; however, we incorporate a number of additions and deletions. Several family names were proposed by Stebbing (1906) to replace older family names in the mistaken belief that a family name must be based on the oldest genus of the family. We have restored these older names, except Orchestiidae, which was unnecessarily placed on the Official Index of Rejected names in favor of the later name Talitridae. The arrangement of the Caprellidea follows Bousfield (1979).

In the Isopoda we have changed four superfamily names that were not based on family names and one family name that was not based on a contained genus. In the Epicaridea we do not feel that morphological differences support the recognition of Bonnier's (1900) seven families of cryptoniscids based primarily on the kinds of hosts, and we list only the Liriopsidae (=Cryptoniscidae, suppressed by Bonnier because he considered *Cryptoniscus* a junior synonym of *Liriopsis*).

We follow Sieg's (1980) rearrangement of the Tanaidacea, which rejects Lang's (1956) widely used system of splitting the group into two suborders, Monokonophora and Dikonophora. Sieg's reconsideration of the Lang classification was made necessary by recent accelerated activity in tanaidacean taxonomy; the number of known species has almost doubled since 1950.

We include in our classification the Maxillopoda, proposed by Dahl (1956) as a group including the classes Mystacocarida, Branchiura, Cirripedia, Copepoda, and possibly the Ostracoda. We believe that there is insufficient evidence to include the Pentastomida in the Crustacea, either separately or as a subgroup of the Branchiura, solely on the basis of the comparative sperm morphology as suggested by Wingstrand (1972; see also Riley *et al*., 1978). Omitted is the class Thoracopoda of Hessler and Newman (1975), comprising the Cephalocarida, Branchiopoda, and Malacostraca. This omission is more or less arbitrary, and based on our feeling that support for the Thoracopoda is quite limited. Even one of its authors no longer advocates its use.

TABLE I

Numbers of Genera and Species and Their General Environmental and Habitat Distribution for the Crustacea

Taxon	Approximate number of species		Environment	Habitat
Class Cephalocarida	4 genera,	9 species	Marine	Fine sediment, intertidal to 1550 m
Class Branchiopoda				
Order Notostraca	2 genera,	11 species	Freshwater	Temporary pools in dry areas
Order Conchostraca	15 genera,	180 species	Freshwater	Shallow areas in lakes, ponds, and temporary pools
Order Cladocera	52 genera,	450 species	Freshwater (95%) Brackish (3%) Marine (2%)	Shallow ponds among vegetation, streams, estuaries
Order Anostraca	25 genera,	180 species	Freshwater Inland saltwater	Temporary pools
Class Remipedia	1 genus,	1 species	Marine	Pelagic in a cave
Class Maxillopoda				
Subclass Mystacocarida	2 genera,	9 species	Marine	Shallow water, interstitial
Subclass Branchiura	4 genera,	150 species	Freshwater (90%) Coastal marine (10%)	Parasitic (almost entirely on fishes)
Subclass Copepoda				
Order Calanoida	265 genera,	2300 species	Marine (75%) Freshwater (25%)	Primarily pelagic
Order Harpacticoida	350 genera,	2800 species	Marine (90%) Freshwater (10%)	Benthic, some pelagic, and a few commensal with invertebrates and vertebrates
Order Cyclopoida	80 genera,	450 species	Freshwater & Marine	Free-living and parasites and commensals of fishes and invertebrates
Order Poecilostomatoida	260 genera,	1320 species	Marine	Parasites and commensals of fishes and invertebrates
Order Siphonostomatoida	200 genera,	1430 species	Marine	Parasites and commensals of fishes and invertebrates

Order Monstrilloida	[illegible] genera,	[illegible] species	Marine	Adults pelagic, young parasitic
Order Misphryioida	2 genera,	3 species	Marine	Pelagic
Order Mormonilloida	1 genus,	2 species	Marine	Pelagic
Subclass Cirripedia				
Order Ascothoracica	10 genera,	45 species	Marine	Parasitic on hexacorals and echinoderms
Order Acrothoracica	12 genera,	50 species	Marine	All bore into calcareous substrates
Order Thoracica	150 genera,	700 species	Marine (98%) Brackish (2%)	Shallow water benthic, some pelagic attached to objects
Order Rhizocephala	31 genera,	230 species	Marine	Parasitic on other crustaceans, especially decapods
Class Ostracoda				
Subclass Myodocopa				
Order Myodocopida	77 genera,	375 species	Marine, few brackish	Most benthic and meroplanktonic, a few bathypelagic and abyssal pelagic
Order Halocyprida				
Suborder Cladocopina	4 genera,	40 species	Marine	0–2600 m, usually < 500 m, benthic, occasionally pelagic
Suborder Halocypridina	10 genera,	160 species	Marine	0–3600 m, most holo- and bathypelagic; few benthic
Subclass Podocopa				
Order Platycopida	2 genera,	73 species	Marine, possibly brackish	Benthic, 0–1234 m, most < 200 m
Order Podocopida	600 genera,	5000 species	Freshwater, marine, brackish, few hypersaline or terrestrial	Mostly benthic; infaunal, epibenthic, on plants; commensal on Crustacea
Subclass Palaeocopa				
Order Palaeocopida	2 genera,	2 species	Marine	Only shells found
Class Malacostraca				
Subclass Phyllocarida	4 genera,	10+ species	Marine	Benthic and pelagic
Subclass Hoplocarida	68 genera,	350 species	Marine	Benthic

(continued)

TABLE I *(Continued)*

Taxon	Approximate number of species	Environment	Habitat
Subclass Eumalacostraca			
Superorder Syncarida			
Order Anaspidacea	10 genera, 15+ species	Freshwater	Benthic, a few interstitial
Order Bathynellacea	23 genera, 100 species	Freshwater (95%) Marine (5%)	Interstitial ground water
Superorder Pancarida			
Order Thermosbaenacea	4 genera, 9+ species	Marine Freshwater Hot Springs	Interstitial, benthic
Superorder Peracarida			
Order Mysidacea	120 genera, 780 species	Marine (90%) Brackish (5%) Freshwater (5%)	Just off the bottom, pelagic, benthic
Order Cumacea	102 genera, 800 species	Marine	Benthic
Order Spelaeogriphacea	1 genus, 1 species	Freshwater	Cavernicolous
Order Amphipoda	840 genera, 6000 species	Marine (82%) Freshwater (15%) Terrestrial (3%)	Benthic, a few parasitic and commensal
Order Isopoda	700 genera, 4000 species	Marine (70%) Freshwater (5%) Terrestrial (25%)	Virtually all habitats; some parasitic
Order Tanaidacea	100 genera, 500 species	Marine Brackish	Benthic to 8300 m
Superorder Eucarida			
Order Euphausiacea	11 genera, 85 species	Marine	Pelagic
Order Amphionidacea	1 genus, 1 species	Marine	Pelagic
Order Decapoda	1200 genera, 10,000 species	Marine (89%) Freshwater (10%) Terrestrial (1%)	Virtually all habitats

The classification of the Eucarida follows Hessler (1969) for the Euphausiacea and Williamson (1973) for the Amphionidacea. The classification of the Decapoda is unsettled and we have followed Glaessner (1969) with some of the changes proposed by Guinot (1977) and De Saint-Laurent (1979). Rice (1980) provides an excellent review of the classification of the Brachyura and some further information is provided by De Saint-Laurent (1980a,b). While we are in agreement with many of the conclusions of the above authors we are constrained by the available taxonomic hierarchy (Infraorder, Section, Superfamily) and by the practical problem of having to include families not yet placed in one major group or another by the current revisions.

The frontispiece very loosely indicates relationships among the various crustacean groups. Table I summarizes numbers of genera and species, and environment and habitat distribution for the Crustacea down to Order.

Phylum, Subphylum, or Superclass Crustacea, Pennant 1777
- Class Cephalocarida Sanders 1955
  - Order Brachypoda Birstein 1960
    - Family Hutchinsoniellidae Sanders 1955
    - Lightiellidae Jones 1961
- Class Branchiopoda Latreille 1817
  - Subclass Calmanostraca Tasch 1969
    - Order Notostraca Sars 1867
      - Family Triopidae Kielhack 1910
  - Subclass Diplostraca Gerstaecker 1866
    - Order Conchostraca Sars 1867
      - Suborder Laeviscauda Linder 1945
        - Family Lyncaeidae Stebbing 1902
      - Suborder Spinicaudata Linder 1945
        - Superfamily Cyzicoidea Stebbing 1910
          - Family Cyzicidae Stebbing 1910
        - Superfamily Limnadioidea Baird 1849
          - Family Cyclestheriidae Sars 1899
          - Leptestheriidae Daday 1923
          - Limnadiidae Baird 1849
    - Order Cladocera Latreille 1829
      - Suborder Haplopoda Sars 1865
        - Family Leptodoridae Lilljeborg 1900
      - Suborder Eucladocera Eriksson 1932
        - Superfamily Sidoidea Baird 1850
          - Family Holopedidae Sars 1865
          - Sididae Baird 1850
        - Superfamily Daphnioidea Straus 1820
          - Family Bosminidae Baird 1845
          - Chydoridae Stebbing 1902
          - Daphniidae Straus 1820
          - Macrotrichidae Norman & Brady 1867
          - Moinidae Goulden 1968

Superfamily Polyphemoidea Baird 1845
Family Cercopagidae Mordukhai-Boltovskoi 1968
Podonidae Mordukhai-Boltovskoi 1968
Polyphemidae Baird 1845
Subclass Sarsostraca Tasch 1969
Order Anostraca Sars 1867
Family Artemiidae Grochowski 1896
Branchinectidae Daday 1910
Branchipodidae Simon 1886
Chirocephalidae Daday 1910
Polyartemiidae Simon 1886
Streptocephalidae Daday 1910
Thamnocephalidae Simon 1886
Class Remipedia Yager 1981
Family Speleonectidae Yager 1981
Class Maxillopoda Dahl 1956
Subclass Mystacocarida Pennak & Zinn 1943
Order Mystacocaridida Pennak & Zinn 1943
Family Derocheilocarididae Pennak & Zinn 1943
Subclass Cirripedia Burmeister 1834
Order Ascothoracica Lacaze-Duthiers 1880
Suborder Synagogoidida Wagin 1976
Family Dendrogastridae Gruvel 1905
Synagogidae Gruvel 1905
Suborder Lauroidida Wagin 1976
Family Lauridae Gruvel 1905
Petrarcidae Gruvel 1905
Order Thoracica Darwin 1854
Suborder Lepadomorpha Pilsbry 1916
Family Heteralepadidae Nilsson-Cantell 1921
Iblidae Leach 1825
Koleolepadidae Hiro 1937
Lepadidae Darwin 1851
Malacolepadidae Hiro 1937
Oxynaspididae Pilsbry 1907
Poecilasmatidae Nilsson-Cantell 1921
Scalpellidae Pilsbry 1916
Suborder Verrucomorpha Pilsbry 1916
Family Verrucidae Darwin 1854
Suborder Balanomorpha Pilsbry 1916
Superfamily Chthamaloidea Darwin 1854
Family Catophragmidae Utinomi 1968
Chthamalidae Darwin 1854
Superfamily Coronuloidea Leach 1817
Family Bathylasmatidae Newman & Ross 1971
Coronulidae Leach 1817
Tetraclitidae Gruvel 1903
Superfamily Balanoidea Leach 1817
Family Archaeobalanidae Newman & Ross 1976
Balanidae Leach 1817
Pyrgomatidae Gray 1825

Order Acrothoracica Gruvel 1905
Suborder Pygophora Berndt 1907
Family Cryptophialidae Gerstaecker 1866
Lithoglyptidae Aurivillius 1892
Suborder Apygophora Berndt 1907
Family Trypetesidae Stebbing 1910
Order Rhizocephala F. Müller 1862
Suborder Kentrogonida Delage 1884
Family Clistosaccidae Boschma 1928
Lernaeodiscidae Boschma 1928
Peltogastridae Lilljeborg 1861
Sacculinidae Lilljeborg 1861
Sylidae Boschma 1928
Suborder Akentrogonida Häfele 1911
Subclass Copepoda Milne-Edwards 1840
Order Calanoida Sars 1903
Superfamily Augaptiloidea Sars 1905
Family Arietellidae Sars 1902
Augaptilidae Sars 1905
Discoidae Gordejeva 1975
Epacteriscidae Fosshagen 1973
Heterorhabdidae Sars 1902
Lucicutiidae Sars 1902
Metridinidae Sars 1902
Phyllopodidae Brodsky 1950
Superfamily Bathypontioidea Brodsky 1950
Family Bathypontiidae Brodsky 1950
Superfamily Centropagoidea Giesbrecht 1892
Family Acartiidae Sars 1903
Candaciidae Giesbrecht 1892
Centropagidae Giesbrecht 1892
Diaptomidae Baird 1850
Parapontellidae Giesbrecht 1892
Pontellidae Dana 1853
Pseudodiaptomidae Sars 1902
Sulcanidae Nicholls 1945
Temoridae Giesbrecht 1892
Tortanidae Sars 1902
Superfamily Clausocalanoidea Giesbrecht 1892
Family Aetideidae Giesbrecht 1892
Clausocalanidae Giesbrecht 1892
Diaixidae Sars 1902
Euchaetidae Giesbrecht 1892
Mesaiokeratidae Matthews 1961
Phaennidae Sars 1902
Pseudocyclopiidae Sars 1902
Scolecitrichidae Giesbrecht 1892
Spinocalanidae Vervoort 1951
Stephidae Sars 1902
Tharybidae Sars 1902

Superfamily Eucalanoidea Giesbrecht 1892
Family Eucalanidae Giesbrecht 1892
Superfamily Megacalanoidea Sewell 1947
Family Calanidae Dana 1849
Calocalanidae Bernard 1958
Mecynoceridae Andronov 1973
Megacalanidae Sewell 1947
Paracalanidae Giesbrecht 1892
Superfamily Platycopioidea Sars 1911
Family Platycopiidae Sars 1911
Superfamily Pseudocyclopoidea Giesbrecht 1893
Family Pseudocyclopidae Giesbrecht 1893
Ridgewayiidae M. S. Wilson 1958
Superfamily Ryocalanoidea Andronov 1974
Family Ryocalanidae Andronov 1974
Order Harpacticoida Sars 1903
Suborder Polyarthra Lang 1948
Family Canuellidae Lang 1948
Longipediidae Sars 1903
Suborder Oligarthra Lang 1948
Infraorder Maxillipedasphalea Lang 1948
Superfamily Cervinioidea (pro Cerviniidimorpha Lang 1948)
Family Aegisthidae Giesbrecht 1892
Cerviniidae Sars 1903
Superfamily Ectinosomatoidea (pro Ectinosomidimorpha Lang 1948)
Family Ectinosomatidae Sars 1903
Superfamily Neobradyoidea (pro Neobradyidimorpha Lang 1948)
Family Chappuisiidae Chappuis 1940
Darcythomsoniidae Lang 1936
Neobradyidae Oloffson 1917
Phyllognathopodidae Gurney 1932
Infraorder Exanechentera Lang 1948
Superfamily Tachidioidea (pro Tachidiidimorpha Lang 1948)
Family Harpacticidae Dana 1846
Tachidiidae Sars 1909
Superfamily Tisboidea (pro Tisbidimorpha Lang 1948)
Family Peltidiidae Sars 1904
Porcellidiidae Sars 1904
Pseudopeltidiidae Poppe 1891
Tegastidae Sars 1904
Tisbidae Stebbing 1910
Infraorder Podogennonta Lang 1948
Superfamily Ameiroidea (pro Ameiridimorpha Lang 1948)
Family Ameiridae Monard 1927
Canthocamptidae Sars 1906
Cylindropsyllidae Sars 1909
Louriniidae Monard 1927
Paramesochridae Lang 1948
Parastenocarididae Chappuis 1933
Tetragonicipitidae Lang 1944

Superfamily Cletodoidea (pro Cletodidimorpha Lang 1948)
Family Ancorabolidae Sars 1909
Cletodidae T. Scott 1904
Laophontidae T. Scott 1904
Superfamily Metoidea (pro Metidimorpha Lang 1948)
Family Metidae Sars 1910
Superfamily Thalestroidea (pro Thalestridimorpha Lang 1948)
Family Balaenophilidae Sars 1910
Diosaccidae Sars 1906
Miraciidae Dana 1846
Parastenheliidae Lang 1936
Thalestridae Sars 1905
Infraorder incertae sedis
Family Gelyellidae Rouch & Lescher-Moutoué 1977
Suborder incertae sedis
Family Latiremidae Bozic 1969
Order Cyclopoida Burmeister 1834
Family Archinotodelphyidae Lang 1949
Ascidicolidae Thorell 1860
Botryllophyllidae Sars 1921
Buproridae Thorell 1859
Cyclopidae Dana 1853
Cyclopinidae Sars 1913
Doropygidae Brady 1878
Enterocolidae Sars 1921
Enteropsidae Aurivillius 1887
Lernaeidae Cobbold 1879
Namakosiramiidae Ho & Perkins 1977
Notodelphyidae Dana 1853
Oithonidae Dana 1853
Schizoproctidae Aurivillius 1887
Order Poecilostomatoida Thorell 1859
Family Anomoclausiidae Gotto 1964
Anomopsyllidae Sars 1921
Bomolochidae Sumpf 1871
Catiniidae Bocquet & Stock 1957
Chondracanthidae Milne Edwards 1840
Clausidiidae Embleton 1901
Clausiidae Giesbrecht 1895
Corallovexiidae Stock 1975
Corycaeidae Dana 1852
Cucumaricolidae Bouligand & Delamare-Deboutteville 1959
Echiurophilidae Delamare-Deboutteville & Nunes-Ruivo 1955
Ergasilidae Nordmann 1832
Eunicicolidae Sars 1918
Gastrodelphyidae List 1890
Lamippidae Joliet 1882
Lichomolgidae Kossmann 1877
Mantridae Leigh-Sharpe 1934
Myicolidae Yamaguti 1936

Mytilicolidae Bocquet & Stock 1957
Nereicolidae Claus 1875
Oncaeidae Giesbrecht 1892
Pharodidae Illg 1948
Philoblennidae Izawa 1976
Philichthyidae Bassett-Smith 1899
Phyllodicolidae Delamare-Deboutteville 1960
Pseudanthessiidae Humes & Stock 1972
Rhynchomolgidae Humes & Ho 1967
Sabelliphilidae Gurney 1927
Sapphirinidae Thorell 1859
Sarcotacidae Yamaguti 1963
Serpulidicolidae Stock 1979
Shiinoidae Cressey 1975
Splanchnotrophidae Norman & Scott 1906
Staurosomatidae Ardeev & Ardeev 1975
Synaptiphilidae Bocquet & Stock 1957
Taeniacanthidae Wilson 1911
Telsidae Ho 1967
Tuccidae Vervoort 1962
Urocopiidae Humes & Stock 1972
Vahiniidae Humes 1966
Xarifidae Humes 1960

Order Siphonostomatoida Thorell 1859

Family Artotrogidae Brady 1880
Ascomyzontidae Thorell 1859
Asterocheridae Giesbrecht 1899
Brychiopontiidae Humes 1974
Caligidae Burmeister 1835
Calvocheridae Stock 1968
Cancerillidae Giesbrecht 1897
Catlaphilidae Tripathi 1960
Cecropidae Dana 1852
Choniostomatidae Hansen 1886
Dichelesthiidae Dana 1853
Dinopontiidae Murnane 1967
Dirivultidae Humes & Dojiri 1980
Dissonidae Yamaguti 1963
Dyspontiidae Giesbrecht 1895
Entomolepidae Brady 1899
Eudactylinidae Yamaguti 1963
Euryphoridae Wilson 1905
Hatschekiidae Kabata 1979
Herpyllobiidae Hansen 1892
Hyponeoidae Heegaard 1962
Kroyeriidae Kabata 1979
Lernaeoceridae Gurney 1933
Lernaeopodidae Olsson 1869
Lernanthropidae Kabata 1979
Megapontiidae Heptner 1968

Micropontiidae Gooding 1957
Myzopontiidae Sars 1915
Nanaspididae Humes & Cressey 1959
Naobranchiidae Yamaguti 1939
Nicothoidae Dana 1852
Pandaridae Milne Edwards 1840
Pennellidae Burmeister 1835
Pontoeciellidae Giesbrecht 1895
Pseudocycnidae Wilson 1922
Rataniidae Giesbrecht 1897
Saccopsidae Lützen 1964
Sphyriidae Wilson 1915
Spongiocnizontidae Stock & Kleeton 1964
Stellicomitidae Humes & Cressey 1958
Tanypleuridae Kabata 1969
Trebiidae Wilson 1932
Ventriculinidae Leigh-Sharpe 1934
Xenocoelomatidae Bresciani & Lützen 1966
Order Monstrilloida Sars 1903
Family Monstrillidae Giesbrecht 1892
Thaumatopsyllidae Sars 1913
Order Misophrioida Gurney 1933
Family Misophriidae Brady 1878
Order Mormonilloida Boxshall 1979
Family Mormonillidae Giesbrecht 1892
Order uncertain
Family Antheacheridae M. Sars 1870
Mesoglicolidae Zulueta 1911
Sponginticolidae Topsent 1928
Staurosomatidae Zulueta 1911
Subclass Branchiura Thorell 1864
Order Arguloida Rafinesque 1815
Family Argulidae Leach 1819
Class Ostracoda Latreille 1806
Subclass Myodocopa Sars 1866
Order Myodocopida Sars 1866
Suborder Myodocopina Sars 1866
Superfamily Cypridinoidea Baird 1850
Family Cylindroleberididae Müller 1906
Cypridinidae Baird 1850
Philomedidae Müller 1906
Rutidermatidae Brady & Norman 1896
Sarsiellidae Brady & Norman 1896
Order Halocyprida Dana 1853
Suborder Cladocopina Sars 1865
Superfamily Polycopoidea Sars 1865
Family Polycopidae Sars 1865
Suborder Halocypridina Dana 1853
Superfamily Halocypridoidea Dana 1852
Family Halocyprididae Dana 1852

Superfamily Thaumatocypridoidea Müller 1906
Family Thaumatocyprididae Müller 1906
Subclass Podocopa Müller 1894
Order Platycopida Sars 1866
Family Cytherellidae Sars 1866
Order Podocopida Sars 1866
Suborder Podocopina Sars 1866
Superfamily Bairdioidea Sars 1865
Family Bairdiidae Sars 1865
Bythocyprididae Maddocks 1969
Superfamily Cytheroidea Baird 1850
Family Australocytherideidae Hartmann 1980
Bonaducecytheridae McKenzie 1977
Bythocytheridae Sars 1866
Cobanocytheridae Schornikov 1975
Cushmanideidae Puri 1973
Cytherettidae Triebel 1952
Cytheridae Baird 1850
Cytherideidae Sars 1925
Cytheromatidae Elofson 1939
Cytheruridae Müller 1894
Entocytheridae Hoff 1942
Eucytheridae Puri 1954
Hemicytheridae Puri 1953
Kliellidae Schäfer 1945
Krithidae Mandelstam 1958
Leptocytheridae Hanai 1957
Limnocytheridae Klie 1938
Loxoconchidae Sars 1925
Microcytheridae Klie 1938
Neocytherideidae Puri 1957
Osticytheridae Hartmann 1980
Paracytheridae Puri 1973
Paracytherideidae Puri 1957
Paradoxostomatidae Brady & Norman 1889
Parvocytheridae Hartmann 1959
Pectocytheridae Hanai 1957
Psammocytheridae Klie 1938
Pseudolimnocytheridae Hartmann & Puri 1974
Trachyleberididae Sylvester-Bradley 1948
Xestoleberididae Sars 1928
Superfamily Terrestricytheroidea Schornikov 1969
Family Terrestricytheridae Schornikov 1969
Suborder Metacopina Sylvester-Bradley 1961
Superfamily Darwinuloidea Brady & Norman 1889
Family Darwinulidae Brady & Norman 1889
Superfamily Cypridoidea Baird 1845
Family Candoniidae Kaufmann 1900
Cyprididae Baird 1845
Cypridopsidae Kaufmann 1910

Ilyocyprididae Kaufmann 1900
Macrocyprididae Müller 1912
Notodromadidae Kaufmann 1900
Paracyprididae Sars 1923
Pontocyprididae Müller 1894
Terrestricypridae Shornikov 1980
Superfamily Healdioidea Harlton 1933
Family Saipanettidae McKenzie 1968
Subclass Palaeocopa Henningsmoen 1953
Order Palaeocopida Henningsmoen 1953
Suborder Beyrichicopina Scott 1961
Superfamily Puncioidea Hornibrook 1949
Family Punciidae Hornibrook 1949
Class Malacostraca Latreille 1806
Subclass Phyllocarida Packard 1879
Order Leptostraca Claus 1880
Family Nebaliidae Baird 1850
Subclass Hoplocarida Calman 1904
Order Stomatopoda Latreille 1817
Suborder Unipeltata Latreille 1825
Superfamily Bathysquilloidea Manning 1967
Family Bathysquillidae Manning 1967
Superfamily Gonodactyloidea Giesbrecht 1910
Family Eurysquillidae Manning 1977
Gonodactylidae Giesbrecht 1910
Hemisquillidae Manning 1980
Odontodactylidae Manning 1980
Protosquillidae Manning 1980
Pseudosquillidae Manning 1967
Superfamily Squilloidea Latreille 1803
Family Harpiosquillidae Manning 1980
Squillidae Latreille 1803
Superfamily Lysiosquilloidea Giesbrecht 1910
Family Coronididae Manning 1980
Lysiosquillidae Giesbrecht 1910
Nannosquillidae Manning 1980
Subclass Eumalacostraca Grobben 1892
Superorder Syncarida Packard 1885
Order Bathynellacea Chappuis 1915
Family Bathynellidae Grobben 1904
Leptobathynellidae Noodt 1965
Parabathynellidae Noodt 1964
Order Anaspidacea Calman 1904
Family Anaspididae Thomson 1893
Koonungidae Sayce 1908
Psammaspididae Schminke 1974
Stygocarididae Noodt 1962
Superorder Pancarida Siewing 1958
Order Thermosbaenacea Monod 1927
Family Thermosbaenidae Monod 1927

Superorder Peracarida Calman 1904
Order Spelaeogriphacea Gordon 1957
Family Spelaeogriphidae Gordon 1957
Order Mysidacea Boas 1883
Suborder Lophogastrida Boas 1883
Family Eucoplidae Sars 1885
Lophogastridae Sars 1870
Suborder Mysida Boas 1883
Family Lepidomysidae Clarke 1961
Mysidae Dana 1850
Petalophthalmidae Czerniavsky 1882
Stygiomysidae Caroli 1937
Order Amphipoda Latreille 1816
Suborder Gammaridea Latreille 1803
Family Ampeliscidae Costa 1857
Amphilochidae Boeck 1871
Ampithoidae Stebbing 1899
Anamixidae Stebbing 1897
Anisogammaridae Bousfield 1977
Argissidae Walker 1904
Artesiidae Holsinger 1980
Bateidae Stebbing 1906
Biancolinidae J. L. Barnard 1972
Bogidiellidae Hertzog 1936
Calliopiidae Sars 1893
Carangoliopsidae Bousfield 1977
Caspicolidae Birstein 1945
Ceinidae J. L. Barnard 1972
Cheluridae Allman 1847
Colomastigidae Stebbing 1899
Corophiidae Dana 1849
Crangonyctidae Bousfield 1973
Cressidae Stebbing 1899
Dexaminidae Leach 1813
Dogielinotidae Gurjanova 1953
Dulichiidae Dana 1849 (pro Podoceridae Stebbing 1906)
Eophliantidae Sheard 1936
Epimeriidae Boeck 1871 (pro Paramphithoidae Stebbing 1906)
Eusiridae Stebbing 1888
Gammaridae Leach 1813
Hadziidae Karaman 1943
Hyalellidae Bulycheva 1957
Hyalidae Bulycheva 1957
Hyperiopsidae Bovallius 1886
Ingolfiellidae Hansen 1903
Iphimedidae Boeck 1871 (pro Acanthonotozomatidae Stebbing 1906)
Ischyoceridae Stebbing 1899
Kuriidae Walker & Scott 1903
Laphystiidae Sars 1893
Laphystiopsidae Stebbing 1899

Lepechinellidae Schellenberg 1926
Leucothoidae Dana 1852
Liljeborgiidae Stebbing 1899
Lysianassidae Dana 1849
Macrohectopidae Sowinsky 1915
Maxillipiidae Ledoyer 1973
Melitidae Bousfield 1973
Melphidippidae Stebbing 1899
Mesogammaridae Bousfield 1977
Metaingolfiellidae Ruffo 1969
Najnidae J. L. Barnard 1972
Nihotungidae J. L. Barnard 1972
Ochlesidae Stebbing 1910
Oedicerotidae Lilljeborg 1865
Pagetinidae K. H. Barnard 1931
Paradaliscidae Boeck 1871
Phliantidae Stebbing 1899
Phoxocephalidae Sars 1891
Platyischnopidae Barnard & Drummond 1979
Pleustidae Buchholz 1874
Plioplateidae J. L. Barnard 1978
Pontoporeiidae Dana 1853 (pro Haustoriidae Stebbing 1906)
Pseudamphilochidae Schellenberg 1931
Salentinellidae Bousfield 1977
Sebidae Walker 1908
Stegocephalidae Dana 1853
Stenothoidae Boeck 1871
Stilipedidae Holmes 1908
Synopiidae Dana 1853
Talitridae Rafinesque 1815
Temnophliantidae Griffiths 1975
Urothoidae Bousfield 1978
Vitjazianidae Birstein & Vinogradov 1955

Suborder Caprellidea Leach 1814
Infraorder Caprellida Bousfield 1979
Superfamily Caprelloidea White 1847
Family Aeginellidae Vassilenko 1968
Caprellidae White 1847
Caprogammaridae Kudrjaschov & Vassilenko 1966
Superfamily Phtisicoidea Vassilenko 1968
Family Dodecadidae Vassilenko 1968
Paracercopidae Vassilenko 1968
Phtisicidae Vassilenko 1968
Infraorder Cyamida Bousfield 1979
Family Cyamidae Rafinesque 1815

Suborder Hyperiidea H. Milne Edwards 1830
Infraorder Physosomata Pirlot 1929
Superfamily Scinoidea Stebbing 1888
Family Archaeoscinidae Stebbing 1904
Mimonectidae Bovallius 1885

Proscinidae Pirlot 1933
Scinidae Stebbing 1888
Superfamily Lanceoloidea Bovallius 1887
Family Chuneolidae Woltereck 1909
Lanceolidae Bovallius 1887
Microphasmatidae Stephensen & Pirlot 1931
Infraorder Physocephalata Bowman & Gruner 1973
Superfamily Vibilioidea Dana 1853
Family Cystisomatidae Willemoes-Suhm 1875
Paraphronimidae Bovallius 1887
Vibiliidae Dana 1853
Superfamily Phronimoidea Rafinesque 1815
Family Dairellidae Bovallius 1887
Hyperiidae Dana 1853
Phronimidae Rafinesque 1815
Phrosinidae Dana 1853
Superfamily Lycaeopsoidae Chevreux 1913
Family Lycaeopsidae Chevreux 1913
Superfamily Platysceloidea Bate 1862
Family Anapronoidae Bowman & Gruner 1973
Lycaeidae Claus 1879
Oxycephalidae Dana 1853
Parascelidae Bate 1862
Platyscelidae Bate 1862
Pronoidae Dana 1853
Order Isopoda Latreille 1817
Suborder Gnathiidea Leach 1814
Family Gnathiidae Harger 1880
Suborder Anthuridea Leach 1814
Family Anthuridae Leach 1814
Hyssuridae Wägele 1981
Paranthuridae Menzies & Glynn 1968
Suborder Microcerberidea Lang 1961
Family Microcerberidae Karaman 1933
Suborder Flabellifera Sars 1882
Family Aegidae Leach 1815
Anuropodidae Stebbing 1893
Argathonidae Stebbing 1905
Bathynataliidae Kensley 1978
Bathynomidae Wood Mason & Alcock 1891
Cirolanidae Dana 1853
Corallanidae Hansen 1890
Cymothoidae Dana 1852
Excorallanidae Stebbing 1904
Keuphyliidae Bruce 1980
Limnoriidae Dana 1853
Phoratopodidae Hale 1925
Plakarthriidae Richardson 1904
Serolidae Dana 1853
Sphaeromatidae Milne Edwards 1840

Suborder Asellota Latreille 1803
Superfamily Aselloidea Rafinesque 1815
Family Asellidae Rafinesque 1815
Atlantasellidae Sket 1980
Stenasellidae Dudich 1924
Superfamily Stenetrioidea Hansen 1905
Family Stenetriidae Hansen 1905
Superfamily Janiroidea Sars 1899
Family Abyssianiridae Menzies 1956
Acanthaspidiidae Menzies 1962
Acanthomunnopsidae Schultz 1978
Dendrotiidae Vanhöffen 1914
Desmosomatidae Sars 1899
Echinothambematidae Menzies 1956
Eurycopidae Hansen 1916
Haplomunnidae Wilson 1976
Haploniscidae Hansen 1916
Ilyarachnidae Hansen 1916
Ischnomesidae Hansen 1916
Jaeropsididae Nordenstam 1933
Janirellidae Menzies 1956
Janiridae Sars 1899
Macrostylidae Hansen 1916
Mesosignidae Schultz 1969
Microparasellidae Karaman 1933
Mictosomatidae Wolff 1965
Munnidae Sara 1899
Munnopsidae Sars 1869
Nannoniscidae Hansen 1916
Paramunnidae Vanhöffen 1914
Pleurocopidae Fresi & Schiecke 1972
Pseudomesidae Hansen 1916
Thambematidae Stebbing 1913
Superfamily Gnathostenetroidoidea Kussakin 1967 (nom. correct. pro Gnathostenetrioidea Kussakin 1967)
Family Gnathostenetroididae Kussakin 1967 (pro Parastenetriidae Amar 1957, nomen nudum)
Superfamily Protallocoxoidea Schultz 1978 (but see Wilson, 1980)
Family Protallocoxidae Schultz 1978
Suborder Valvifera Sars 1882
Family Amesopodidae Stebbing 1905
Arcturidae Sars 1899
Chaetiliidae Dana 1853
Holognathidae Thomson 1904
Idoteidae Milne Edwards 1840
Pseudidotheidae Ohlin 1901
Xenarcturidae Sheppard 1957
Suborder Phreatoicidea Stebbing 1893
Family Amphisopodidae Nicholls 1943
Nichollsiidae Tiwari 1958

Phreatoicidae Chilton 1891
Suborder Epicaridea Latreille 1831
Family Bopyridae Rafinesque 1815
Dajidae Sars 1882
Entoniscidae F. Müller 1871
Liriopsidae Bonnier 1900
Suborder Oniscidea Latreille 1803
Infraorder Tylomorpha Vandel 1943
Family Tylidae Milne Edwards 1840
Infraorder Ligiamorpha Vandel 1943
Section Diplocheta Vandel 1957
Family Ligiidae Brandt 1883
Mesoniscidae Verhoeff 1908
Section Synocheta Legrand 1946
Superfamily Trichoniscoidea Sars 1899
Family Buddelundiellidae Verhoeff 1930
Trichoniscidae Sars 1899
Superfamily Styloniscoidea Vandel 1952
Family Schoebliidae Verhoeff 1938
Styloniscidae Vandel 1952
Titaniidae Verhoeff 1938
Tunanoniscidae Borutskii 1969
Section Crinocheta Legrand 1946
Superfamily Oniscoidea Dana 1852 (pro Atracheata)
Family Bathytropidae Vandel 1973
Berytoniscidae Vandel 1973
Detonidae Budde-Lund 1906
Halophilosciidae Vandel 1973
Olibrinidae Vandel 1973
Oniscidae Brandt 1851
Philosciidae Vandel 1952
Platyarthridae Vandel 1946
Pudeoniscidae Lemos de Castro 1973
Rhyscotidae Arcangeli 1947
Scyphacidae Dana 1853
Speleoniscidae Vandel 1948
Sphaeroniscidae Vandel 1964
Stenoniscidae Budde-Lund 1904
Tendosphaeridae Verhoeff 1930
Superfamily Armadilloidea (pro Pseudotracheata Verhoeff)
Family Actaeciidae Vandel 1952
Armadillidae Verhoeff 1917
Armadillidiidae Brandt 1833
Atlantidiidae Arcangeli 1954
Balloniscidae Vandel 1963
Buddelundiellidae Verhoeff 1930
Cylisticidae Verhoeff 1949
Eubelidae Budde-Lund 1904
Periscyphicidae Ferrara 1973
Porcellionidae Verhoeff 1918
Trachelipidae Strouhal 1953

Order Tanaidacea Hansen 1895
Suborder Tanaidomorpha Sieg 1980
Superfamily Tanaoidea Dana 1849
Family Tanaidae Dana 1849
Superfamily Paratanaoidea Lang 1949
Family Agathotanaidae Lang 1971
Anarthruridae Lang 1971
Leptognathiidae Sieg 1973
Nototanaidae Sieg 1973
Paratanaidae Lang 1949
Pseudotanaidae Sieg 1973
Suborder Neotanaidomorpha Sieg 1980
Family Neotanaidae Lang 1956
Suborder Apseudomorpha Sieg 1980
Superfamily Apseudoidea Leach 1814
Anuropodidae Bacescu 1980 (homonym of Anuropodidae Stebbing 1893, Isopoda)
Family Apseudellidae Gutu 1972
Apseudidae Leach 1814
Cirratodactylidae Gardiner 1972
Gigantapseudidae Kudinova-Pasternak 1978
Kalliapseudidae Lang 1956
Leviapseudidae Sieg 1980
Metapseudidae Lang 1970
Pagurapseudidae Lang 1970
Sphyrapidae Gutu 1980
Tanapseudidae Bacescu 1978
Order Cumacea Kryer 1846
Family Archaeocumatidae Bacescu 1972
Bodotriidae T. Scott 1901
Ceratocumatidae Calman 1905
Diastylidae Bate 1856
Lampropidae Sars 1878
Leuconiidae Sars 1878
Nannastacidae Bate 1866
Pseudocumatidae Sars 1878
Superorder Eucarida Calman, 1904
Order Euphausiacea Dana, 1852
Family Bentheuphausiidae Colosi, 1917
Euphausiidae Dana, 1852
Order Amphionidacea Williamson, 1973
Family Amphionididae Holthuis, 1955
Order Decapoda Latreille, 1803
Suborder Dendrobranchiata Bate, 1888
Superfamily Penaeoidea Rafinesque, 1815
Family Aristeidae Wood-Mason, 1891
Penaeidae Rafinesque, 1815
Solenoceridae Wood-Mason and Alcock, 1891
Sicyoniidae Ortmann, 1898
Superfamily Sergestoidea Dana, 1852
Family Sergestidae Dana, 1852

Suborder Pleocyemata Burkenroad, 1963
  Infraorder Stenopodidea Claus, 1872
      Family Stenopodidae Claus, 1872
  Infraorder Caridea Dana, 1852
    Superfamily Procaridoidea Chace and Manning, 1972
      Family Procarididae Chace and Manning, 1972
    Superfamily Atyoidea De Haan, 1849
      Family Nematocarcinidae Smith, 1884
        Atyidae De Haan, 1849
        Oplophoridae Dana, 1852
    Superfamily Stylodactyloidea Bate, 1888
      Family Stylodactylidae Bate, 1888
    Superfamily Pasiphaeoidea Dana, 1852
      Family Pasiphaeidae Dana, 1852
    Superfamily Rhynchocinetoidea Ortmann, 1890
      Family Bresiliidae Calman, 1896
        Eugonatonotidae Chace, 1936
        Rhynchocinetidae Ortmann, 1890
    Superfamily Palaemonoidea Rafinesque, 1815
      Family Campylonotidae Sollaud, 1913
        Gnathophyllidae Dana, 1852
        Palaemonidae Rafinesque, 1815
    Superfamily Psalidopodoidea Wood-Mason and Alcock, 1892
      Family Psalidopodidae Wood-Mason and Alcock, 1892
    Superfamily Alpheoidea Rafinesque, 1815
      Family Alpheidae Rafinesque, 1815
        Hippolytidae Dana, 1852
        Ogyrididae Hay and Shore, 1918
        Processidae Ortmann, 1896
    Superfamily Pandaloidea Haworth, 1825
      Family Pandalidae Haworth, 1825
        Thalassocarididae Bate, 1888
    Superfamily Physetocaridoidea Chace, 1940
      Family Physetocarididae Chace, 1940
    Superfamily Crangonoidea Haworth, 1825
      Family Crangonidae Haworth, 1825
        Glyphocrangonidae Smith, 1884
  Infraorder Astacidea Latreille, 1803
    Superfamily Nephropoidea Dana, 1852
      Family Nephropidae Dana, 1852
        Thaumastochelidae Bate, 1888
    Superfamily Astacoidea Latreille, 1803
      Family Astacidae Latreille, 1803
        Cambaridae Hobbs, 1942
    Superfamily Parastacoidea Huxley, 1879
      Family Parastacidae Huxley, 1879
  Infraorder Thalassinidea Latreille, 1831
      Superfamily Thalassinoidea Latreille, 1831
      Family Axianassidae Schmitt, 1924
        Axiidae Huxley, 1879

Callianassidae Dana, 1852
Callianideidae Kossmann, 1880
Laomediidae Borradaile, 1903
Thalassinidae Latreille, 1831
Upogebiidae Borradaile, 1903
Infraorder Palinura Latreille, 1903
Superfamily Glypheoidea Zittel, 1885
Family Glypheidae Zittel, 1885
Superfamily Eryonoidea De Haan, 1841
Family Polychelidae Wood-Mason, 1874
Superfamily Palinuroidea Latreille, 1803
Family Palinuridae Latreille, 1803
Scyllaridae Latreille, 1825
Synaxidae Bate, 1881
Infraorder Anomura H. Milne Edwards, 1832
Superfamily Coenobitoidea Dana 1851
Family Coenobitidae Dana, 1851
Diogenidae Ortmann, 1892
Lomisidae Bouvier, 1895
Pomatochelidae Miers, 1879
Superfamily Paguroidea Latreille, 1803
Family Lithodidae Samouelle, 1819
Paguridae Latreille, 1803
Parapaguridae Smith, 1882
Superfamily Galatheoidea Samouelle, 1819
Family Aeglidae Dana, 1852
Chirostylidae Ortmann, 1892
Galatheidae Samouelle, 1819
Porcellanidae Haworth, 1825
Superfamily Hippoidea Latreille, 1825
Family Albuneidae Stimpson, 1858
Hippidae Latreille, 1825
Infraorder Brachyura Latreille, 1803
Section Dromiacea De Haan, 1833
Superfamily Dromioidea De Haan, 1833
Family Dromiidae De Haan, 1833
Dynomenidae Ortmann, 1892
Homolodromiidae Alcock, 1899
Section Archaeobrachyura Guinot, 1977
Superfamily Tymoloidea Alcock, 1896
Family Cymonomidae Bouvier, 1897
Tymolidae Alcock, 1896
Superfamily Homoloidea De Haan, 1839
Family Homolidae De Haan, 1839
Latreilliidae Stimpson, 1858
Superfamily Raninoidea De Haan, 1839
Family Raninidae De Haan, 1839
Section Oxystomata H. Milne Edwards, 1834
Superfamily Dorippoidea MacLeay, 1838
Family Dorippidae MacLeay, 1838

Superfamily Leucosioidea Samouelle, 1819
Family Calappidae De Haan, 1833
Leucosiidae Samouelle, 1819
Section Oxyrhyncha Latreille, 1803
Superfamily Majoidea Samouelle, 1819
Family Majidae Samouelle, 1819
Superfamily Hymenosomatoidea MacLeay, 1838
Family Hymenosomatidae MacLeay, 1838
Superfamily Mimilambroidea Williams, 1979
Family Mimilambridae Williams, 1979
Superfamily Parthenopoidea MacLeay, 1838
Family Parthenopidae MacLeay, 1838
Section Cancridea Latreille, 1803
Superfamily Cancroidea Latreille, 1803
Family Atelecyclidae Ortmann, 1893
Cancridae Latreille, 1803
Corystidae Samouelle, 1819
Pirimelidae Alcock, 1899
Thiidae Dana, 1852
Section Brachyrhyncha Borradaile, 1907
Superfamily Portunoidea Rafinesque, 1815
Family Geryonidae Colosi, 1923
Portunidae Rafinesque, 1815
Superfamily Bythograeoidea Williams, 1980
Family Bythograeidae Williams, 1980
Superfamily Xanthoidea MacLeay, 1838
Family Goneplacidae MacLeay, 1838
Hexapodidae Miers, 1886
Platyxanthidae Guinot, 1977
Xanthidae MacLeay 1838
Superfamily Bellioidea Dana, 1852
Family Belliidae, 1852
Superfamily Grapsidoidea MacLeay, 1838
Family Gecarcinidae MacLeay, 1838
Grapsidae MacLeay, 1838
Mictyridae Dana, 1851
Superfamily Pinnotheroidea De Haan, 1833
Family Pinnotheridae De Haan, 1833
Superfamily Potamoidea Ortmann, 1896
Family Deckeniidae Bott, 1970
Gecarcinucidae Rathbun, 1904
Isolapotamidae Bott, 1970
Parathelphusidae Alcock, 1910
Potamidae Ortmann, 1896
Potamocarcinidae Ortmann, 1899
Potamonautidae Bott, 1970
Pseudothelphusidae Ortmann, 1893
Sinopotamidae Bott, 1970
Sundathelphusidae Bott, 1969
Trichodactylidae H. Milne Edwards, 1853

Superfamily Ocypodoidea Rafinesque, 1815
Family Ocypodidae Rafinesque, 1815
Palicidae Rathbun, 1898
Retroplumidae Gill, 1894
Superfamily Hapalocarcinoidea Calman, 1900
Family Hapalocarcinidae Calman, 1900

ACKNOWLEDGMENTS

We are most grateful to those colleagues who have responded so generously to our requests for information and assistance. Without their aid we would have had great difficulty in compiling the classification contained in this chapter. Any errors and omissions are, however, our responsibility. Our thanks go to J. L. Barnard, P. Bodin, A. Cohen, R. F. Cressey, E. Dahl, F. D. Ferrari, D. G. Frey, R. R. Hessler, J.-S. Ho, L. B. Holthuis, A. G. Humes, Z. Kabata, B. F. Kensley, L. S. Kornicker, R. B. Manning, W. A. Newman, H. K. Schminke, F. R. Schram, G. A. Schultz, J. Sieg, G. C. Steyskal, A. B. Williams, and G. D. Wilson.

REFERENCES

Andronov, V. N. (1974). Phylogenetic relation of large taxa within the suborder Calanoida (Crustacea, Copepoda). *Zool. Zh.* **53,** 1002-1012.

Barnard, J. L. (1969). The families and genera of marine gammaridean Amphipoda. *Bull. U.S. Natl. Mus.* **271,** 1-535.

Barnard, J. L., and Karaman, G. S. (1980). Classification of gammarid Amphipoda. *Crustaceana Suppl.* **6,** 5-16.

Bocquet-Védrine, J. (1972). Suppression de l'ordre des Apodes (Crustacea Cirripèdes) et rattachement de son unique représentant, Proteolepas bivincta, à la famille des Crinoniscidae (Crustacés Isopodes, Cryptonisciens). *C. R. Hebd. Séances Acad. Sci.* **275,** 2145-2148.

Bodin, P. (1979). "Catalogue des nouveaux Copépodes Harpacticoïdes marins (Nouvelle édition)." Université de Bretagne Occidentale, Laboratoire d'Ócéanographie Biologique, Brest.

Bonnier, J. (1900). Contribution à l'étude des Épicarides; les Bopyridae. *Trav. Stn. Zool. Wimereux* **6,** 1-395, pls. 1-41.

Bousfield, E. L. (1977). A new look at the systematics of gammaroidean amphipods of the world. *Crustaceana, Suppl.* **4,** 282-316.

Bousfield, E. L. (1979). A revised classification and phylogeny of amphipod crustaceans. *Trans. R. Soc. Can.* **(4)16,** 343-390.

Bowman, T. E., and Gruner, H.-E. (1973). The families and genera of Hyperiidea (Crustacea: Amphipoda). *Smithson. Contrib. Zool.* **46,** 1-64.

Boxshall, G. A. (1979). The planktonic copepods of the northeastern Atlantic Ocean: Harpacticoida, Siphonostomatoida and Mormonilloida. *Bull. Br. Mus. (Nat. Hist.), Zool.* **35,** 201-264.

Dahl, E. (1956). Some crustacean relationships. *In* "Bertil Hanström: Zoological Papers in Honour of his Sixty-fifth Birthday, November 20, 1956" (K. G. Wingstrand, ed.), pp. 138-147. Lund Zool. Inst., Lund, Sweden.

Dahl, E. (1977). The amphipod functional model and its bearing upon systematics and phylogeny. *Zool. Scr.* **6,** 221-228.

De Saint-Laurent, M. (1979). Vers une nouvelle classification des Crustacés Décapodes Reptantia. *Bull. Off. Natl. Pêches Tunisie,* **3**(1), 15-31.

De Saint-Laurent, M. (1980a). Sur la classification et la phylogénie des Crustacés Décapodes Brachyoures. I. Podotremata Guinot, 1977, et Eubrachyura sect. nov. *C.R. Hebd. Séances Acad. Sci., Ser. D* **290,** 1265-1268.

De Saint-Laurent, M. (1980b). Sur la classification et la phylogénie des Crustacés Décapodes Brachyoures. II. Heterotremata et Thoracotremata Guinot, 1977. *C.R. Hebd. Séances Acad. Sci., Ser. D* **290,** 1317-1320.

Glaessner, M. F. (1969). Decapoda. *In* "Treatise on Invertebrate Paleontology" (R. C. Moore, ed.), Part R, Arthropoda 4, Vol. II, pp. R400-R533. Geol. Soc. Am., Boulder, Colorado, and the Univ. of Kansas Press, Lawrence.

Guinot, D. (1977). Propositions pour une nouvelle classification des Crustacés Décapodes Brachyoures. *C. R. Hebd. Séances Acad. Sci., Ser. D* **285,** 1049-1052.

Gupta, A. D., ed. (1979). "Arthropod Phylogeny." Van Nostrand-Reinhold, Princeton, New Jersey.

Hartmann, G., and Puri, H. S. (1974). Summary of neontological and paleontological classification of Ostracoda. *Mitt. Hamburg Zool. Mus. Inst.* **70,** 7-73.

Hessler, R. R. (1969). Euphausiacea. *In* "Treatise on Invertebrate Paleontology" (R. C. Moore, ed.), Part R, Arthropoda 4, Vol. II, pp. R394-R398. Geol. Soc. Am., Boulder, Colorado, and Univ. of Kansas Press, Lawrence.

Hessler, R. R., and Newman, W. A. (1975). A trilobitomorph origin for Crustacea. *Fossils Strata* **4,** 437-459.

Hornibrook, N. de B. (1949). A new family of living Ostracoda with striking resemblances to some Palaeozoic Beyrichiidae. *Trans. R. Soc. N. Z.* **77,** 469-471, pls. 50-51.

Hornibrook, N. de B. (1963). The New Zealand ostracode family Punciidae. *Micropaleontology* **9,** 318-320.

Kabata, Z. (1979). "Parasitic Copepoda of British Fishes." Ray Society, London.

Karaman, G., and Barnard, J. L. (1979). Classificatory revisions in gammaridean Amphipoda (Crustacea). Part 1. *Proc. Biol. Soc. Wash.* **92(1),** 106-165.

Kornicker, L. S. (1975). Antarctic Ostracoda (Myodocopina). *Smithson. Contrib. Zool.* **163(1-2),** 1-720.

Kornicker, L. S., and Sohn, I. G. (1976). Phylogeny, ontogeny, and morphology of living and fossil Thaumatocypridacea (Myodocopa: Ostracoda). *Smithson. Contrib. Zool.* **219,** 1-124.

Kunze, J. C. (1981). The functional morphology of stomatopod Crustacea. *Philos. Trans. R. Soc. London, Ser. B.* **292,** 255-328.

Lang, K. (1948a). "Monographie der Harpacticiden." A-B. Nordiska Bokhandeln, Stockholm.

Lang, K. (1948b). Copepoda "Notodelphyoida" from the Swedish west-coast with an outline on the systematics of the copepods. *Ark. Zool.* **40A,** 1-36, pl. 1.

Lang, K. (1956). Neotanaidae, nov. fam., with some remarks on the phylogeny of the Tanaidacea. *Ark. Zool.* [2] **9**(21), 469-475.

McKenzie, K. G. (1977). Bonaducecytheridae, a new family of cytheracean Ostracoda, and its phylogenetic significance. *Proc. Biol. Soc. Wash.* **90,** 263-273.

Manton, S. M. (1977). "The Arthropoda: Habits, Functional Morphology and Evolution." Oxford Univ. Press (Clarendon), London and New York.

Moore, R. C., and McCormick, L. (1969). General features of Crustacea. *In* "Treatise on Invertebrate Paleontology" (R. C. Moore, ed.), Part R, Arthropoda 4, Vol. I, pp. R57-R120. Geol. Soc. Am., Boulder, Colorado, and Univ. of Kansas Press, Lawrence.

Rice, A. L. (1980). Crab zoeal morphology and its bearing on the classification of the Brachyura. *Trans. Zool. Soc. London* **35,** 271-424.

Riley, J., Banaja, A. A., and James, J. L. (1978). The phylogenetic relationships of the Pentastomida: The case for their inclusion within the Crustacea. *Int. J. Parasitol.* **8,** 245-254.

Schminke, H. K. (1978). Die phylogenetische Stellung der Stygocarididae (Crustacea, Syncarida)—unter besonderer Berüchsichtigung morphologischer Ähnlichkeiten mit Larvenformen der Eucarida. *Z. Zool. Syst. Evol.* **16,** 225-239.

Schram, F. R. (1969a). Some Middle Pennsylvanian Hoplocarida and their phylogenetic significance. *Fieldiana, Geol.* **12,** 235-289.

Schram, F. R. (1969b). Polyphyly in the Eumalacostraca? *Crustaceana* **16,** 243-250.

Schram, F. R. (1978). Arthropods: A convergent phenomenon. *Fieldiana, Geol.* **39,** 61-108.

Serban, E. (1970). "À propos du genre Bathynella Vejdovsky (Crustacea Syncarida). Livre du Centenaire Émile G. Racovitza 1868-1968," pp. 265-273. Éditions de l'Académie de la République Socialiste de Roumanie.

Sieg, J. (1980). Sind die Dikonophora eine Polyphyletische Gruppe? *Zool. Anz.* **205**(5/6), 401-416.

Stebbing, T. R. R. (1906). Amphipoda Gammaridea. *Tierreich* **21,** xxxix, 1-806.

Wagin, W. L. (1976). "Meshkogrydye Raki (Ascothoracida)." Kazan University, Kazan (in Russian).

Williamson, D. I. (1973). *Amphionides reynaudii* (H. Milne Edwards), representative of a proposed new order of eucaridean Malacostraca. *Crusataceana* **25,** 35-50.

Wilson, C. B. (1924). New North American parasitic copepods, new hosts, and notes on copepod nomenclature. *Proc. U.S. Nat. Mus.* **64**(2507), 1-22.

Wilson, G. D. (1980). Superfamilies of the Asellota (Isopoda) and the systematic position of *Stenetrium weddellense* (Schultz). *Crustaceana* **38,** 219-221.

Wingstrand, K. G. (1972). Comparative spermatology of a pentastomid, *Raillietiella hemidactyli*, and a branchiuran, *Argulus foliaceus*, with a discussion of pentastomid relationships. *Dansk. Vidensk. Selskab. Biol. Skrifter* **19,** 1-72.

2

Systematic Methods in Research

PATSY A. McLAUGHLIN, GEORGE T. TAYLOR,
AND MARTIN L. TRACEY

I. INTRODUCTION

Systematic methods in crustacean research encompass several levels, from basic species definitions to considerations of phylogenetic relationships among major taxa. The term systematics has its root in the latinized Greek word, *systema,* and as employed by early naturalists, such as Lin-

THE BIOLOGY OF CRUSTACEA, VOL. I

ISBN 0-12-106401-8

naeus (1758, 1767) and Fabricius (1775, 1792–1794, 1795), it referred to a system of classification. Modern interpretations have been broadened to include considerations of relationships and diversities among organisms (Simpson, 1961; Mayr, 1969). As befitted the philosphy of the times and the relatively limited scope of their collections, the specific descriptions of crustaceans given by Linnaeus, Fabricius, and their contemporaries consisted of brief Latin phrases describing one or more particular morphological traits of a type of specimen or series. Advances in optical equipment gave nineteenth and early twentieth century carcinologists the opportunity to examine morphological characters more critically; moreover, the numerous major oceanographic and exploratory expeditions of the period provided greatly expanded collections with which to work. Darwin's theory of evolution released systematists from the creationist or strictly typological interpretation of morphological characters and permitted the inclusion of some degree of variation in species definitions. Although descriptive morphology remains the cornerstone of crustacean systematic research, in recent years it has been complemented by several new approaches, e.g., statistical analyses, numerical taxonomy, cladistic analysis, electron microscopy, and biochemistry, particularly comparisons of electrophoretically separated proteins.

Crustaceans typically all have similar body plans, the notable exceptions being the Cirripedia and the various parasitic groups. However, this basic similarity does not necessarily make systematic techniques or morphologically significant characters the same from one major taxon to another. The size range of the study organisms or the particular diagnostic characters under consideration will determine the specific techniques needed. The significance imparted to structural differences, armature, and ornamentation vary greatly among the major taxa. In free-living Crustacea, obvious differences such as the absence or presence and form of the carapace, number of body somites and appendages, tagmata of the body, and the position(s) of the genital apertures delineate the principal subdivisions. At progressively subordinate levels, structural differences become much more subtle, and armature and ornamentation often may become the criteria upon which taxonomic decisions are based. Only a general overview of techniques and methodology can be presented in this chapter; however, references to more detailed accounts have been included.

II. COLLECTION

Freshwater and marine planktonic crustaceans typically are collected by means of a variety of qualitative and quantitative plankton nets, pumps, high-speed oceanographic plankton samplers, and midwater trawls (e.g., Welch, 1948; Wiborg, 1948; Banse, 1955, 1962; Aron, 1958, 1962;

Tranter, 1968). Gear selectivity and mesh size must be considered in choosing a specific sampling apparatus. Hansen and Andersen (1962) have reported that an 8 liter plankton bottle is effective in quantitatively sampling zooplankters, at least up to the size of the copepod *Calanus finmarchicus* (Gunnerus). Benthic and infaunal crustaceans usually are sampled by a wide range of trawls, dredges, sledges, grabs, cores, and suction pumps (e.g., Welch, 1948; Barnes, 1959; Gilat, 1963; Tait and de Santo, 1972). Dredge, core, grab, and pump samples all require subsequent sieving; also sledge and trawl samples collected by relatively fine mesh nets may be most effectively processed by sieving. A series of U.S. standard graded sieves, or larger specially built sieves of equivalent mesh sizes are the most often used. Mesh sizes of 0.5 mm and finer are required to retain small copepods, cephalocarids, and mystacocarids. A variety of push-nets have been found effective for collecting shallow-water crustaceans, particularly those living attached to vegetation (e.g., Strawn, 1954; Allen and Inglis, 1968; Manning, 1975). Freshwater crustaceans, particularly decapods, are most effectively collected by the use of a fine-meshed dipnet, seine, or pull-trap (Higer and Kolipinski, 1967; Hobbs, 1975). Burrowing forms, such as crayfishes and callianassids, require special techniques. Hobbs (1975) recommends hand collecting for the most part, especially when collecting species that rarely, if ever, leave their burrows. Haistone and Stephenson (1961) and Manning (1975) have found a yabby pump highly successful for removing callianassids and upogebids from their burrows, and Hobbs (1975) has reported success with this pump for extracting crayfish species that construct vertical burrows with only one or two openings. Although rotenone and most similar fish poisons have been reported generally to have no effect on crustaceans, in shallow-water habitats, particularly in the tropics, excessive use of these toxins can have devastating effects on entire communities, including crustaceans (P. A. McLaughlin, personal observations). Manning (1960), while recommending caution in the use of fish poisons, has reported that the commercial fish toxin "Pro-noxfish" dispensed from a polyethylene squeeze bottle into restricted areas can be used effectively to incapacitate decapods and stomatopods without adverse effects to the overall watermass. Undoubtedly one of the most difficult groups of crustaceans to sample adequately is the meiofaunal group. Hulings and Gray (1971) have thoroughly reviewed the numerous collecting devices that have been developed for sampling this particular habitat, as well as the specialized techniques that are required for meiofaunal separations.

A. Narcotization and Preservation

Standard narcotizing agents such as magnesium chloride, magnesium sulfate, menthol, and chloral hydrate have little or no effect on most crustaceans. A few drops of clove oil in a finger bowl of milieu water have been found to successfully narcotize phyllopods, copepods, cladocerans, some amphipods, and decapods, but have no visible effect on isopods. Chloretone (chlorobutanol) also has been found to narcotize phyllopods, and occasionally other small crustaceans (J. E. Lynch, personal communication; Knudsen, 1966). Dilute (1–2%), unbuffered, seawater formalin will narcotize some reef-dwelling amphipods (J. Thomas, personal communication). Provenzano and Rice (1964) have narcotized pagurid larvae with propylene phenoxytol. Often, the best narcotizing effect can be achieved by allowing the milieu water to come to ambient temperature (particularly for subtidal crustaceans). Once the animals become unresponsive, they must be preserved as they will die quickly and begin to disintegrate. Specimens narcotized with clove oil should be changed from the original preservative after a day or two to remove the clove oil, which tends to adhere to body surfaces.

Formaldehyde is still the best general fixative for most crustaceans, but ethyl alcohol (95%) is preferable for cirripeds, except acrothoracicans. For this group, Tomlinson (1969) has recommended alcohol (70%) with hydrochloric acid (2%) added, Bouin's solution, or formalin (10%). The terms formalin, formol, and Formal commonly are used to denote diluted formaldehyde, and the dilution percentage is based on 100% formaldehyde; e.g., 90 parts milieu water and 10 parts commercial formaldehyde would be stated as 10% formalin. However, commercial formaldehyde actually contains 36–50% by weight dissolved formaldehyde (CH_2O); the standard U.S.P. formaldehyde solution is 37% (Walker, 1964). Because confusion has arisen concerning the computation of the percentage of concentrated formaldehyde required for a "10%" diluted solution, the SCOR (Scientific Committee for Oceanic Research) Working Group 23 has recommended that a 4% solution contain 10 ml of commercial formaldehyde and 90 ml of milieu water. It has further recommended that the term formaldehyde be adopted for all concentrations and that the terms formalin, formol, and Formal be abandoned entirely (Steedman, 1976). Formaldehyde concentrations referred to in this chapter will follow the SCOR Group's recommendation. A disadvantage to formaldehyde as a preservative is its tendency to become acidic. For short-term (weeks) storage of marine crustaceans, seawater provides an adequate buffer; for long-term buffering, chemical buffers are required. Among those most frequently recommended are sodium tetraborate (Borax), basic magnesium carbonate, calcium carbonate, hexamethylenetetramine (hexamine), and sodium-β-glycerophosphate; however, hexamethylenetetramine has been shown to be unsatisfactory for the preservation of marine plankton (McGowan, 1967; Fleminger, 1968).

Steedman (1976) reviews an array of formaldehyde buffers and points out the advantages and disadvantages of each. Not considered by Steedman, but recommended by Mahoney (1966) as a buffer for dilute formaldehyde, is a combination of sodium dihydrogen phosphate (anhydrous) and disodium hydrogen phosphate (anhydrous). Borax is probably the most often used buffer for generalized preservation. For preservation of most crustaceans, supersaturation of concentrated formaldehyde with 2% Borax will provide sufficient buffering to provide the diluted preservative with an adequate pH level (6.5-7.5) for short periods of time. If samples are to be stored in formaldehyde over long periods (years), periodic checks of pH must be made every 3-6 months. If the pH of the preservative is found to be falling, 0.2 g of Borax should be added for every 100 ml of preservative (Steedman, 1976). Actually it is far more satisfactory that long-term collections be transferred to 70% ethyl alcohol (ethanol). And Hobbs (1975) has reported that adequate fixation, as well as long-term preservation, of freshwater isopods, amphipods, and mysids is obtained by placing specimens directly into 70-80% ethyl alcohol. Isopropyl alcohol also is used as a preservative, but decapods may become very brittle with long-term storage. In recent years, several new preservatives have been suggested, e.g., polyethylene glycol (Fraser, 1961), ethylene glycol (Williamson and Russell, 1965), Ionol (bulated hydroxytoluene) (Waller and Eschmeyer, 1965), Phenoxetol (ethyleneglycol-monophenylester) (Mahoney, 1966), and proplyene phenoxetol (cf. Steedman, 1976). Although Provenzano (1968, 1971) has reported good results from the storage of pagurid larvae in ethylene glycol, laboratory experiences have shown that long-term storage of adult pagurids in solutions of Ionol, formaldehyde plus Ionol, ethylene glycol, or alcohol plus ethylene glycol is very unsatisfactory. With time, Ionol produces a crystalline precipitate; ethylene glycol produces a "cotton-like" precipitate that adhers to body surfaces, spines, and setae. Tests on individual crustacean groups should be made before any unfamiliar preservative is put into extensive use. A number of more specialized fixatives and preservatives are available if the specimens are to be used in histological studies (e.g., Galigher and Kozloff, 1964; Steedman, 1976; electron microscopy section, this chapter).

B. Specimen Preparation

With the exception of certain parasites that require histological examination for identification, taxonomic evaluations and/or identifications typically are based on external morphological characters. Examination of these characters often requires that the animal be cleared, stained, dissected, and mounted. Standard clearing and staining techniques using toluene, xylene, or terpineoltoluene are time-consuming, tend to make the specimens more brittle, and frequently are difficult to adapt to many crustaceans because of the dehydration requirement. Some carcinologists recommend clearing and mounting specimens in glycerin; others rely on lactic acid, lactophenol, or dehydration with glacial acetic acid followed by wintergreen oil with mounting in balsam. Provenzano (1967) has used a 3-5% solution of hot potassium hydroxide for clearing decapod larvae. Steedman (1966) suggests the use of dimethyl hydantoin formaldehyde dissolved in ethanol or water for clearing and mounting virtually all arthropods. McHardy (1966), Owre and Foyo (1967), Fryer (1968), Henry and McLaughlin (1975), and McLaughlin (1976, 1980), among others, have used polyvinyl (alcohol) lactophenol for a variety of crustacean groups. The major advantage to the use of dimethyl hydantoin formaldehyde or polyvinyl lactophenol (the authors prefer the latter) is that dissections and mounting can be made on a single slide without problems of rapid hardening of the mounting medium or the need for transferring dissected parts to a second slide and medium; however, as Hulings and Gray (1971) have pointed out, the latter can be used only for the appendages of ostracods as it causes decalcification of the valves. Dissection in lactic acid with mounting in Hoyer's solution (e.g., Provenzano, 1962a,b) or Turtox CMC-10 or CMC-5 (Knudsen, 1966; Gore, 1970; Yang, 1971) appears to work well with decapod larvae and other microcrustaceans. Stock (1976) has found mounting in Reyne's modification of Faure's successful for thermobaenaceans.

The stains most commonly used in mounts of crustacean appendages or whole specimens are lignin pink, Mallory's acid fuchsin red, fast green, and chorazol black. Hall and Hessler (1971), Hulings and Gray (1971), and Ritchie (1975) use methyl blue in lactic acid for staining mystacocarids and copepods; Monod and Cals (1970) and McLaughlin (1976) have used heated chorazol black, B and E, respectively, in lactophenol to stain planktonic decapod larvae and cephalocarids. Perhaps the major drawback to mounting either in dimethyl hydantoin formaldehyde or in polyvinyl lactophenol is that stains ultimately are leached from the specimens. Consequently, when using either of these media, it is advisable always to ring small structures with ink on the lower surface of the slide for ease in locating them later.

An alternative method that might be considered by those who make sections of paraffin-embedded specimens are the plastic embedding (Araldite or Epon or one of the combinations) and glass-knife sectioning techniques of the transmission electron microscopists. With this method it is possible to produce 0.5-2 nm sections with few of the problems inherent to paraffin-embedded material, and generally to obtain better resolution with the light microscope. However, serial sections are difficult to produce, and the staining methods for plastic-embedded material are limited (Lewis and Knight, 1977). The recently introduced DuPont/Sorvall JB-4 microtome, which uses a larger glass knife than the standard ultramicrotome and permits the sectioning of larger blocks, takes advantage of the dramatically superior embedding that is possible with plastic media and allows even a relatively unskilled operator to produce high quality sections. Since the plastic-embedding media used with the JB-4 are permeable to aqueous stains, a much wider range of staining procedures can be used.

III. DESCRIPTIVE TAXONOMY

In major taxa such as birds and mammals, most of the alpha level taxonomic problems have been resolved. In the majority of crustacean classes, hosts of new species remain to be described. Despite the debates on species concepts (e.g., Mayr, 1969, 1976; Sneath and Sokal, 1973; Ghiselin, 1974; Hull, 1976; Slobodchikoff, 1976; White, 1978; Wiley, 1978, 1980), to the practicing carcinologist the species is still defined primarily by its morphological characters. As the number of recognized species has increased, so has the need for more definitive descriptions. No longer will a few characters suffice for the description of a new species, nor are the characters that have routinely been used in earlier species descriptions of a particular group necessarily the most diagnostic. Among early carcinologists, frequently characters were selected simply because they were the most obvious or were the easiest to examine. But often these are the characters most influenced by environmental factors, sexual dimorphism, and allometry, or altered by preservation. Modern carcinologists usually rely not only on gross morphological structures, but also on mouthparts, sexual specializations, and color patterns as diagnostic characters.

Although the typological concept of Linnaeus has been appreciably broadened by the recognition that wide ranges of intraspecific variation in characters do occur, the taxonomist's species ultimately is defined by its type specimen (holotype, lectotype, neotype). However, type specimens may be inaccessible to other investigators because of deposition in obscure repositories, or as frequently has happened in the past, types are lost or destroyed. Thus it is imperative that the taxonomist provide accurate and complete descriptions of new taxa, including known ranges of intraspecific variation in characters. Descriptive morphology is hampered, however, by the subjectivity of descriptive terms. For example, there is no universal definition of a spine, a tooth, or a tubercle. Similarly, the meaning of a descriptive adjective may be lost totally in its subtlety or in its translation from another language. Expressions of size, length, and breadth without actual measurements are relative and may not impart to the reader the impressions intended by the describer. What might be considered a short rostrum, for example, in a taxon in which rostra usually are long, may be a long rostrum to the observer who has never seen a truly long rostrum.

The successful use of descriptive morphology depends on several things. The first is the taxonomist's knowledge of the morphology of the organisms under study and the inherent variation that can be expected in particular structures. The second is the careful selection of conservative diagnostic characters. The third is the definition of descriptive terms, and as much as possible the avoidance of relative terms. A meaningful species description then must be based on a detailed description of the animal's morphology, with particular emphasis on conservative diagnostic characters. Descriptions of behavior patterns, habitat, bathymetric or geographical ranges, and feeding habits, when known, also are of considerable importance. And lastly, each original description must be independent. A species description that is based, for meaning, on a comparison with another species, is virtually meaningless unless the second species also is available.

Often almost as important to the correct interpretation of a species as its actual description are illustrations of the animal and its diagnostic characters. Care must be taken to ensure that such illustrations are scientifically accurate rather than artistically pleasing. The use of a camera lucida can provide even the less artistically inclined carcinologist with the means for producing good line drawings. With many crustaceans, photographs also have provided an excellent method of depicting diagnostic characters (e.g., Owre and Foyo, 1967; Newman and Ross, 1971; Crane, 1975; Henry and McLaughlin, 1975; McLaughlin, 1981).

The step by step procedures for presenting species descriptions, diagnoses, and synonymies for publication have been presented most recently by Blackwelder (1967) and Mayr (1969). Equally important to the carcinologist is a working familiarity with the International Code of Zoological Nomenclature (1964 edition). It is the Code, together with subsequent Opinions and Declarations of the International Commission (cf. Bulletin of Zoological Nomenclature), that provides the framework within which systematists must work when dealing with taxonomic matters at the species, genus, and family levels. Any carcinologist who has dealt with complex taxonomic problems is well aware of the havoc that can result when uninformed taxonomists ignore the nomenclatorial rules outlined in the Code.

IV. CLASSIFICATION AND PHYLOGENY

Above the species level, the systematist is faced with the determination of relationships, i.e., the grouping or clustering of related species in genera, and the arrangement of genera into higher taxa. These higher taxa are then ranked in a taxonomic hierarchy. Once again, the basis for determining relationships usually is morphological similarity. How that similarity is defined and determined has been the source of many heated debates and has led to the fragmentation of the systematic community into three philosophical "schools" (cf. Simpson, 1961; Sokal and Sneath, 1963; Hennig, 1966; Mayr, 1969). To proponents of the phenetic school, often also referred to as the school of numerical taxonomy, overall similarity between any two entities is a function of their similarities in each of the many characters in which they are being compared. Classification is based on a matrix of these resemblances (Sneath and Sokal, 1973). Proponents of the phylogenetic school, or cladists, are concerned primarily with genealogical relationships. Implicit in developing a phylogeny for a group is the concept that two (or more) taxa which share and immediate common ancestor are more closely related to each other than either is to another group. The most important kind of intrinsic data for indicating phylogenetic relationships is the shared derived character (Hecht and Edwards, 1977). The classical or evolutionary school bases its evaluation of relationship on the degree of genetical similarity between organisms as judged by the degree of phenotypical similarity. Greater phenotypical similarity implies greater genetical similarity and hence closer relationships (Bock, 1973). Although Ghiselin and Jaffe (1973) have argued that Darwin's (1851, 1854) classification of the Cirripedia was based, at least in part, on genealogical considerations, most crustacean classifications of the ninteenth and early twentieth centuries were based exclusively on similarities in morphological characters, the significance of which were often intuitively determined.

A major criticism of the classical approach to classification has been its perceived lack of objectivity. Many systematists have attempted to overcome this weakness by quantification of their data and application of mathematical analyses. Most notable in their efforts have been the proponents of the school of numerical taxonomy. They have eliminated intuitive character weighting by giving all characters *a priori* equal weight. For each character, two or more character states are defined and assigned numerical values. Estimates of resemblance between operational taxonomic units (OTU's) can then be computed through a variety of mathematical procedures. Although widely adopted by systematists in a number of vertebrate and invertebrate groups, numerical taxonomic methodology has had relatively limited application in studies of crustaceans (e.g., Kaesler, 1966; 1967, 1969, 1970; Moulton, 1973). However, a number of statistical techniques, not all necessarily encompassing aspects of numerical taxonomy as defined by Sneath and Sokal (1973), have been successfully applied to crustacean systematic studies. Techniques such as factor analysis, canonical variate analysis, discriminant function, and distance function analyses are those most frequently employed.

Multiple factor analysis examines a complex set of phenomena stated in terms of correlations among the variables under consideration and attempts to express these phenomena as functions of a small number of new variables (factors) (Sokal and Sneath, 1963; Seal, 1964). Factor analyses based on matrices of correlation coefficients have been used by Matthews (1972) and Lawson (1973) to evaluate species within two families of copepods. In the latter study the author used 24 characters obtained from the right swimming appendage of adult females from species of the carnivorous family Candaciidae. Using Q-mode factor analysis, which measures the relationship between taxonomic units in terms of Euclidean distance in multidimensional space, Lawson has evaluated the interspecific relationships among the species and has compared the results with a phenogram based on the weighted variable group method of cluster analysis (Sokal and Sneath, 1963). Lawson has found considerable support for his proposed clusters from previous interpretations made by copepidologists using more classical approaches. Moulton (1973) has used principal component (R-mode) factor analysis, as well as Q-mode analysis, in analyzing variation in another copepod species. The former type of factor analysis, which describes the interrelationships among taxonomic units in terms of an arbitrary orthogonal or unrelated system, has the advantage of relating trends in variability to the actual characters that cause them. A mathematical account of principal component analysis has been given by Lawley and Maxwell (1963) and Seal (1964).

Another multivariate technique that is employed more frequently is the discriminant function, mathematically derived by Fisher (1936) and Rao (1952), the biological applications of which Healy (1965) has reviewed. This function, linear in observation, has the property of discrimination between any normally distributed pairs of classes. It provides a measure or index of generalized distance (D^2), or difference, between any pair. Not only may the discriminant function be used to measure relationships between pairs of taxa, but it also can be used to classify individuals of unknown identity belonging to one member of a pair of taxa. Saila and Flowers (1969) have used discriminant function analysis successfully both to distinguish between populations of the American lobster *Homarus americanus* H. Milne Edwards and to classify specimens of unknown origin into their respective populations.

Canonical variate analysis most typically is used as a statistical tool in ecology rather than systematics, but it can be used in conjunction with the discriminant function. For many species, canonical variate analysis allows the comparison of the means of species abundances (variates) by a transformation which reduces numerous original variates to a few canonical variates. This technique cannot be used to define groups or taxa, but rather is a method of statistical comparison of already defined groups. A conceptual and computational outline of canonical analysis is give by Buzas (1967) and a detailed account by Seal (1964). Barnes and Healy (1965, 1969, 1971) have combined regression and canonical variate analyses with discriminant function analysis to evaluate morphological changes in several species of *Balanus* over a broad geographical range and have found clinal changes in these barnacles that correlate with environmental factors.

Henry and McLaughlin (1975) have used a combination of the discriminant function and the generalized distance function to analyze intra- and interspecific relationships of 16 taxa of closely related barnacles. Whereas, the discriminant function is designed to separate pairs of taxa, the generalized distance function establishes constellations of groups or taxa based on similarities of group characters. This resemblance also is expressed in terms of Euclidian geometry as the square of the distance (D^2) between groups in multidimensional space. Rao (1952) has outlined the mathematical computations of the generalized distance function. Computed D^2 values and percents of overlap between pairs of taxa reveal the general pattern of relationships. When the distance between certain pairs of taxa is much less than the distance between either member of the pair and the next closest taxon, the pairs are termed clusters.

There are no formal rules for clustering taxa determined through the use of the general distance function. Henry and McLaughlin (1975), following Rao's (1952) suggestion, utilized a method that began with a pair of closely related groups, to which a third with the smallest average D^2 from the first two was added (Table I). This procedure was followed until the average D^2 obtained by adding a group was particularly high. The latter group was therefore considered to lie outside the cluster. This method was repeated until all groups had been clustered. Constellations, representing intercluster relationships, could then be developed in terms of the square root of the average square of the distance (D^2), e.g., Figs. 1 and 2.

Phenetic methods of clustering generally are based on similarity coefficients. Sokal and Sneath (1963) describe several techniques for clustering when numerical taxonomic methods of classification are employed. Clusters are based on bi- or multimodality in the distribution of association, correlation, or distance (Δ) coefficients. In its simplest form this type of clustering consists of selecting, arbitrarily, a level on a scale of similarity coefficients. This type of clustering is not usually satisfactory, because the selection of a

TABLE I

Computational Procedures for Determining Clusters[a]

Taxon added to cluster	Sum of corrected $D''(n)$	terms	Increase in D'' / increase in n	Average corrected D'' ($\Sigma D''/n$)	Resulting cluster
Balanus variegatus–*B. reticulatus*	4.06	1	—	4.06	*variegatus*–*reticulatus*
B. kondakovi	29.00	3	8.75	7.19	*kondakovi*
B. peruvianus	65.25	6	14.56	10.88	
B. c. pacificus–*B. c. mexicanus*	3.50	1	—	3.50	*c. pacificus*–*c. mexicanus*
B. variegatus	29.00	3	12.75	9.66	*variegatus*[b]
B. kondakovi	70.36	6	13.79	11.73	
B. venustus–*B. inexpectatus*	8.31	1	—	8.31	*venustus*–*inexpectatus*
B. peruvianus	34.41	3	13.05	11.47	
B. peruvianus–*B. inexpectatus*	12.08	1	—	12.08	*peruvianus*
B. suturaltus–*B. c. pacificus*	12.30	1	—	12.30	*suturaltus*
B. citerosum–*B. suturaltus*	13.91	1	—	13.91	*citerosum*
B. dentivarians–*B. suturaltus*	87.06	1	—	87.06	*dentivarians*
B. a. amphitrite–*B. a. saltonensis*	9.73	1	—	9.73	*a. amphitrite*–*a. saltonensis*
B. subalbidus	49.24	3	19.76	16.41	
B. subalbidus–*B. a. amphitrite*	16.49	1	—	16.49	*subalbidus*
B. eburneus–*B. subalbidus*	24.04	1	—	24.04	*eburneus*
B. improvisus–*B. eburneus*	28.15	1	—	28.15	*improvisus*

[a] After Henry and McLaughlin, 1975.
[b] Not included in second cluster.

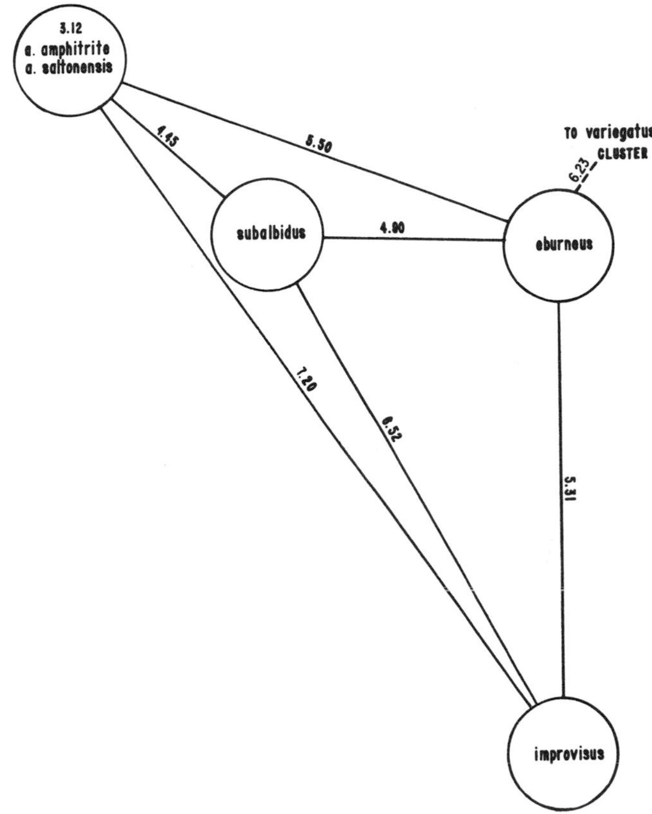

Fig. 1. Constellation representing intercluster relationships of species of the *Balanus amphitrite* Darwin group of barnacles. Relationships are expressed in terms of the square root of the average square of the distance (D^2). (From Henry and McLaughlin, 1975.)

very high similarity coefficient would result in very few clusters, yet the selection of a low coefficient would result in considerable overlap of clusters.

More complex clustering techniques do not group units or taxa related above an arbitrarily fixed level of similarity coefficient. The single linkage technique calls first for clustering those units most closely related, i.e., with the highest similarity coefficients, and then successively lowering the level of admitting to the cluster by steps of equal magnitude. The technique of average linkage clustering bases admission to a cluster on the average of the similarities of the unit with members of the cluster. With this method, the size of the clusters at any level is likely to vary and the number of units joining a new cluster also may vary. If no units join a new cluster in any one

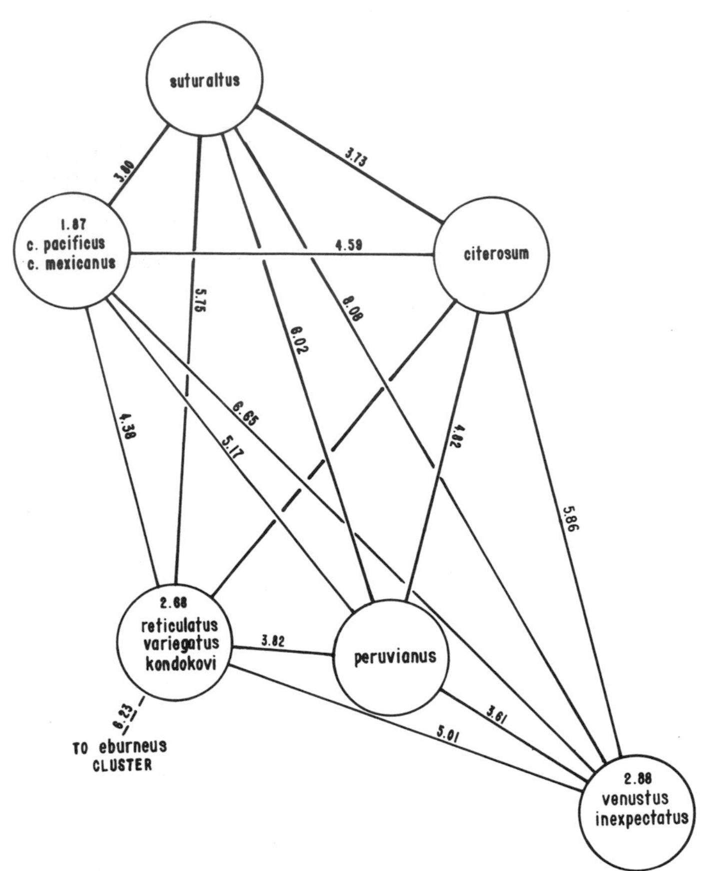

Fig. 2. Constellation representing intercluster relationships of species of the *Balanus amphitrite* complex of barnacles. Relationships are expressed in terms of the square root of the average square of the distance (D^2). (From Henry and McLaughlin, 1975.)

computational cycle, it implies that all prospective members have higher relationships to other clusters than to the one under consideration. Clustering of these types usually are pictorially represented in phenograms. The method of clustering used is dependent on the comparison matrix utilized by the investigator. Sneath and Sokal (1973) and McNeill (1979) have described various techniques for clustering. When expanded, under certain assumptions, similarity clustered phenograms presumably give reasonable estimations of evolutionary trees (Colless, 1970; Goodman and Moore, 1971).

From the phylogenetic viewpoint, related species are clustered into lineages on the basis of shared derived (synapomorphic) characters (Cracraft, 1974); a classification includes in each taxon all descendants, but only the decendants, of a presumed ancestral species of that taxon (Boudreaux, 1979). Means by which primitive (plesiomorphic) and derived (apormorphic) characters are determined have been reviewed by Nelson (1969, 1972, 1973), Schaeffer *et al.* (1972), Hecht (1976), and Hecht and Edwards (1976, 1977), and a comprehensive overview of cladistic theory has been presented by Bonde (1977). Cladistic analyses frequently are presented in the forms of cladograms or phylogenetic trees (cf. Hennig, 1966; Kluge and Farris, 1969; Bonde, 1977). A cladogram is a graphic presentation of interrelationships among taxa, wherein a taxon may be represented either as a tip of a branch (terminal taxon), or as a branch point (inclusive taxon). Two types of cladograms are recognized: (1) a fundamental cladogram in which data regarding the interrelationships of terminal taxa are analyzed, and (2) a derivative cladogram in which the cladistic aspects of hierarchical classification are specified (Nelson, 1979). As originally proposed, cladistic analyses were nonquantitative and, as they have been applied to carcinological classifications, have remained so (e.g., Bousfield, 1977; Watling, 1979). However, a number of methods have been proposed for quantitative character weighting of phylogenetic data (e.g., Cavalli-Sforza and Edwards, 1967; Kluge and Farris, 1969; Farris, 1970, 1972, 1977, 1979; Estabrook, 1972; Hecht, 1976; Hecht and Edwards, 1976). Kluge and Farris' (1969) and Farris' (1970) phylogenetic tree (Wagner tree) clustering, although proposed originally for attaining a quantitative phyletic taxonomy, was modified by Farris (1972) to reflect electrophoretic and immunological distance data.

The construction of evolutionary trees is based on Wagner's (1961) method for producing a most parsimonious tree. Farris' Wagner tree is a network of minimum length in the space in which "length of the tree" is defined as the sum of the lengths of all the intervals of the tree, where an interval is the connection between an OTU and its most recent ancestor, and the length of the interval is the difference between the OTU and its most recent ancestor. It is assumed throughout that the character states of the OTU's and the differences computed from them are weighted values; however, the choice of weighting coefficients can usually affect the parsimoniousness of the tree. The tree of minimum length is defined as the most parsimonious.

V. PRACTICAL APPLICATIONS

With the modern-day emphasis on ecological and environmental assessment studies, the systematist provides critical assistance to members of other biological disciplines through the publication of monographic studies of regional crustacean faunas (e.g., Hobbs, 1942; Williams, 1965; Menzies and Glynn, 1968; Manning, 1969; Chace and Hobbs, 1969; Barnard, 1970, 1972, 1974; Chace, 1972; McLaughlin, 1974; Powers, 1977). Perhaps the most useful aids to the nonspecialist have been the illustrated keys to major crustacean groups that have appeared in recent years (e.g., Barnard, 1971; Bousfield, 1973; Felder, 1973; Williams, 1974; Hobbs, 1975). Keys, of course, cannot incorporate full ranges of morphological characters and, therefore, must be used with caution. However, they do provide invaluable references for the identifications of unfamiliar crustaceans. And when keys are used in conjunction with well-developed species descriptions, accurate identifications usually can be made.

Another very useful tool that has been provided by systematists for the benefit of the nonspecialist is the species identification sheet. Developed for fishery purposes under the auspices of the Food and Agriculture Organization of the United Nations, these identification sheets provide illustrations of the species, diagnostic characters, and distinguishing characters of similar species, as well as data on geographical distributions, color patterns, and known fishing groups (e.g., Pérez-Farfante, 1978; Manning, 1978; Williams, 1978).

VI. ELECTRON MICROSCOPY

It has now been more than 25 years since biologists were introduced to the use of the transmission electron microscope (TEM) (Reid, 1974) and to satisfactory fixatives for tissues (Palade, 1952) and the design of microtomes for sectioning (Porter and Blum, 1953). These developments and later modifications have been pursued avidly, much to the benefit of histologists and cell biologists. The scanning electron microscope (SEM), which developed almost concurrently with the TEM, but at a slower pace because of early technical problems, was not available to biologists until much later; the first commercial model was introduced in 1965. Progress has been rapid and today the biologist is presented with a multitude of preparative techniques for use in scanning electron microscopy (Hayat, 1978).

Notwithstanding the fact that the TEM is used primarily for the visualization of thin sections of tissues and the SEM is used primarily for the visualization of the surfaces of cells and whole structures, the techniques of preparation are not dramatically different. Indeed, the criteria for good preservation, which apply to TEM, also apply to SEM. Simply stated, the appearance of material examined in the TEM or SEM should be as close to that of the living specimen as is possible. Shrinkage or swelling should be minimized, structural damage should be absent, and loss of material should not be apparent.

To the systematist, the chief advantages of SEM are greater resolution and depth of field as compared with that available with conventional light microscopy. In addition, SEM permits the greater magnification needed for the visualization of minute surface structures. The SEM should prove to be especially valuable in the examination of larvae, small specimens, and isolated small structures of larger crustaceans. To date, TEM has not been employed in crustacean systematics; however, the potential for its use in the study of parasitic copepods and rhizocephalans, where sections are required for identification, should be explored.

VII. SCANNING ELECTRON MICROSCOPY

In the preparation of a specimen for SEM there are six fundamental steps: (1) fixation, (2) cleaning, (3) dehydration, (4) drying, (5) mounting (on a metal stub), and (6) coating with an electrical conductor. Each of these steps should be carried out with care, since distortion or damage of the specimen is possible in each step. Ideally, cleaning of the specimen should be done prior to fixation, but since this is rarely possible except under the best of conditions, our discussion will concentrate on the cleaning of fixed material.

A. Fixation

The principles and methods of fixation have been reviewed extensively by Hayat (1978), although his discussion is directed to the preparation of soft tissues. For whole animals with firm exoskeletons, successful fixation has not presented as many problems as it has for soft tissues. Specimens collected, killed, and fixed in the field in formaldehyde diluted either with seawater or freshwater (from the appropriate habitat) have produced adequate SEM pictures. However, we suspect that some level of distortion results from this treatment, particularly of small crustaceans. Shrinkage of the carapace and compression of the thoracic region are the most apparent forms of distortion produced by this type of fixation. We recently had the occasion to examine reef amphipods that had been fixed in 2% seawater formaldehyde, and this fixation appeared to give satisfactory results. If more specialized fixatives are not available, or collecting conditions will not permit specialized techniques, fixation in 2–4% buffered formaldehyde is probably the best compromise. However, if this fixation technique is used, specimens should be washed with several changes of distilled water or with the buffer solution used with the formaldehyde before the SEM processing is continued.

For critical work, the method of choice would be a modification of the paraformaldehyde-glutaraldehyde fixative of Karnovsky (1965). As it is now generally used, the concentration of paraformaldehyde is reduced to 2% and that of glutaraldehyde to 3% from the original formulation. In fixing marine Crustacea, the vehicle for preparation of paraformaldehyde solution and the cacodylate buffer should be filtered seawater.

In our laboratory, the Karnovsky fixative routinely is prepared by adding 2 g of paraformaldehyde to 50 ml of seawater and heating to 60°C with constant stirring, followed by the addition of three to six drops of 1 *N* sodium hydroxide when the temperature reaches 60°C (there is usually a small residue of undissolved paraformaldehyde, but this will eventually go into solution). Following cooling to ambient temperature, 6 ml of 50% glutaraldehyde (biological grade) is added and the resulting solution is diluted to 100 ml with 0.2 *M* sodium cacodylate (or Sorenson's phosphate buffer) in seawater. If stored at 5°C, this solution should last for several weeks, but it is preferable that it be made up just before use. At ambient temperature, most small animals should be fixed satisfactorily within 2 hr. Caution should be used when fixing animals with a body thickness greater than 3-5 mm, since this fixative penetrates somewhat slowly. In preparing larger animals it is advisable to fix for longer than 2 hr, and, since for SEM observation it is the surface structures which are of interest, 24-72 hr in the fixative should do no harm. Following fixation, the specimens should be washed in several changes of 0.1 *M* buffer solution to remove excess fixative.

B. Cleaning

One of the most persistent problems is that of contamination by adherent particles and mucous, and no cleaning method has proven to be consistently effective. Various methods, such as washing fixed animals in dilute KOH or HCl and the use of ultrasonic waves (Pulsifier, 1975), have produced inconsistent results. Such methods also introduce the possibility of structural damage to the specimen. If adherent mucous or gelatinous substances are a problem, it may be necessary to clean the animals prior to fixation by repeated washings with filtered seawater (for marine animals) or distilled water (all others). An alternative that apparently has not been tried with crustaceans but has proven successful in the preparation of acanthocephalans for SEM (D. Miller, personal communication) might be used. In this method, the living animals are placed on a millipore filter and washed rapidly with short jets of increasing concentrations (10%, 20%, 30%, etc.) of sodium chloride solution.

A novel and rather effective cleaning method for the removal of adherent particulate matter is to gently brush the fixed specimen with a fine sable brush, and to follow with sonication in several changes of distilled water or buffer (J. Thomas, personal communication). This method seems to be successful in removing most particulate matter.

C. Dehydration

In most instances, dehydration through a graded series of concentrations of ethanol (25%, 50%, 60%, etc.) finishing with several changes in 100% is the best procedure. The usual sequence is 10-15 min in each concentration, with at least three changes of 100% ethanol. Specimens thicker than 3 mm should receive 15 to 20 min or longer in each concentration. Acetone may be substituted for ethanol with equally good results.

In some laboratories an intermediate fluid, such as amyl acetate, is used prior to critical point drying, although this is not absolutely essential (Hayat, 1978). For the preparation of whole animals, this step does not seem necessary unless it is dictated by the drying method to be used.

D. Drying

Following dehydration the specimen must be dried. The easiest and most rapid method is air drying, but, for biological material, this is the least desirable. The forces of surface tension associated with air drying usually result in distortions, such as flattening, shrinkage, surface wrinkling, and contraction. For all biological material the methods of choice are critical point drying and freeze drying. The latter method has a number of disadvantages, chief of which is the damage introduced by the formation of ice crystals. However, the critical point method is also not without hazards, although the proliferation of commercially available critical point drying instruments, and the availability of one model or another in most SEM laboratories, makes this the easiest method to use. The techniques of critical point drying and the ways of handling specimens are treated extensively by Hayat (1978) and are beyond the scope of this brief discussion. Much of the technique required will be dictated by the particular commercial model of dryer available to the investigator.

E. Mounting

Immediately after drying, the specimen should be mounted on an aluminum SEM stub. This can be done with a number of types of adhesives, such as conductive silver paint and Duco cement. In our laboratory we routinely use a conductive silver paint and find this to be very satisfactory. If Duco cement is used, it is essential to ensure electrical contact between the specimen and the stub by making a thin connection of silver paint between the two. Use of the paint as the mounting medium makes this extra step unnecessary.

One should be cautioned that critical point-dried specimens are extremely fragile and must be handled with great care during mounting. After the specimen is firmly attached to the stub, it is possible to take advantage of this fragility in dissecting away appendages, mouthparts, etc., with fine steel needles. Dissections are limited only by the size of the specimen and the skill of the investigator.

F. Coating

All biological specimens should be coated with a heavy metal to prevent a build-up of electric charge in the electron beam of the SEM. Again, there exist a variety of methods and instruments for coating, and the investigator may be limited by the equipment available for this purpose. In our experience, a coating of gold-palladium alloy produced by a sputter-coating instrument has proven quite adequate.

Once coated, specimens should be protected from dust contamination and rehydration until they can be observed in the SEM. A simple vacuum dessicator is adequate for this purpose.

To date, SEM has had relatively limited use in crustacean systematics since the store of comparative data is still small. However, it has been shown to be very useful in critical comparisons of diagnostic characters such as shell surfaces and structures in ostracods (e.g., Kornicker, 1975), cirriped mouthparts (e.g., Yamaguchi, 1973, 1977), and decapod gonopods (e.g., Abele, 1971).

VIII. ELECTROPHORESIS

Electrophoretic techniques may be used to separate proteins according to their relative mobilities in an electric field, where mobility is determined by molecular weight, conformation, and net charge. Since polypeptides are the most easily identified molecular products of genes, and since they reflect mutational changes in the DNA more or less directly, comparisons of proteins among different animals may be assumed to reflect genetic differences. To date, electrophoretic techniques have been employed by very few carcinologists working in systematics. Indeed, the techniques of molecular biology have not been widely employed by systematists working on any group of plants or animals. Electrophoresis does, however, offer several theoretical advantages in generating data bases on which systematic judgements might be made: (1) It makes possible the comparison of as many as 50 protein characters, all specified by different genes, among individual specimens. Once the appropriate assays have been determined for a group of species, scoring 40 or 50 protein characters is far less time-consuming than scoring an equivalent number of morphological characters. (2) Determination of protein differences is reduced to a reasonably straightforward, objective exercise in measuring the relative mobility of various proteins. (3) Interspecific and higher taxonomic classifications, in groups that are difficult to separate morphologically, frequently show unambiguous differences at the protein level (e.g., Murphy, 1978). (4) Where a large number of proteins have been examined in a sample of several closely related species, the data may be summarized in a phylogenetic tree or dendrogram (e.g., Nemeth and Tracey, 1979).

Before employing electrophoretic techniques, the following disadvantages should also be considered: (1) living or freshly frozen specimens must be used; (2) identical protein mobility does not necessarily imply that the proteins are identical (Johnson, 1973); (3) the genes sampled electrophoretically specify soluble proteins—we do not know if this represents a random sample; (4) protein mobility, or even the presence or absence of a particular protein phenotype may not be constant either spatially or temporally: e.g., the mobility of some proteins may change through various physiological cycles. M. L. Tracey, S. T. Nemeth, and S. A. Espinet (unpublished data) have recorded changes in the enzyme leucine aminopeptidase that correlated with the molt cycle in the crayfish *Orconectes propinquus*. Khan *et al.* (1977) have reported similar changes during the ovarian cycle of the hermit crab *Clibanarius longitarsis*. An interesting discussion of the advantages and disadvantages of electrophoretic data in systematics may be found in Avise (1975).

A. Electrophoretic Techniques

Living organisms must be collected in order to extract proteins for electrophoretic separation and protein-specific or enzyme-specific staining. In many cases, samples may be frozen in the field or specimens may be homogenized (see below) and held in an ultracold (−80°C) unit for later use. The proteins of some organisms do not, however, respond well to such treatment, and pilot studies should be run before large samples are needlessly collected and frozen.

Smaller animals usually are prepared for electrophoresis by grinding the whole animal in either distilled water or the buffer used in the electrophoretic separation. Since volume prevents utilization of this simple expedient with larger animals, various organs or tissues are removed and homogenized in tissue grinders (e.g., Tracey *et al.,* 1975). In some cases the homogenized supernatant is used directly; in others the homogenate is centrifuged to remove debris before it is used. The additional effort involved in preparing organ or tissue samples from larger animals is rewarded by the availability of additional information. For example, many proteins are produced only in specific cells; these can be diluted out to the point of loss if large animals are homogenized without dissection. Thus the use of specific organs or tissues may increase the gene sample size; if one aim of the study is the construction of a phylogenetic tree, a large gene sample size is critical (Nei, 1975). Moreover, positive identification of proteins, where many are assayed simultaneously, is frequently facilitated by preparing different organs for assay (Tracey *et al.,* 1975).

Once the samples have been homogenized, they may be frozen for future use (see cautions cited above) or they may be loaded onto electrophoretic gels for separation. Samples are loaded onto gels either by absorbing the sample onto a piece of filter paper that is inserted into the gel, or by loading a small amount of fluid directly into a well in the gel.

The type of gel selected will be dictated by the needs of the particular investigator. Generally, better resolution is achieved with acrylamide (Hebert and Ward, 1972) than with starch gels; however, starch is less expensive and, because starch gels may be sliced and stained for as many as five different proteins, starch gel electrophoresis is more efficient (Tracey *et al.,* 1975). Once the samples are applied to the gel, the gel is placed into a support containing electrolytic buffers and current is applied. Separation time varies with gel type and buffer. After separation has been achieved, the current is switched off, gels are removed, and various histochemical stains adapted for electrophoretic detection of specific proteins are applied (Brewer, 1970; Shaw and Prasad, 1970; Selander *et al.,* 1971). Since the proteins separated are not pure, it is not possible to select, *a priori,* the best buffers, gel type, or protein stains to be used in studying a particular group. Operationally one prepares a number of buffers, gels, and stains; these are tried in various combinations and the best combination is determined empirically. Good introductions to electrophoretic techniques are given by Brewer (1970) and Gordon (1975).

The electropherogram or zymogram that is produced following staining of the gel must be interpreted. A number of different, idealized patterns are presented in Fig. 3; they may be interpreted genetically. The first four sets of bands are labeled Locus 1, denoting a genetic locus exhibiting electrophoretically separable protein variation. The first band is labeled 100. By convention, the most common allele in a sample is labeled 100, and alleles are identified by adding or subtracting their millimeter differences in migration distance. The 90 band is homozygous 90/90, and the multiple bands in

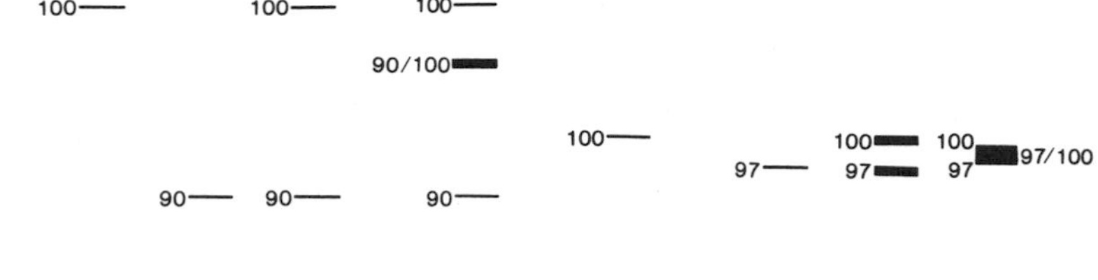

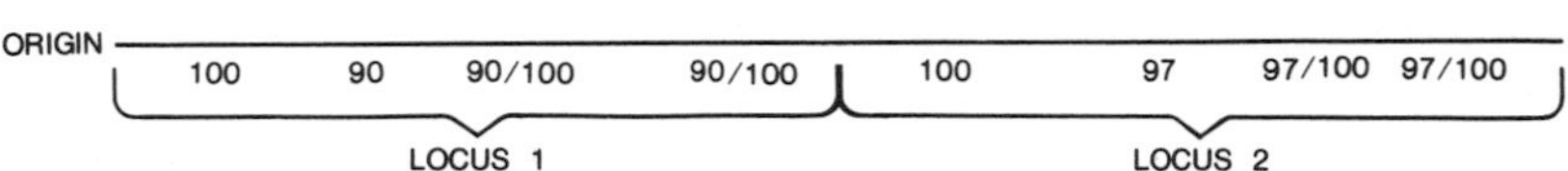

Fig. 3. A hypothetical zymogram or electropherogram. Genotypes are identified below the origin for each animal and allelic constituents of each band are written next to the bands.

columns three and four are formed by 90/100 heterozygotes. The two bands of column three, one 100 and one 90, are produced where the protein assayed is monomeric or a homeric multimer. The additional 90/100 band of column four is seen in dimeric or other multimeric proteins; heteromeres (90/100) of this type usually produce darker and wider bands, since there is twice the amount of protein present. The patterns presented for Locus 2 are similarly interpreted. Single band homozygotes are labeled 100 and 97; 97/100 are heterozygotes. In interpreting heterozygotes, where alleles differ by only one or a few millimeters, as in column seven, it is often necessary to interpret thick bands as heterozygotes. Hedgecock *et al.* (1975) provide additional examples of band interpretation and genetic tests of these interpretations. Where genetic crosses can be done, genetic interpretations of band patterns should be checked experimentally. If ovigerous animals are available, genetic interpretations may be tested by predicting the frequency of various genotypes among the progeny from knowledge of the maternal genotype and various hypothetical paternal genotypes (Hedgecock *et al.*, 1975). In many cases, neither of these verifications is feasible, and although genetic interpretations may still be made, they should be made with caution. Electrophoresis permits genetic interpretations in the absence of breeding studies. This is an advantage where evolutionary interpretations are sought, as the data may be compared with various genetic models of divergence (Nei, 1977a). The value of these comparisons depends, obviously, on the accuracy of the genetic interpretations, which are not always simple and straightforward. If breeding tests are impossible, genetic interpretations may be checked by calculating allele frequencies and comparing the observed genotype frequencies with the Hardy-Weinberg expectations (Li, 1976). As allozymes are codominantly expressed, genotype frequencies are calculated by counting the genotypic classes. Allele frequencies are calculated, in diploids, by counting each allele twice for homozygotes and once for heterozygotes. These sums are then divided by twice the total in diploids.

B. Applications in Systematics

The first systematic use of electrophoretically separated protein characters was in a study of nine species in the *Drosophila virilis* group (Hubby and Throckmorton, 1965). In that study, preliminary ammonium sulfate fractionation of proteins was employed since the majority of the proteins studied were visualized by staining with analine blue black. Only α-glycerophosphate dehydrogenase and malic dehydrogenase were specifically stained. Consequently most differences could not be interpreted genetically, and only general estimates of the number of loci differing among species could be obtained.

In studies where genetic interpretations are not possible, protein differences may be used in describing taxa by specifying the mobilities of protein bands. If many bands are being studied, mathematical techniques (see Numerical Methods) should be employed in summarizing the data.

Specific enzyme stains were employed by Hubby and Throckmorton (1968) in their next study of six species groups in *Drosophila*. These differences were genetically interpretable: for the first time, species similarities as well as differences could be identified and quantified in genetic terms, and the advent of multi-locus electrophoretic studies provided the data over which the selectionist-neutralist controversy raged (cf. Dobzhansky, 1970; Kimura and Ohta, 1971; Lewontin, 1974; Nei, 1975; Ayala, 1976; Dobzhansky *et al.*, 1977, for recent reviews). In addition, it led to the formulation of genetically interpretable identity or similarity coefficients, calculated from electrophoretic allele, allozyme, and genotype frequencies. A number of similar coefficients have been devised (cf. Nei, 1977b; Smith, 1977; Wright, 1978); however, Nei's coefficient (Nei, 1975, 1978) is readily interpretable in genetic terms and as a result is widely used. Since neither the approach nor the coefficient differ substantially when various measures of distance or similarity are employed (Avise, 1975), only Nei's indices of genetic identity I and genetic distance D will be further discussed.

If organisms from two populations (R and S of Table II) are sampled, the single locus genetic identity (I) between the two populations is calculated as follows:

$$I = j_{RS} \, (j_R \cdot j_S)^{-\frac{1}{2}} \tag{1}$$

TABLE II

Allele Frequencies for Two Loci in Five Populations

	Populations sampled				
Allele	*P*	*Q*	*R*	*S*	*T*
Locus 1					
85	1.00	1.00	0.10	0.15	
90			0.80	0.80	
98			0.10	0.05	
100					0.95
104					0.05
n	22	46	84	20	100
Locus 2					
97			1.00	1.00	
100	1.00	1.00			0.90
104					0.10
n	22	46	84	20	100

where $j_R = \Sigma x_i^2$, $j_S = \Sigma y_i^2$, x_i is the allele frequency of population *R*, y_i is the allele frequency of population *S*, *j* is the expected homozygosity or probability of identity, and *I* is the identity index. Similarly, the probability of identity of two alleles, chosen at random from each of the populations, is $j_{RS} = \Sigma x_i y_i$. For example, $j_R = 0.66$, $j_S = 0.665$, and $j_{RS} = 0.66$ when populations *R* and *S* (Table II) are compared at locus 1, and $I = 0.996$. As *R* and *S* differ only with respect to allele frequencies, this high value of *I* seems intuitively reasonable. When *I* is calculated for locus 1 between populations *P* and *Q*, it is seen that $I = 1.00$. The populations are identical: all organisms examined show the same electrophoretic band, 85, which migrates 15 mm less than the standard 100 that is seen only in population *T*. The computed *I* between *P* and *R* and between *Q* and *R* is equal to 0.123. The identity between population *T* and all others clearly is 0.00 since no alleles are shared between these populations. Similar single locus *I* values may be calculated for all loci sampled. Where this has been done for fifteen or more loci, a locus by locus picture of near total identity or near total difference emerges, i.e., loci fall into two classes: $I > 0.90$ or $I < 0.10$. As expected in intraspecific populations, more than 95% of all loci compared fall into the $I > 0.90$ class. In interspecific comparisons and those involving higher taxonomic categories, the $I < 0.10$ class becomes progressively larger (Ayala, 1975; Avise, 1976).

This bimodality makes the use of a large number of loci imperative in calculating average identies or similarities (see below). Indeed, the number of loci sampled is far more important than the organismal sample size (Nei and Roychoudhary, 1974; Nei, 1977b). In most electrophoretic studies it has been found that comparisons of populations within a species detect few differences at or below the $I = 0.85$ level. Species may exhibit varying degrees of protein polymorphism, but when the I values that normalize this polymorphism are calculated, population to population differentiation is very small. At the subspecific level, I values less than 0.10 are seen for a few proteins. The majority of the proteins, however, remain in the $I > 0.90$ class. Thus, once the appropriate proteins are identified, it is possible to classify individuals even at the subspecies level by assaying only those proteins which are diagnostic. If three or four such proteins are identified, the probability of misclassification may be extremely small (e.g., 10^{-5}) (Ayala and Powell, 1972).

Allozyme classifications have been employed in only a few studies of crustaceans, but the utility of the technique already has been demonstrated. Eight populations of the American lobster, *Homarus americanus,* were surveyed for genetic variation at 44 loci encoding electrophoretically detectable proteins. Differentiation between populations was found only at the *Malic enzyme* locus; however, the degree of differentiation ($I = 0.03$ and $I = 0.06$) supports the suggestion from previous morphological and migration studies that *H. americanus* is subdivided into populations which do not interbreed (Tracey *et al.,* 1975). Flowerdew and Crisp (1975) were able to separate eastern and western Atlantic races of *Balanus balanoides* on the basis of cholinesterase and arylesterase differences. Formal genetic confirmation of their band interpretations was impossible; they did, however, compare genotype frequencies with Hardy-Weinberg expectations. Although the Welsh population appeared to be in Hardy-Weinberg equilibrium, the Canadian population showed an excess of homozygotes at the locus for which gel interpretations appeared relatively straightforward. They did not comment on this discrepancy. Sywula and Bartkowiak (1978) studied esterases and acid phosphatases in six species of ostracodes. The species were readily separable, and dendrograms, based on these enzyme differences, are presented in their paper.

The bimodal distribution of single locus I values suggests that allozyme data should only be used as a base for constructing phylogenetic trees or dendrograms when a large number, i.e., 15 or more, of loci have been sampled. Clearly with I values falling into either $I > 0.90$ or $I < 0.10$ classes, the use of only two or three protein loci may give misleading impressions of the overall similarity or difference (Nei and Raychoudhury, 1974). If sufficient loci have been assayed, average I values may be clacu-lated and transformed to D for use in the construction of a dendrogram.

With reference to Table II, average genetic identity over loci may be estimated by:

$$\bar{I} = J_{RS}(J_R \cdot J_S)^{-\frac{1}{2}} \quad (2)$$

where J_R, J_S, and J_{RS} are the means of j_R, j_S, and j_{RS}, and $\bar{I}$ is the average genetic identity. From this, the distance D can be computed as:

$$D = -\log_e \bar{I} \quad (3)$$

Since 0 and 1.00 are the limits of $\bar{I}$, a matrix of D values between taxa is generally used to construct the dendrograms.

Many methods of constructing dendrograms based on matrices of similarity coefficients are available. They may be categorized mathematically as either maximum likelihood or parsimony methods (for further references and comparisons, see Felsenstein, 1973, 1978; C. H. Nelson, 1978). The methods and assumptions of the statistical techniques vary; they should be carefully considered before sweeping conclusions are drawn. The technique is useful when judiciously applied, (Avise, 1975), and results in many plant and animal groups exhibit close agreement with morphological, behavioral, or chromosomal phylogenies. As yet, very few studies of genetic distance in crustaceans have sampled enough loci (>15) to provide adequate estimates of D.

In a large study of twenty-six loci in six species of decapods, i.e., three species of *Orconectes* and three species of *Cambarus,* good agreement between established relationships and the allozyme-based dendrogram were observed (Nemeth and Tracey, 1978). However, when *Procambarus clarkii* and *Procambarus acutus acutus* were studied, an interesting and as yet unexplained discrepancy was noted. The two most similar species ($I = 0.84 \pm 0.04$) were *P. a. acutus* and *Orconectes propinquus* (Tracey and Nemeth, 1979). In attempting to explain this and other discrepancies to come, it will be necessary to examine the assumptions underlying both the biochemical and the classical systematic approaches.

In the first case it should be emphasized that the molecular techniques sample only a small fraction of the genome. Gene substitutions are subject to stochastic errors, and D is a measure of the "expected" number of mutant substitutions (Nei, 1975, 1977b). Moreover, electrophoretic separation of proteins does not lead to unequivocal separation of all proteins in which amino acid substitutions have taken place (Johnson, 1973; Coyne and Felton, 1977). Clearly, in resolving discrepancies of this type, the preferred approach is further study of the loci exhibiting apparent identity. If the identity turns out to be spurious, the problem is solved. If the identity is not further explainable, it may be real or it may represent molecular convergence (C. H. Nelson, 1978). On the other hand, there is no *a priori* reason for suspecting that a dendrogram or classification based on morphological characters is inherently more accurate than a dendrogram or classification based on biochemical data. Where discrepancies remain unresolved between the two approaches, the morphological classification also should be re-examined.

Classifications based on molecular techniques other than electrophoretic separation of proteins also have been used, e.g., DNA hybridization, immunological cross reaction, and amino acid sequencing (Dobzhansky *et al.*, 1977). More recently, restriction endonucleases have been employed at the species level to decipher population structure (Avise *et al.*, 1979; Lansman *et al.*, 1981). Molecular confirmation of established classifications is reassuring, but the detection of discrepancies is more interesting and probably more informative. Molecular-level studies of evolutionary or phenetic relationships offer readily quantifiable data and the ability to compare even distantly related extant organisms. These techniques will, hopefully, be more widely and wisely employed in the future. They are valuable tools to complement morphological studies.

REFERENCES

Abele, L. G. (1971). Scanning electron photomicrographs of brachyuran gonopods. *Crustaceana* **21,** 218-219.

Allen, D. M., and Inglis, A. (1968). A pushnet for quantitative sampling in shallow estuaries. *Limnol. Oceanogr.* **3,** 239-241.

Aron, W. (1958). The use of a large capacity portable pump for plankton sampling, with notes on plankton patchiness. *J. Mar. Res.* **16,** 158-173.

Aron, W. (1962). Some aspects of sampling the macroplankton. *Rapp. Proc.-Verb.* **153,** No. 5, 29-38.

Avise, J. C. (1975). Systematic value of electrophoretic data. *Syst. Zool.* **23,** 465-481.

Avise, J. C. (1976). Genetic differentiation during speciation. *In* "Molecular Evolution" (F. J. Ayala, ed.), pp. 106-122. Sinauer Assoc., Sunderland, Massachusetts.

Avise, J. C., Lansman, R., and Shade, R. O. (1979). The use of restriction endonucleases to measure mitochondrial DNA sequence relatedness in natural populations. I. Population structure and evolution in the genus *Peromyscus*. *Genetics* **92,** 279-295.

Ayala, F. J. (1975). Genetic differentiation during the speciation process. *Evol. Biol.* **8,** 1-78.

Ayala, F. J., ed. (1976). "Molecular Evolution." Sinauer Assoc., Sunderland, Massachusetts.

Ayala, F. J., and Powell, J. R. (1972). Allozymes as diagnostic characters of sibling species of *Drosophila*. *Proc. Natl. Acad. Sci. U.S.A.* **69,** 1094-1096.

Banse, K. (1955). Über das Verhalten von meroplanktischen Larven in geschichtetem Wasser. *Kiel. Meeresforsch.* **11,** 188-200.

Banse, K. (1962). Net zooplankton and total zooplankton. *Rapp. Proc.-Verb.* **153,** No. 36, 211-214.

Barnard, J. L. (1970). Sublittoral Gammaridea (Amphipoda) of the Hawaiian Islands. *Smithson. Contrib. Zool.* **34,** 1-286.

Barnard, J. L. (1971). Keys to the Hawaiian marine Gammaridea, 0-30 meters. *Smithson. Contrib. Zool.* **58,** 1-135.

Barnard, J. L. (1972). Gammaridean Amphipoda of Australia. Part I. *Smithson. Contrib. Zool.* **103,** 1-333.

Barnard, J. L. (1974). Gammaridean Amphipoda of Australia. Part II. *Smithson. Contrib. Zool.* **139,** 1-148.

Barnes, H. (1959). "Oceanography and Marine Ecology." Allen & Unwin, London.

Barnes, H., and Healy, M.J.R. (1965). Biometrical studies on some common cirripedes. I. *Balanus balanoides:* Measurements of the scuta and terga of animals from a wide geographical range. *J. Mar. Biol. Assoc. U.K.* **45,** 779-789.

Barnes, H., and Healy, M.J.R. (1969). Biometrical studies on some common cirripedes. II. Discriminant analysis of measurements on the scuta and terga of *Balanus balanus* (L.), *B. crenatus* Brug., *B. improvisus* Darwin, *B. glandula* Darwin, and *B. amphitrite stutsburi* Darwin (*B. pallidus stutsburi*). *J. Exp. Mar. Biol. Ecol.* **4,** 51-70.

Barnes, H., and Healy, M.J.R. (1971). Biometrical studies on some common cirripedes. III. Discriminant analysis of measurements on the scuta and terga of *Balanus eburneus* Gould. *J. Exp. Mar. Biol. Ecol.* **6,** 83-90.

Blackwelder, R. E. (1967). "Taxonomy." Wiley, New York.

Bock, W. J. (1973). Philosophical foundations of classical evolutionary classification. *Syst. Zool.* **22,** 375-392.

Bonde, N. (1977). Cladistic classification as applied to vertebrates. *In* "Major Patterns in Vertebrate Evolution" (M. K. Hecht, P. C. Goody, and B. M. Hecht, eds.), pp. 741-803. Plenum, New York.

Bourdeaux, H. B. (1979). "Arthropod Phylogeny with Special Reference to Insects." Wiley (Interscience), New York.

Bousfield, E. L. (1973). "Shallow-water Gammaridean Amphipoda of New England." Cornell Univ. Press, Ithaca, New York.

Bousfield, E. L. (1977). A new look at the systematics of gammaroidean amphipods of the world. *Crustaceana, Suppl.* **4,** 281-316.

Brewer, G. J. (1970). "An Introduction to Isozyme Techniques." Academic Press, New York.

Buzas, M. A. (1967). An application of canonical analysis as a method for comparing faunal areas. *J. Anim. Ecol.* **36,** 563-577.

Cavalli-Sforza, L. L., and Edwards, A. W. F. (1967). Phylogenetic analysis: Models and estimating procedures. *Evolution* **21,** 550-570.

Chace, F. A., Jr. (1972). The shrimps of the Smithsonian-Bredin Caribbean expeditions with a summary of the West Indian shallow-water species (Crustacea: Decapoda: Natantia). *Smithson. Contrib. Zool.* **98,** 1-179.

Chace, F. A., Jr., and Hobbs, H. H., Jr. (1969). The freshwater and terrestrial decapod crustaceans of the West Indies with special reference to Dominica. *Bull.—U.S. Natl. Mus.* **292,** 1-258.

Colless, D. H. (1970). The phenogram as an estimate of phylogeny. *Syst. Zool.* **19,** 352-362.

Coyne, J. A., and Felton, A. A. (1977). Genic heterogeneity at two alcohol dehydrogenase loci in *Drosphila pseudoobscura* and *Drosophila persimilis. Genetics* **87,** 285-304.

Cracraft, J. (1974). Phylogenetic models and classification. *Syst. Zool.* **23,** 71-90.

Crane, J. (1975). "Fiddler Crabs of the World." Princeton Univ. Press, Princeton, New Jersey.

Darwin, C. (1851). "A Monograph on the Subclass Cirripedia, with Figures of all Species. The Lepadidae; or, Pedunculated Cirripedes." Ray Society, London.

Darwin, C. (1854). "A Monograph on the Subclass Cirripedia with Figures of all the species. The Balanidae, the Verrucidae, etc." Ray Society, London.

Dobzhansky, T. (1970). "Genetics of the Evolutionary Process." Columbia Univ. Press, New York.

Dobzhansky, T., Ayala, F. J., Stebbins, G. L., and Valentine, J. W. (1977). "Evolution." Freeman, San Francisco, California.

Estabrook, G. F. (1972). Cladistic methodology: A discussion of the theoretical basis for the induction of evolutionary history. *Annu. Rev. Ecol. Syst.* **3,** 427–456.

Fabricius, J. C. (1775). "Systema Entomologiae." Flensburgi and Lipsiae.

Fabricius, J. C. (1792-1794). "Systema Entomologiae II." Flensburgi and Lipsiae.

Fabricius, J. C. (1795). "Entomologiae Systema," Supplement.

Farris, J. S. (1970). Methods for computing Wagner trees. *Syst. Zool.* **19,** 83-92.

Farris, J. S. (1972). Estimating phylogenetic trees from distance matrices. *Am. Nat.* **106,** 645-668.

Farris, J. S. (1977). Phylogenetic analysis under Dollo's Law. *Syst. Zool.* **26,** 77-88.

Farris, J. S. (1979). The information content of the phylogenetic system. *Syst. Zool.* **28,** 483-519.

Felder, D. L. (1973). An annotated key to crabs and lobsters (Decapoda, Reptantia) from coastal waters of the northwestern Gulf of Mexico. Cent. Wetland Res. **LSU-SG-73-02,** 1-103.

Felsenstein, J. (1973). Maximum likelihood and minimum-steps methods for estimating evolutionary trees from data on discrete characters. *Syst. Zool.* **22,** 240-249.

Felsenstein, J. (1978). The number of evolutionary trees. *Syst. Zool.* **27,** 27-33.

Fisher, R. A. (1936). The use of multiple measurements in taxonomic problems. *Ann. Eugen. (London)* **7,** 179-188.

Fleminger, A. (1968). Effects of high concentration of hexamine during fixation and preservation of copepods. SCOR Working Group (unpublished).

Flowerdew, W. W., and Crisp, D. J. (1975). Esterase heterogeneity and an investigation into racial differences in the cirripede *Balanus balanoides* using acrylamide gel electrophoresis. *Mar. Biol.* **33,** 33-39.

Fraser, J. H. (1961). Use of polyethylene glycol in biology. *Nature (London)* **189,** 241-242.

Fryer, G. (1968). Evolution and adaptive radiation in the Chydoridae (Crustacea: Cladocera): A study in comparative functional morphology. *Philos. Trans. R. Soc. London, Ser. B* **254,** 221-385.

Galigher, A. E., and Kozloff, E. N. (1964). "Essentials of Practical Microtechnique." Lea & Febiger, Philadelphia, Pennsylvania.

Ghiselin, M. T. (1974). A radical solution to the species problem. *Syst. Zool.* **23,** 536-544.

Ghiselin, M. T., and Jaffe, L. (1973). Phylogenetic classification in Darwin's monofgraph on the sub-class Cirripedia. *Syst. Zool.* **22,** 132-140.

Gilat, E. (1963). Methods of study in marine benthonic ecology. Comm. Int. Explor. Sci. Mer Mediterr. *Colloque Com. Benthos* 7-13.

Goodman, M., and Moore, G. W. (1971). Immunodiffusion systematics of primates. I. The Catarrhini. *Syst. Zool.* **20,** 19-62.

Gordon, A. H. (1975). "Electrophoresis of Proteins in Polyacrylamide and Starch Gels." Am. Elsevier, New York.

Gore, R. H. (1970). *Petrolisthes armatus:* A redescription of larval development under laboratory comidtions (Decapoda, Porcellanidae). *Crustaceana* **18,** 75-89.

Haistone, T. S., and Stephenson, W. (1961). The biology of *Callianassa (Trypaea) australiensis* Dana 1852 (Crustacea, Thalassinidea). *Univ. Queensl. Pap.* **1,** 259-285.

Hall, J. R., and Hessler, R. R. (1971). Aspects of the population dynamics of *Derocheilocaris typica* (Mystacocarida, Crustacea). *Vie Milieu* **22,** 305-326.

Hansen, V., Jr., and Andersen, K. P. (1962). Sampling the smaller zooplankton. *Rapp. Proc.-Verb* **152,** No. 6, 39-47.

Hayat, M. A. (1978). "Introduction to Biological Scanning Electron Microscopy." University Park Press, Baltimore, Maryland.
Healy, M. J. R. (1965). *In* "Mathematics and Computer Science in Biology and Medicine." HM Stationery Office, London.
Hebert, P.D.N., and Ward, R. D. (1972). Inheritance during parthenogenesis in *Daphnia magna*. *Genetics* **71,** 639-642.
Hecht, M. K. (1976). Phylogenetic inference and methodology as applied to the vertebrate record. *Evol. Biol.* **9,** 335-363.
Hecht, M. K., and Edwards, J. L. (1976). The determination of parallel or monophyletic relationships: The proteid salamanders—a test case. *Am. Nat.* **110,** 653-677.
Hecht, M. K., and Edwards, J. L. (1977). The methodology of phylogenetic inference above the species level. *In* "Major Patterns in Vertebrate Evoltuion" (M. K. Hecht, P. C. Goody, and B. M. Hecht, eds.), pp. 3-51. Plenum, New York.
Hedgecock, D., Nelson, K., Shleser, R. A., and Tracey, M. L. (1975). Biochemical genetics of lobsters (*Homarus*). II. Inheritance of allozymes in *H. americanus*. *J. Hered.* **66,** 114-118.
Hennig, W. (1966). "Phylogenetic Systematics." Illinois Univ. Press, Urbana.
Henry, D. P., and McLaughlin, P. A. (1975). The barnacles of the *Balanus amphitrite* complex (Cirripedia, Thoracica). *Zool. Verh.* **141,** 1-254.
Higer, A. L., and Kolipinski, M. C. (1967). Pull-up trap: A quantitative devide for sampling shallow-water animals. *Ecology* **48,** 1008-1009.
Hobbs, H. H., Jr. (1942). The crayfishes of Florida. *Univ. Fla. Publ., Biol. Sci. Ser.* **3,** 1-179.
Hobbs, H. H., Jr. (1975). Crustacea: Malacostraca. *In* "Keys to Water Quality Indicative Organisms of the Southeastern United States" (F. K. Parrish, ed.), 2nd ed., pp. 87-122. Environmental Monitoring and Support Laboratory, USEPA, Cincinnati, Ohio.
Hubby, J. L., and Throckmorton, L. H. (1965). Protein differences in *Drosphila*. II. Comparative species genetics and evolutionary problems. *Genetics* **52,** 203-215.
Hubby, J. L., and Throckmorton, L. H. (1968). Protein differences in *Drosphila*. IV. A study of sibling species. *Am. Nat.* **102,** 193-205.
Hulings, N. C., and Gray, J. S. (1971). A manual for the study of meiofauna. *Smithson. Contrib. Zool.* **78,** 1-83.
Hull, D. L. (1976). Are species really individuals? *Syst. Zool.* **25,** 174-191.
Johnson, G. B. (1973). Enzyme polymorphism and biosystematics: The hypothesis of selective neutrality. *Annu. Rev. Ecol. Syst.* **4,** 93-116.
Kaesler, R. L. (1966). Quantitative re-evaluation of ecology and distribution of Recent Foraminifera and Ostracoda of Todos Santos Bay, Baja California, Mexico. *Univ. Kans. Paleontol. Contrib., Pap.* **10,** 1-50.
Kaesler, R. L. (1967). Numerical taxonomy in invertebrate paleontology. *In* "Essays in Paleontology and Stratigraphy: Raymond C. Moore Commemorative Volume" (C. Teichert and E. L. Yochelson, eds.), pp. 63-81. Univ. of Kansas Press, Lawrence.
Kaesler, R. L. (1969). Numerical taxonomy of selected Recent British Ostracoda. *In* "Symposium on the Taxonomy, Morphology and Ecology of Recent Ostracoda" (J. W. Neale, ed.), pp. 21-47. Oliver & Boyd, Edinburgh.
Keasler, R. L. (1970). Numerical taxonomy in paleontology: Classification, ordination and reconstruction of phylogenies. *In* "Proceedings of the North American Paleontological Convention, Chicago 1969" (E. L. Yochelson, ed.), Vol. 1, Part B, pp. 84-100. Allen Press, Lawrence, Kansas.
Karnovsky, M. J. (1965). A formaldehyde-glutaraldehyde fixative of high osmolarity for use in electron microscopy. *J. Cell Biol.* **27,** 137A.
Khan, S. A., Reddy, P. S., and Natarjan, R. (1977). Electro phoretic studies of protein patterns in the hermit crab, *Cl b narius longitarsis* (Dettaan) during the Ovarian cycle. *In* "Advances

in Invertebrate Reproduction" (K. G. Adiyodi and R. G. Adiyodi, eds.), Vol. I, pp. 179-184. Peralam-Kenoth, Karivellur, Kerala, India.

Kimura, M., and Ohta, T. (1971). "Theoretical Aspects of Population Genetics." Princeton Univ. Press, Princeton, New Jersey.

Kinzelback, R. K. (1971). Morphologische Befunde an Facherflugern und ihre Phylogenetische Bedeutung (Insecta: Strepsiptera). *Zoologica* **41,** 1-256.

Kluge, A. G., and Farris, J. S. (1969). Quantitative phyletics and the evolution of anurans. *Syst. Zool.* **18,** 1-32.

Knudsen, J. W. (1966). "Biological Techniques." Harper, New York.

Korniker, L. S. (1975). Antarctic ostracods (myodocopina) [in two parts] Part II. Smithsonian contributions to *Zoology* **63,** 375-720.

Lansman, R. Q., Shade, R. O., Shapira, J. F., and Avise, J. C. (1981). The use of restriction endonucleases to measure mitochondrial DNA sequence relatedness in natural populations. III. Techniques and potential applications. *Mol. Evol.* **17,** 214-225.

Lawley, D. N., and Maxwell, A. E. (1963). "Factor Analysis as a Statistical Method." Butterworth, London.

Lawson, T. J. (1973). Factor analysis of variation in the first pair of swimming feet of the Candaciidae (Copepoda: Calanoida). *Syst. Zool.* **22,** 302-309.

Lewis, P. R., and Knight, D. P. (1977). Staining methods for sectional material. *Pract. Methods Electron Microsc.* **5.**

Lewontin, R. C. (1974). "The Genetic Basis of Evolutionary Change." Columbia Univ. Press, New York.

Li, Ching Chun (1976). "First Course in Population Genetics." Boxwood Press, Pacific Grove, California.

Linnaeus, C. (1758). "Systema Naturae," 10th ed. Holmiae.

Linnaeus, C. (1767). "Systema Naturae," 12th ed. Holmiae.

McGowan, J. A. (1967). Comments on zooplankton curation at IOBC. Rep. Consultative Comm. IOBC (unpublished).

McHardy, R. A. (1966). Polyvinyl Lactophenol microscope preparations made permanent with Turtox CMC-10. *Turtox News* **44,** 63.

McLaughlin, P. A. (1974). The hermit crabs (Crustacea Decapoda, Paguridea) of northwestern North America. *Zool. Verh.* **130,** 1-396.

McLaughlin, P. A. (1976). A new species of *Lightiella* (Crustacea: Cephalocarida) from the west coast of Florida. *Bull. Mar. Sci.* **26,** 593-599.

McLaughlin, P. A. (1980). "Comparative Morphology of Recent Crustacea." Freeman, San Francisco, California.

McLaughlin, P. A. (1981). Revision of *Pylopagurus* and *Tomopagurus* (Crustacea: Decapoda: Paguridae), with the descriptions of new genera and species: Part I. Ten new genera of the Paguridae and a redescription of *Tomopagurus* A. Milne Edwards and Bouvier. *Bull. Mar. Sci.* **31,** 1-30.

McNeill, J. (1979). Purposeful phenetics. *Syst. Zool.* **28,** 465-482.

Mahoney, R. (1966). "Laboratory Techniques in Zoology." Butterworth, London.

Manning, R. B. (1960). A useful method for collecting Crustacea. *Crustaceana* **1,** 372.

Manning, R. B. (1969). Stomatopod Crustacea of the western Atlantic. *Stud. Trop. Oceanogr.* **8,** 1-380.

Manning, R. B. (1975). Two methods for collecting decapods in shallow water. *Crustaceana* **29,** 317-319.

Manning, R. B. (1978). Lobsters. *In* "FAO Species Identification Sheets for Fishery Purposes." (W. Fischer, ed.), pp. 1-47. Food and Agricultural Organization of the United Nations, Rome.

Matthews, J.B.L. (1972). The genus *Euaugaptilus* (Crustacea, Copepoda). New descriptions and a review of the genus in relation to *Augaptilus, Haloptilus* and *Pseudaugaptilus*. *Bull. Br. Mus. (Nat. Hist.), Zool.* **24,** 1-71.

Mayr, E. (1969). "Principles of Systematic Zoology." McGraw-Hill, New York.

Mayr, E. (1976). "Evolution and the Diversity of Life." Belknap Press, Cambridge, Massachusetts.

Menzies, R. J., and Glynn, P. W. (1968). The common marine isopod Crustacea of Puerto Rico. *Stud. Fauna Curacao Carib. Is.* **27,** 1-133.

Monod, T., and Cals, P. (1970). Mission zoologique Belge aux iles Galapagos et en Ecuador. **2,** 57-103.

Moulton, T. P. (1973). Principal component analysis of variation in form within *Oncaea conifera* Giesbrecht, 1891, a species of Copepoda (Crustacea). *Syst. Zool.* **22,** 141-156.

Murphy, P. G. (1978). *Collisella austrodigitalis* sp. nov.: A sibling species of limpet (Acmaeidae) discovered by electrophoresis. *Biol. Bull. (Woods Hole, Mass.)* **155,** 193-205.

Nei, M. (1975). "Molecular Population Genetics and Evolution." Am. Elsevier, New York.

Nei, M. (1977a). Standard error of immunological dating of evolutionary time. *J. Mol. Evol.* **9,** 203-211.

Nei, M. (1977b). Genetic distance. *In* "Genetics" (E. Matsunaga and K. Omoto, eds.), pp. 29-62. Yuzankaku Publ, Tokyo.

Nei, M. (1978). The theory of genetic distance and evolution of human races. *Jpn. J. Hum. Genet.* **23,** 341-369.

Nei, M., and Roychoudhury, A. K. (1974). Sampling variances of heterozygosity and genetic distance. *Genetics* **76,** 379-390.

Nelson, C. H. (1978). Recognition of covergence and parallelism on Wagner Trees. *Syst. Zool.* **27,** 122-124.

Nelson, G. (1969). Gill arches and the phylogeny of fishes, with notes on the classification of vertebrates. *Bull. Am. Mus. Nat. Hist.* **141,** 475-552.

Nelson, G. (1972). Comments on Hennig's "phylogenetic systematics" and its influence on ichthyology. *Syst. Zool.* **21,** 364-373.

Nelson, G. (1973). Classification as an expression of phylogenetic relationships. *Syst. Zool.* **22,** 344-359.

Nelson, G. (1979). Cladistic analysis and synthesis: Principles and definitions, with a historical note on Adanson's Families des Plantes (1763-1764). *Syst. Zool.* **28,** 1-21.

Nemeth, S. T., and Tracey, M. L. (1979). Allozyme variability and relatedness in six crayfish species. *J. Hered.* **70,** 37-43.

Newman, W. A., and Ross, A. (1971). "Antarctic Cirripedia," Antarct. Res. Ser. 14. Am. Geophys. Union, Washington, D.C.

Owre, H. B., and Foyo, M. (1967). Copepods of the Florida current. *Fauna Caribaea* **1,** 1-137.

Palade, G. E. (1952). A study of fixation for electron microscopy. *J. Exp. Med.* **95,** 285.

Pérez-Farfante, I. (1978). Shrimps and prawns. *In* "FAO Identification Sheets for Fishery Purposes" (W. Fischer, ed.), pp. 1-44. Food and Agriculture Organization of the United Nations, Rome.

Porter, K. R., and J. BLum (1953). A study in microtomy for electron microscopy. *Anat. Rec.* **117,** 685.

Powers, L. W. (1977). A catalogue and bibliography to the crabs (Brachyura) of the Gulf of Mexico. *Contrib. Mar. Sci.* **20,** Suppl. 1-190.

Provenzano, A. J., Jr. (1962a). The larval development of *Calcinus tibicen* (Herbst) (Crustacea, Anomura) in the laboratory. *Biol. Bull. (Woods Hole, Mass.)* **123,** 179-202.

Provenzano, A. J., Jr. (1962b). The larvel development of the tropical land hermit *Coenobita clypeatus* (Herbst) in the laboratory. *Crustaceana* **4,** 207-228.

Provenzano, A. J., Jr. (1967). The zoeal stages and glaucothoe of the tropical eastern Pacific hermit crab *Trizopagurus magnificu* (Bouvier, 1898) (Decapoda; Diogenidae), reared in the laboratory. *Pac. Sci.* **21,** 457–473.

Provenzano, A. J., Jr. (1968). The complete larval development of the West Indian hermit crab *Petrochirus diogenes* (L.) (Decapoda, Diogenidae) reared in the laboratory. *Bull. Mar. Sci.* **18,** 143-181.

Provenzano, A. J., Jr. (1971). Zoeal development of *Pylopaguropsis atlantica* Wass, 1963, and evidence from larval characters of some generic relationships within the Paguridae. *Bull. Mar. Sci.* **21,** 237-255.

Provenzano, A. J., Jr., and Rice, A. L. (1964). The larval stages of *Pagurus marshi* (Decapoda, Anomura) reared in the laboratory. *Crustaceana* **7,** 217-235.

Pulsifer, J. (1975). Some techniques for mounting copepods for examination in a scanning electron microscope. *Crustaceana* **28,** 101.

Rao, C. R. (1952). "Advanced Statistical Methods in Biometric Research." Wiley, New York.

Reid, N. (1974). Ultramicrotomy. *Pract. Methods Electron Microsc.* **3,** 224.

Ritchie, L. (1975). A new genus and two new species of Choniostomatidae (Copepoda) parasitic on two deep sea isopods. *J. Linn. Soc. London, Zool.* **57,** 155-178.

Saila, S. B., and Flowers, J. M. (1969). Geographic morphometric variation in the American lobster. *Syst. Zool.* **18,** 330-338.

Salmon, J. T. (1949). New methods in microscopy for the study of small insects and arthropods. *Trans. R. Soc. N.Z.* **77,** 250-253.

Schaeffer, B., Hecht, M. K., and Eldredge, N. (1972). Phylogeny and paleontology. *Evol. Biol.* **6,** 31–46.

Seal, H. L. (1964). "Multivariate Statistical Analysis for Biologists." Wiley, New York.

Selander, R. K., Smith, M. H., Yang, S. Y., Johnson, W. E., and Gentry, J. B. (1971). Biochemical polymorphism and systematics in the genus *Peromyscus.* I. Variation in the old-field mouse (*Peromyscus polionotus*). *Univ. Tex. Publ.* **7103,** 49-90.

Shaw, C. R., and Prasad, R. (1970). Starch gel electrophoresis of enzymes—a compilation of recipes. *Biochem. Genet.* **4,** 297-320.

Simpson, G. G. (1961). "Principles of Animal Taxonomy." Columbia Univ. Press, New York.

Slobodchikoff, C. N. (ed.) (1976). "Concepts of Species." Wiley, New York.

Smith, C.A.B. (1977). A note on genetic distance. *Ann. Hum. Genet.* **40,** 463–479.

Sneath, P.H.A., and Sokal, R. R. (1973). "Numerical Taxonomy." Freeman, San Francisco, California.

Sokal, R. R., and Sneath, P.H.A. (1963). "Principles of Numerical Taxonomy." Freeman, San Francisco, California.

Steedman, H. F. (1966). Dimethyl hydantoin formaldehyde: A new water-soluble resin for use as a mounting medium. *Q. J. Microsc. Sci.* [N.S.] **99,** 451–452.

Steedman, H. F. (1976). "Zooplankton Fixation and Preservation." UNESCO Press, Paris.

Stock, J. (1976). A new genus and two new species of the crustacean order Thermobaenacea from the West Indies. *Bijdr. Dierkd.* **46,** 47-70.

Strawn, K. (1954). The pushnet, a one-man net for collecting in attached vegetation. *Copeia* pp. 195-197.

Sywula, T., and Bartkowiak, S. (1978). Preliminary study of the isoenzymes of Ostracoda. *Crustaceana* **35,** 265-272.

Tait, R. V., and de Santo, R. S. (1972). "Elements of Marine Ecology." Springer-Verlag, Berlin and New York.

Tomilinson, J. T. (1969). The burrowing barnacles (Cirripedia: Order Acrothoracica). *Bull.—U.S. Natl. Mus.* **296,** 1-161.

Tracey, M. L., and Nemeth, S. T. (1978). Genic variability and speciation in crayfish. *Genetics* **88,** s99-s100.

Tracey, M. L., Nelson, K., Hedgecock, D., Shleser, R. A., and Pressick, M. L. (1975). Biochemical genetic of lobsters: Genetic variation and structure of American lobster (*Homarus americanus*) populations. *J. Fish. Res. Board Can.* **32,** 2091-2101.

Tranter, D. J. (1968). "Reviews on Zooplankton Sampling Methods." UNESCO, Switzerland.

Wagner, W. H. (1961). Problems in the classification of ferns. *Recent Adv. Bot.* pp. 841-844.

Walker, J. F. (1964). "Formaldehyde," 3rd ed. van Nostrand-Reinhold, Princeton, New Jersey.

Waller, R. A., and Eschmeyer, W. N. (1965). A method for preserving color in biological specimens. *BioScience* **15,** 361.

Watling, L. (1979). The phylogeny of the superorder Peracarida (Crustacea: Malacostraca). *Am. Zool.* **19,** 870.

Welch, P. S. (1948). "Limnological Methods." McGraw-Hill (Blakiston), New York.

White, M.J.D. (1978). "Modes of Speciation." Freeman, San Francisco, California.

Wiborg, K. F. (1948). Experiments with the Clarke-Bumpus plankton sampler and with a plankton pump in the Lofoten area in norther Norway. *Rep. Norw. Fish. Mar. Invest.* **9,** 1-32.

Wiley, E. O. (1978). The evolutionary species concept reconsidered. *Syst. Zool.* **27,** 17-26.

Wiley, E. O. (1980). Is the evolutionary species fiction?—A consideration of classes, individuals and historical entities. *Syst. Zool.* **29,** 76-80.

Williams, A. B. (1965). Marine decapod crustaceans of the Carolinas. *Fish. Bull.* **65,** 1-298.

Williams, A. B. (1974). Marine flora and fauna of the northeastern United States. Crustacea: Decapoda. *NOAA Tech. Rep., NMFS Circ.* **389,** 1-50.

Williams, A. B. (1978). True crabs. *In* "FAO Species Identification Sheets for Fishery Purposes" (W. Fischer, ed.), pp. 1-34. Food Agriculture Organization of the United Nations, Rome.

Williamson, D. I., and Russell, G. (1965). Ethylene glycol as a preservative for marine organisms. *Nature (London)* **206,** 1370-1371.

Wright, S. (1978). Variability within and among natural populations. *In* "Evolution and the Genetics of Populations," pp. 89-92. Univ. of Chicago Press, Chicago, Illinois.

Yamaguchi, T. (1973). On megabalanus (Cirripedia, Thoracia) of Japan. *Pub. Seto Mar. Biol. Lab.* **21,** 115-140.

Yamaguchi, T. (1977). Taxonomic studies on some fossil and Recent Japanese Balanoidea. *Trans. Proc. Paleontol. Soc. Jpn.* **107/108,** 135-201.

Yang, W. T. (1971). The larval and post larval development of *Panthenope serrata* reared in the laboratory and the systematic position of the Parthenopinae (Crustacea, Brachyura). *Biol. Bull. (Woods Hole, Mass.)* **140,** 166-189.

3

Origin of the Crustacea

JOHN L. CISNE

I. INTRODUCTION

Despite continually growing knowledge of Arthropoda and ever more refined approaches to phylogenetic analysis, there is still no general agreement concerning the origin of Crustacea and their relationships to the other major arthropod groups. The problem of crustacean origins and relationships is part of a still larger and more complex one concerning the unity of the Arthropoda as a natural taxonomic group. In reviewing earlier thinking on arthropod classification and phylogeny, Tiegs and Manton (1958) raised the possibility that Arthropoda are polyphyletic, not monophyletic, as widely believed at that time (e.g., Snodgrass, 1938). Surveys and syntheses of comparative functional morphology of arthropod limb mechanisms (Manton, 1950, 1963, 1964a,b, 1965, 1966, 1967, 1972, 1973, 1977) and comparative annelid-arthropod embryology (Anderson, 1973) have led to the

THE BIOLOGY OF CRUSTACEA, VOL. 1

ISBN 0-12-106401-8

conclusion, at last fully stated by Manton (1977), that the three major modern groups—Crustacea, Chelicerata, and Uniramia—represent distinct phyla, no two of which share a common ancestor that was itself an arthropod. This conclusion is controversial; many workers find what they consider to be strong evidence for a monophyletic Arthropoda (e.g., papers in Gupta, 1979a). It will probably be some time before any general agreement is reached concerning the major features of arthropod phylogeny.

This review briefly surveys progress toward resolving the origin and relationships of Crustacea. Specifically, it considers these questions: What are the primitive unifying characteristics of Crustacea? How might an ancestral crustacean have embodied them? What were the anatomical, functional, morphological, developmental, life-historical, and ecological characteristics of the ancestral crustacean? What sort of organism was the ancestral crustacean's ancestor? What do comparisons of primitive representatives of Crustacea and other groups indicate about their mutual origins and relationships? How might Crustacea have arisen and diversified?

II. CRUSTACEA, THE ANCESTRAL CRUSTACEAN, AND THE CRUSTACEAN ANCESTOR

A. Characterization of Crustacea

Crustacea evidently represent a single, well-defined group. Its characteristics of interest include the following:

(1) The adult head includes five segments that typically bear limbs: two preoral segments that bear the first and second antennae, and three postoral segments that bear the mandibles and first and second maxillae. In the nauplius larva, the second antenna is postoral. Because the segmental nature of the one or two pre-antennal somites found in some forms is unclear (see Anderson, 1973), the exact segmental composition of the head remains unclear.

(2) The mandibular mechanism is gnathobasic. The adult mandible is formed of the basal part of the limb; the endopod and exopod are reduced or absent in the adult. In feeding nauplius larvae, the naupliar enditic process near the base of the second antenna serves a mandibular function. Typically, preoral limbs are developed for sensory operations and postoral head limbs for trophic ones.

(3) Development proceeds on a modified spiral cleavage pattern in which a yolk-rich 4D cell comprises the presumptive midgut and the 3A, 3B, and 3C cells comprise the presumptive mesoderm (Anderson, 1973). In forms with early-hatching larvae, the larva is a nauplius.

B. The Ancestral Crustacean

In addition to uniformity in the characters listed above, similarities in limb structure (Sanders, 1957, 1963; Hessler and Newman, 1975), skeletomusculature (Hessler, 1964), and feeding mechanisms (Sanders, 1963; Hessler, 1964; Hessler and Newman, 1975) among more primitive groups justify the idea that there was a single crustacean ancestor that was itself a crustacean. However, dichotomies between branchiopods and non-branchiopods on larval feeding mechanism (Gauld, 1959; Sanders, 1963) and between malacostracans and non-malacostracans on the organization of the nervous system, the eye in particular, and the urogenital system (Dahl, 1963) indicate that divergences between the major crustacean groups extend back to the most primitive levels of crustacean organization. Though each of the most primitive modern groups—notostracan, anostracan, and conchostracan branchiopods, phyllocarid malacostracans, and, most primitive of all, cephalocarids—is specialized in its own characteristic ways, together they form a sound basis for reconstructing the ancestral crustacean's essential characteristics.

1. LIFE CYCLE AND DEVELOPMENT

The ancestral crustacean was probably marine and free-living throughout its life cycle. To judge from benthic cephalocarids and phyllocarids, and from the particle-adapted trunk limb feeding mechanisms in these forms and the similar mechanism expected in ancestors of pond-living branchiopods (Cannon, 1933; Sanders, 1963), at least the adult was probably benthic. Reproduction was probably sexual. The various reproductive specializations among hermaphroditic cephalocarids (Hessler *et al.*, 1970) and the usually dioecious branchiopods give no clear guidelines as to the number of eggs laid at a time nor as to the possible brooding of eggs. Development was probably very gradual, beginning with a nauplius that hatched at a very early stage (i.e., a true nauplius) from a small egg and commenced feeding soon after hatching. It is unclear whether the nauplius was primitively a pelagic, dispersal-adapted larva, as de Beer (1958) suggested; benthic and not dispersal-adapted, as Sanders (1963) suggested on the basis of cephalocarid development; or demersal and dispersal-adapted (see Mileikovsky, 1971, on this newly recognized larval type). Sanders' (1963) comparative evaluation of pelagic and benthic copepod nauplii indicated that the nauplius is adaptively more plastic than had been thought previously. Metamorphosis between larval and juvenile stages was probably marked by little more than the loss of the naupliar enditic process. No more than one or two segments and fully formed segmental limb pairs were probably added at each molt (see Sanders, 1963). The transition from the naupliar net-sweep locomotory and feeding mechanism to the adult trunk limb mechanism was probably smooth and gradual, as in cephalocarids and extinct lipostracan branchiopods (see Sanders, 1963).

The size of the ancestral crustacean is important because many features associated with the larval and/or phyletically primitive condition actually may be more directly related to small size and developmental state than to phyletic primitiveness. Because developmental trends (e.g., forward movement of segmental organs around the oral region and progressive differentiation of segments) parallel evolutionary trends [respectively, shift of anterior head segments from postoral to preoral position in the adult and progressive tagmosis (Lankester, 1904)], simplistic attempts at understanding arthropod evolution in terms of direct transition from one adult form to another can and have led to confusing and erroneous conclusions. It is the entire developmental pattern that evolves, and it evolves in concert. Müller's (1864) misunderstanding of this fundamental point in his still-fascinating book on Crustacea provided one of the inspirations for Haeckel's Biogenic Law. Misunderstanding of this point persists even in current work, and particularly in some efforts on comparative functional morphology for which developmental studies provide no background (see Section III,B).

In line with the general evolutionary trend toward evolutionary increase in adult size, one would expect the ancestral crustacean to have been relatively small. At 3 mm length, adult caphalocarids fall in this size range. Still tinier mystacocarids are probably secondarily specialized for meiofaunal life, with miniaturization having perhaps been achieved through progenesis (in the sense of Gould, 1977).

A critical question is the extent to which the apparently very primitive features of Cephalocarida are in fact expressions of secondary specialization through progenesis. Because early development has apparently been very conservative within Crustacea, progenesis and coincident miniaturization of the adult can result in a body organization that is only apparently primitive. Copepoda are a case in point. As Calman (1909) pointed out, retention of the mandibular palp in the adult, a phyletically primitive character in line with the evolutionary trend toward limb specialization by simplification (Sanders, 1963), is in fact more directly related to retention of a typically larval feature in an adult that is in effect ecologically specialized as a professional larva.

All evidence suggests that Cephalocarida are not secondarily modified in this way. Progenesis, as it seems to have taken place in the origins of Cladocera and Copepoda and perhaps other maxillopod groups, appears to have been coupled with the appearance of other adaptations for ecological opportunism. A short generation time, achieved through progenesis, is perhaps the most effective way of dramatically increasing and maintaining a high intrinsic rate of natural increase—a *sine qua non* for ecological opportunism (Lewontin, 1965). Cephalocarids show, if anything, the extreme opposite ecology and life history design from opportunists. They are evidently sparce but cosmopolitan in the world ocean, including the deep sea (Hessler and Sanders, 1973). The benthic larva is not adapted for long distance dispersal (Sanders, 1963). Though never measured, the mean generation time is probably relatively long. *Hutchinsoniella* occurs commonly throughout the year in Buzzards Bay, Massachusetts (Sanders, 1960), but bears eggs there only during the summer (Sanders and Hessler, 1964), suggesting that the generation time is on the order of a year, i.e., many times longer than generation times characteristic of comparably small, opportunistic crustaceans. Considering that *Hutchinsoniella* bears only two eggs at a time, the population's intrinsic rate of increase must be relatively very low in Buzzards Bay. Full development of the trunk in Cephalocarida also argues against progenesis, for in other groups, progenesis has generally been accompanied by substantial trunk reduction.

Despite apparent specializations in their reproductive biology, Cephalocarida probably preserve the essential features of development in the ancestral crustacean: very gradual development to an adult that is not greatly different from the larva in size or other characteristics as are much larger adults in more derived groups. Hermaphroditism (Hessler, *et al.*, 1970), retention of the eggs on the parent's body, and hatching as a metanauplius are all probably phyletic specializations, but ones of minor nature. Hermaphroditism is perhaps an accomodation for reproduction at the low population densities at which cephalocarids normally seem to occur.

2. ADULT ANATOMY AND FUNCTIONAL MORPHOLOGY

Hessler and Newman (1975) synthesized previous work on the problem of reconstructing the ancestral crustacean in unprecedented detail (Figs. 1-3). Beyond the basic crustacean features, the reconstruction's essential points are as follows:

(1) The preoral region includes a stalked eye and a long, sensory-type first antenna (Figs. 1 and 2).

(2) The second antenna, situated beside the mouth, is essentially postoral, much as in the nauplius. It shows a high degree of serial homology with trunk limbs, retaining an ambulatory-type endopod, but is differentiated from them in having the exopod developed as a flagelliform, sensory-type antenna and in having the epipod reduced (Figs. 1 and 2).

(3) All postoral segments show a very high degree of serial homology in external and internal anatomy (Figs. 1 and 2). Postantennal head limbs and trunk segments all bear serially very similar limbs. There is no sharp distinction between head and trunk or between thorax and abdomen.

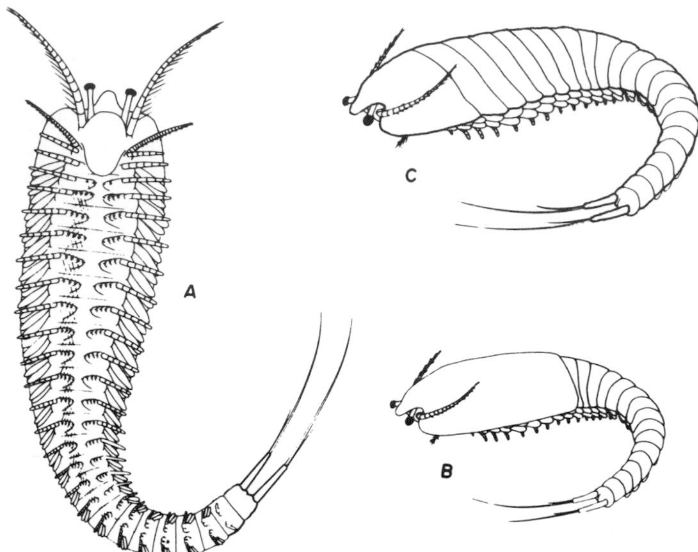

Fig. 1. The ancestral crustacean as reconstructed by Hessler and Newman (1975): (A) ventral view and (B and C) oblique dorsal views showing the crustacean (B) with and (C) without the carapace it may or may not have had. (From Hessler and Newman, 1975.)

(4) Postantennal limbs conform to a triramous pattern (Figs. 1 and 3). The telson bears a caudal furca (Fig. 1).

(5) A carapace may or may not have covered the anterior part of the trunk (Fig. 1).

(6) It possessed a trunk limb locomotory and feeding mechanism of the cephalocarid type: As in cephalocarids, the food collection mechanism lacked elaborate filtration devices [cephalocarids are non-selective deposit-feeders (Sanders, 1963)], and food material was transported forward in the food groove borne on endites.

The one truly new feature of Hessler's and Newman's reconstruction is the triramous limb (Fig. 3). While the constancy of the endopod and the exopod had long been noted in what was thought to be the typically biramous crustacean limb, the constancy of the epipod (the third ramus) in primitive groups had simply been overlooked, perhaps because earlier comparative studies had emphasized higher malacostracans at the expense of the now much better known entomostracans. Otherwise, earlier workers independently came to similar reconstructions of the ancestral crustacean. Calman (1909) verbally sketched a basically similar ancestor that, understandably, emphasized Notostraca, the most primitive Crustacea known at the time. Based on his comparative studies of branchiopod feeding mechanisms,

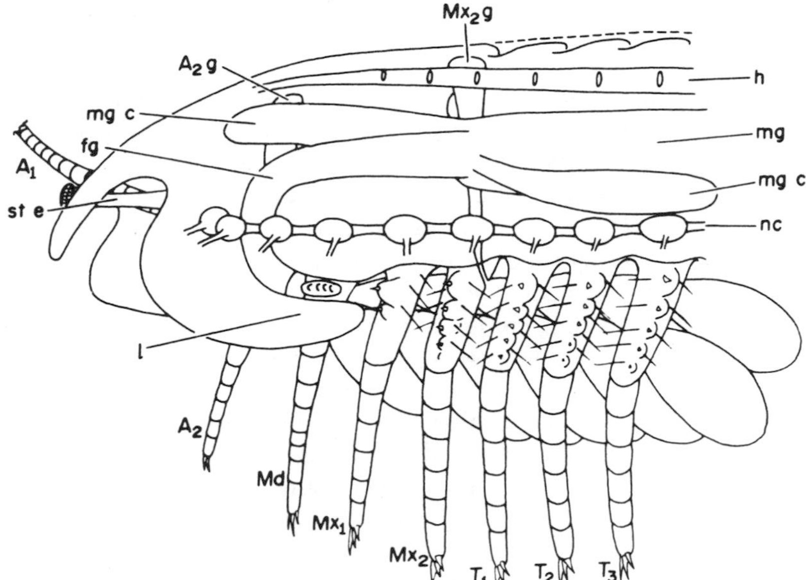

Fig. 2. Midsaggital section through the ancestral crustacean's head and first three thoracic segments as reconstructed by Hessler and Newman (1975). Symbols indicate the first (A_1) and second (A_2) antennae, mandible (Md), first (Mx_1) and second (Mx_2) maxillae, first (T_1) through third (T_3) thoracopods, antennal gland (A_2g), foregut (fg), heart (h), labrum (1), midgut (mg), midgut caecum (mg c), maxillary gland (Mx_2g), nerve cord (nc), and stalked compound eye (st e). (From Hessler and Newman, 1975.)

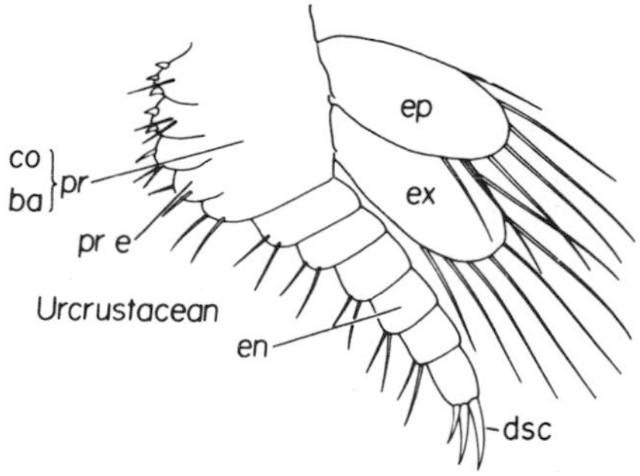

Fig. 3. The ancestral crustacean's trunk limb according to Hessler and Newman (1975). Symbols indicate the coxa (co) and basis (ba), distal setal claws (dsc), endopod (en), epipod (ep), exopod (ex), protopod (pr), and protopodal endites (pr e). (From Hessler and Newman, 1975.)

Cannon (1933) most insightfully proposed that the branchiopod ancestor had a setose basal endite developed for mechanical transport of material in the food groove—a condition unknown in branchiopods and discovered by Sanders (1963) in cephalocarids. Whether or not the ancestral crustacean had a carapace has long been controversial, as it was between Hessler (Fig. 1,B) and Newman (Fig. 1,C). My own opinion is that the ancestral crustacean probably lacked one. The carapace seems to be an adaptively plastic feature that is developed in a variety of contexts. It probably has no great phylogenetic significance. I envision an extensive carapace first developing among large descendents of the crustacean ancestor in connection with encasement of the feeding current and the limbs that created it.

Briggs' (1978) reconstruction of the Middle Cambrian phyllocarid *Canadaspis* (Figs. 4 and 5) confirms several of the less extensively documented primitive features that Hessler and Newman (1975) give the ancestral crustacean. All postmandibular limbs show a very high degree of serial homology, higher, and thus phyletically more primitive, than in Cephalocarida (Fig. 4). Even the first maxilla has the basic form of thoracic limbs (Fig. 5). Setose endites at the bases of postmandibular limbs indicate that the food groove was developed for mechanical transport of food, as in cephalocarids and modern leptostracan phyllocarids. The fact that all postmandibular limbs are so similar, and that they all bear ambulatory-type endopods, suggests that food collection and locomotion were shared more or less equally among these limbs, not neatly divided between thorax and abdomen as they are in leptostracans. There are no indications of a thoracic brood pouch, suggesting that *Canadaspis,* unlike the leptostracans, did not brood eggs and young, and that eggs hatched at an early stage, not as

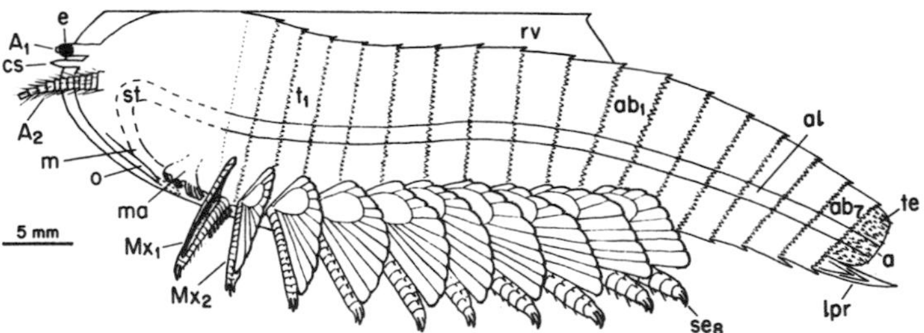

Fig. 4. Lateral view of the Middle Cambrian phyllocarid crustacean *Canadaspis perfecta* as reconstructed by Briggs (1978). Symbols indicate the anus (a), first (ab_1) and seventh (ab_7) abdominal segments, first (A_1) and second (A_2) antennae, intestine (al), cephalic spine (cs), stalked eye (e), lateral projection of the seventh abdominal segment (lpr), mouth region (m), mandible (ma), first (Mx_1) and second (Mx_2) maxillae, labrum (o), carapace (rv), eighth thoracopod (se_8), stomach (st), first thoracic segment (t_1), and telson (te). (From Briggs, 1978.)

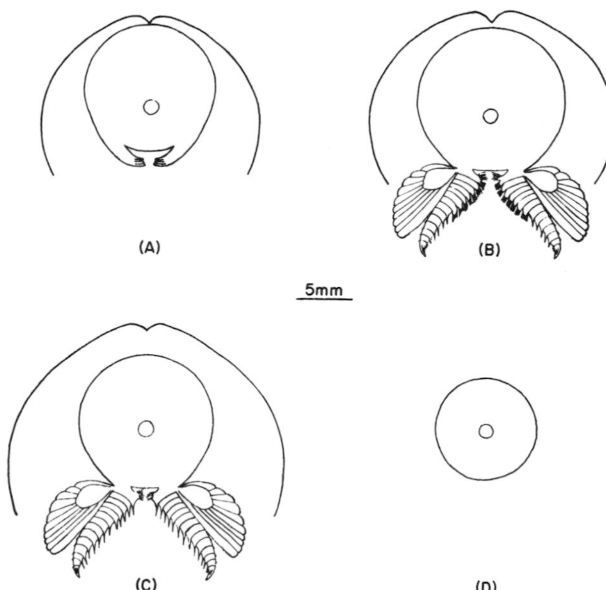

Fig. 5. Cross sections through the Middle Cambrian phyllocarid crustacean *Canadaspis perfecta* as reconstructed by Briggs (1978): (A) mandibular segment (B) first maxillary segment, (C) second thoracic segment, and (D) fourth abdominal segment. (From Briggs, 1978.)

leptostracan-type mancas. Thus one cannot agree with Manton's assessment of *Canadaspis* as "a near leptostracan" (Manton, 1977, p. 269). It is very much more primitive in most, but not all, respects. Its specializations may relate to its large size (10 cm length). Among these are the reduced relative size and loss of trophic function of the second antenna. The much smaller modern phyllocarids have relatively much larger second antennae with both sensory and trophic functions in the adult (Cannon, 1927). As the best known of the earliest Crustacea, *Canadaspis* indicates that divergences and specializations, such as malacostracan tagmatization, took place rapidly and at very primitive organizational grades not long after the appearance of Crustacea near the beginning of the Cambrian.

Hessler and Newman (1975) recount strong evidence from comparative functional morphology (Sanders, 1963) and comparative skeletomusculature (Hessler, 1964) that the basic trunk limb feeding mechanism is an evolutionarily conservative, unifying characteristic of the primitive Crustacea, and they counter Cannon's and Manton's (1927; Cannon, 1927; Manton, 1977) arguments to the contrary. To summarize, Cannon and Manton were more or less forced to conclude that the trunk limb mechanism evolved independently in Phyllocarida and Branchiopoda from the then widely accepted but not untenable belief that Mysidacea are phyletically more primitive than Leptostraca. Briggs (1978) rather dramatically reconfirms the conclusion from a great deal of evidence now available that in fact phyllocarids represent an organizational grade much more primitive than mysids. Though the argument would have seemed daringly speculative until discovery of the Cephalocarida, the cephalocarid-like leptostracan food groove mechanism falls into place beautifully with the ancestral mechanism that Cannon (1933), himself, postulated from comparative study of branchiopods, as Sanders (1963) showed. It is interesting to note how Cannon's and Manton's evidently mistaken ranking of two groups by degree of phyletic primitiveness could, through its far-reaching implications, create so distorted a picture of evolution among many more groups.

Because developmental and evolutionary trends parallel one another in a potentially confusing way (see Section II,B,1), there will probably always be disagreement as to whether any one reconstruction of the ancestral crustacean represents the desired organism, its ancestor, or its larva. The one feature of Hessler's and Newman's (1975) reconstruction that I, myself, find difficult to accept is the high degree of similarity between the second antenna and postantennal limbs. This high similarity would suggest that the limb did not have a much different larval developmental history from others, which it certainly would have had in the ancestral crustacean's nauplius larva. I would expect the second antenna to lack an ambulatory type endopod, i.e., to retain more the form and trophic functions it presumably had in the naupliar net-sweep feeding mechanism, much as in Cephalocarida and Lipostraca, though I would also expect it to have the same components as postoral limbs. Thus I wonder whether Hessler and Newman have in fact reconstructed the ancestral crustacean's ancestor on this one point—a crustacean-like arthropod that did not have a nauplius larva.

C. The Ancestral Crustacean's Ancestors

Certain characteristics of the ancestral crustacean's immediate ancestors can be reconstructed by extrapolating back from the cephalocarid condition along evolutionary trends within Crustacea, as done by Sanders (1957, 1963), Cisne (1974, 1975), and Hessler and Newman (1975). These characteristics include the following:

(1) From the trend toward progressive cephalization of segments (Lankester, 1904), the adult head probably included only four segments that typically bear limbs; the second maxillary segment, which is little-cephalized in cephalocarids (Sanders, 1955, 1963; Hessler, 1964) and which is embryologically not part of the primitive crustacean head (Anderson, 1973), was probably a trunk segment.

(2) From the evolutionary trend toward progressive movement of anterior head segments to a preoral position (Lankester, 1904), the second antenna was probably postoral in the adult as well as in the larva.

(3) From the evolutionary trend toward progressive specialization and tagmatization of segments (Lankester, 1904), all postoral head segments were probably less strongly differentiated among themselves. In the extreme, postoral head segments were probably scarcely differentiated among themselves, and scarcely differentiated from trunk segments. The larva may not have had the naupliar net-sweep feeding mechanism with its accompanying specialization of the second antennal segment.

(4) From the trend toward specialization of limbs by simplification of a primitively complex primitive pattern (Sanders, 1957), the postoral limb must have been at least as complex as the primitive triramous limb (Fig. 3).

Such an anthropod would probably be called a trilobitoid if known, as Hessler and Newman (1975) point out, which is to say that it would have been a trilobite-like but basically unclassifiable (see Section III,C,1) marine anthropod. The four limb-bearing head segments, the one preoral limb-bearing segment, and the serially similar postoral limbs point toward trilobites as crustacean ancestors; but the multiramous limb, more complex and hence phyletically more primitive than the biramous trilobite limb (Fig. 6), points back to some still more primitive, pre-trilobite organizational grade.

III. RELATIONSHIPS OF CRUSTACEA

A. Problems of Arthropod Classification and Phylogeny

For Arthropoda, as for any major group, developing a classification and a phylogeny that seem to satisfactorily depict the relationships among the constituent natural groups involves finding the one logically most satisfactory solution to a very complicated set of simultaneous logical problems—the ordination of characters of taxonomic importance, the orientation of evolutionary trends, and the organization of constituent taxa that, together, give the most plausible phylogeny. There is no general agreement as to which among several possible solutions is best, and no agreement as to whether Crustacea share a common ancestry with other groups. In recent reviews, authors have endeavored to make the evidence best fit schemes in which Arthropoda exclusive of Onychophora comprise a single, natural phylum (Snodgrass, 1938; Baccetti, 1979; Boudreaux, 1979; Clark, 1979; Paulus, 1979; Weygoldt, 1979), in which Arthropoda inclusive of Onychophora comprise two phyla (Cisne, 1974), at least two phyla (Anderson, 1973), or at least three phyla (Manton, 1977; Schram, 1978). I suspect that attempting to determine relationships among Arthropoda from available evidence rather resembles attempting to solve a set of simultaneous equations in which unknowns outnumber equations: there is more than one logically satisfactory solution, and no one of them can be proved to be the unique solution desired. Discovery of one phylogenetic scheme that, from a certain taxonomic viewpoint, seems to satisfy evidence on certain aspects of arthropod biology does not necessarily signify discovery of the scheme that best satisfies all evidence as seen from all taxonomic viewpoints.

Historically, three interrelated complications have bedeviled phylogenetic studies on Arthropoda:

(1) Commonness of close convergence between otherwise quite different, seemingly unrelated arthropods: Tiegs and Manton (1958) and Manton (1977) showed the logical impossibility of constructing a phylogenetic scheme that did not presuppose close convergences between at least some members of different major groups, and pointed out the likelihood of convergent development of gross anatomical characters used in earlier classifications relative to apparently more conservative functional morphological and embryological characters. The taxonomically most conservative characters, it seems, are to be found in the operation of limb mechanisms and in embryonic development.

(2) Nonpreservation of these most desirable high-level taxonomic characters in fossils: By and large, the fossil record preserves only the gross anatomy of the most durable parts of the cuticle. Embryonic development is intrinsically nonpreservable. Larval development is difficult to study owing to the great rarity of fossil larvae, and to problems of identifying them surely enough to place them in a developmental series. Fine anatomical details from which the workings of limb mechanisms might be surmised with any degree of confidence are very rarely preserved in adults, much less in larvae. Tiny head limbs that, in the absence of embryological evidence, might be the only clues as to head structure, and thus to identification as to major group, may be either not preserved or difficult to interpret even in the rare cases in which they are preserved.

(3) Incompleteness of modern arthropods as a representation of all arthropods that have ever existed: Unknown "missing links" implied by gaps within major modern groups (e.g., between merostome and arachnid chelicerates) abound. The fossil record not only fails to fill in the missing links but, in its great diversity of distinctive forms (e.g., trilobites and the miscellany of trilobitoids; see Section III,C,1), it suggests that the modern representation of Arthropoda is still less complete than modern forms alone would imply.

These factors have the following implications for phylogenetic analysis and classification:

(1) Classifications and phylogenies must be based primarily on modern forms, the only ones for which good information on taxonomic characters is available.

(2) The ordination of characters and the orientation of evolutionary trends developed from comparative study of modern forms generally cannot be tested against the fossil record.

(3) Extinct forms without obvious similarities to modern major groups—i.e., the very forms that would give the most new information about phylogeny—cannot be integrated into the comparative, synthetic process through which phylogenies and classifications are arrived at. These forms are basically unclassifiable and unidentifiable except as singular oddities.

(4) The exact phylogenetic significance of similarities and differences among major modern groups is unclear. The distinctness of the major groups as represented at present has in all likelihood been accentuated through extinction of primitive forms and through secondary specialization among survivors. The extent of this bias is impossible to determine without better information from the fossil record.

(5) Necessarily based primarily on the incomplete modern representation of Arthropoda, a phylogeny is likely to be lacking as a depiction of relationships among all groups that have ever existed. Some major groups have perhaps gone undiscovered (see Section III,C,1).

(6) Such a phylogeny cannot be adequately tested against the fossil record.

In light of these problems, it is easy to understand how available evidence leaves so much room for the traditional differences between taxonomic "lumpers" and "splitters" to express themselves in the controversy over arthropod phylogeny.

B. Uniramia, Mandibulata, and Crustacea

Though Manton's (1964, 1950, 1963, 1964a, 1965, 1966, 1967, 1972, 1973) and Anderson's (1973) work at first seemed to settle the issue of the unity of Uniramia (Onychophora-Myriapoda-Hexapoda) versus the unity of Mandibulata (Crustacea-Myriapoda-Hexapoda), recent reviews (e.g., Hessler and Newman, 1975; Baccetti, 1979, on sperm ultrastructure; Gupta, 1979b, on hemocytes; Paulus, 1979, on eye structure; Weygoldt, 1979, on embryology) have brought up new evidence and interpretations to support resurrecting Snodgrass' (1938) Mandibulata. This complex matter appears to be a side issue as regards crustacean origins. The possibility of descent of

Crustacea from onychophoran-like forms, as Snodgrass (1938) hypothesized, is of little interest here because no unequivocal evidence bears on this speculation. The possibility of descent of Crustacea from Myriapoda or Hexapoda has seemed implausible enough not to have been worth considering. If any direct relationship exists, the line of descent is the reverse.

The Uniramia-Mandibulata question will not be resolved until considerably more evidence has been brought to bear on it. Thus far, evidence has been less than overwhelming that the hexapod mandible represents either a whole limb, as Manton (1964b, 1972) contended in making the whole limb jaw a diagnostic character of Uniramia, or a crustacean-type gnathobasic jaw, as advocates of Mandibulata would have it (see Boudreaux, 1979). Both Manton's comparative functional morphology and Anderson's comparative fate map embryology are as yet relatively new approaches to arthropod phylogeny and classification. There are large gaps in the coverage of Arthropoda among the relatively few arthropods that have been studied and evaluated by these methods. Among the most obvious and important ones are the embryology of Cephalocarida, and, at least for the perspective it might give, the functional morphology of limb mechanisms in Tardigrada, whether or not their little-known embryology indicates that they are arthropods. Attention needs to be given to the ontogenetic development of limb mechanisms (e.g., Sanders, 1963), particularly as regards transformations in mechanisms during development, in order that transformations during evolution can be better understood (see Section II,B,1). Therein may lie the keys to problems such as the one Cannon and Manton (1927; Cannon, 1927) confronted in assessing the evolution of feeding mechanisms from comparison of adult forms alone (see Section II,B,2). More perhaps needs to be known about the adaptive significance of developmental patterns in relation to habitat and life history design. One wonders to what extent the differences in developmental patterns that Anderson (1973) records for Uniramia and marine arthropods merely reflect adjustments for reproduction under physiologically quite different circumstances.

C. Trilobitomorpha, Chelicerata, and Crustacea

Though modern Chelicerata and Crustacea show no evidence of direct relationship, both show suggestive indications of relationship to Trilobitomorpha. Whether relationships among Trilobitomorpha, Chelicerata, and Crustacea bind the three together as parts of a natural group is now controversial (see Cisne, 1974, 1975; Hessler and Newman, 1975; Manton, 1977; Schram, 1978). The question devolves onto the ones discussed below.

1. NATURE OF TRILOBITOMORPHA

Once thought to be united as a natural group by possession of Størmer's (1939, 1944) "trilobitan limb," the Trilobitomorpha now appear to be no more than a heterogeneous, unnatural collection of forms united by no taxonomically important characters. Though Trilobita appear to be well defined as a major group by their head structure (Whittington, 1957; Cisne, 1974, 1975), not even they are united by Størmer's "trilobitan limb" (Bergström, 1969, 1973; Cisne, 1974, 1975; Whittington, 1975a, 1977, 1979, 1980). Forms have been included in Trilobitoidea (i.e., the non-trilobite Trilobitomorpha) that are now considered to be crustaceans (e.g., *Canadaspis:* Briggs, 1978, trilobites (*Naraoia:* Whittington, 1977), altogether distinctive arthropods (e.g., *Marrella:* Whittington, 1971), and problematical non-arthropods (e.g., *Opabinia:* Whittington, 1975b). Past assertions that some particular organism is a trilobitoid amount to little more than assertions of ignorance about its true affinities. Persistence of the names Trilobitomorpha and Trilobitoidea in the literature is justified only by their utility as taxonomic wastebaskets.

Trilobitomorpha may contain representatives of major groups other than Trilobita, undiscovered major groups that cannot be adequately characterized for want of reliable information on head structure (see Manton, 1969). Though it is difficult to conceive of a jawless arthropod with a head comprised of two limb-bearing segments (both preoral and bearing antenniform limbs), this is what Whittington's (1971) reconstruction indicates head structure to be in *Marrella.* Following criteria applied to modern Arthropoda, one could propose a new major group for *Marrella* and for numerous other distinctive Paleozoic forms [e.g., *Mimetaster* and *Vachonisia* (Stürmer and Bergström, 1976), and as yet undescribed forms from Mazon Creek]. What greater understanding would be achieved by so labeling these possible "evolutionary experiments" is debatable.

2. CHELICERATA AND TRILOBITA

The degree of relationship between Chelicerata and other groups is difficult to assess owing to the specialized nature of known chelicerates. Truly primitive chelicerates are evidently unknown things of the past, all the more so now that the extinct, trilobite-like Aglaspida, once candidates for such forms, have been shown to be probably non-chelicerate (Briggs *et al.,* 1979). The chelicerate prosoma represents a degree of cephalization paralleled only in more derived Malacostraca among the Crustacea. Chelicerate development, which is almost completely embryonized, is apparently specialized to the point that no hints of relationship to other groups remain (Anderson, 1973). Perhaps this specialization has to do with adaptation for terrestrial embryonic development, not only in terrestrial arachnids, but in marine horseshoe crabs. Modern *Limulus* lays its eggs in high intertidal, though not fully terrestrial, habitats. Fisher (1979) found abundant indications that the Carboniferous horseshoe crab *Euproops* actually lived subaerially amid coal swamp vegetation.

Størmer (1933, 1939, 1942, 1944, 1951) popularized the idea of a close relationship between Trilobita and Chelicerata. Some of the similarities he pointed out have stood the test of time as valid indications of relationship while others have not (see Hessler and Newman, 1975), and some new ones have been discovered. The apparently valid indications include the following:

(1) Flexible coxa-body articulation in trilobite trunk limbs (Manton, 1964b; Cisne, 1974) and merostome prosomal limbs (Manton, 1964b): The condition in merostomes probably represents a somewhat specialized version of a very primitive condition in which limb movements about this joint were controlled not by a two point hinge and an appropriately simplified musculature, as in larger crustaceans, but by an appropriately more complicated musculature—a condition probably found in trilobites (Cisne, 1975, 1981). Persistence of this condition in merostomes is evidently related to the flexibility required for promotor-remotor ambulatory movements, transverse biting movements, and a combination of both movements during feeding (see Manton, 1964b).

(2) Structure of the ventral endoskeleton: The ladderlike structure of the primitive chelicerate endoskeleton reconstructed by Firstman (1973) is also found in trilobites (Cisne, 1975, 1981).

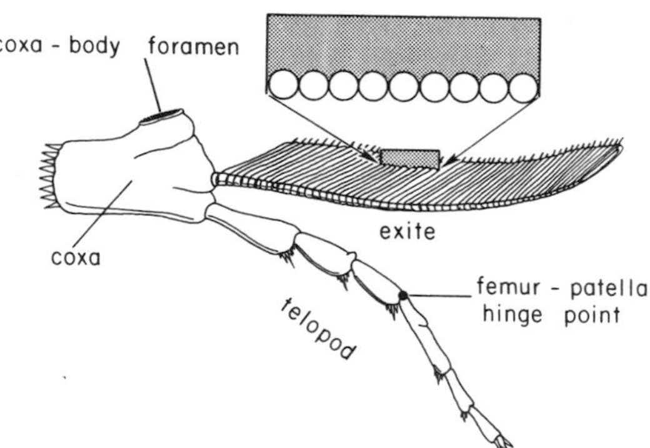

Fig. 6. Anterior view of the fifth thoracic limb in the Middle Ordovician trilobite *Triarthrus eatoni* showing the circular shape of exite filaments in cross section (see Fig. 7). (Redrawn from Cisne, 1975, 1981, with additions.)

(3) Gnathobasic food-handling in trilobites (Cisne, 1975; Whittington, 1975a, 1977, 1980) and in merostomes and some arachnids (Manton, 1964b): The trilobite *Naraoia* is particularly merostome-like in having relatively massive, spinose coxae suited for transverse biting (see Whittington, 1977, Fig. 99).

(4) Telopod structure in trilobites, merostomes, and other chelicerates (Størmer, 1933, 1939, 1944, 1951): The femur-patella joint may be homologous between trilobites (Fig. 6) and chelicerates. Location of the femur-patella hinge point at the top of the limb in *Triarthrus* (Fig. 6) indicates that extension of the limb across this joint could not have been accomplished by the musculature and must have been accomplished by means of internal hydrostatic pressure or cuticular elasticity, as in the corresponding chelicerate joint. The femur-patella hinge points and more distal intersegmental hinge points in *Olenoides* are likewise located at the top of the limb (Whittington, 1980). It is difficult to see how intrinsic muscles could have extended the limb, as Whittington (1980) claimed. Since intrinsic muscles could only have been situated ventral to the hinge, it would seem that they could only have flexed the limb.

The first three points above indicate only that Trilobita and Chelicerata share a common ground plan and, evidently, a common arthropod ancestor at some very primitive organizational grade; the fourth point may represent a shared derived character indicative of more direct relationship.

Based on Størmer's (1939, 1944) work, it has been widely accepted that filaments on the trilobite exite categorically have a bladelike shape, and that this structural similarity and the implied similarity as to respiratory function with opisthosomal gills in merostomes imply an uniquely close relationship between Trilobita and Chelicerata (e.g., Manton, 1964b, 1977). However, recent discoveries on *Triarthrus* (Figs. 6 and 7) indicate that Størmer generalized too broadly from too little evidence on trilobite exite filaments, and that their structure provides no compelling evidence as to a close relationship with chelicerates. To be sure, a few trilobites belonging to more derived groups have blade-shaped exite filaments (e.g., *Ceraurus:* Størmer, 1939). But it has been difficult to determine the original shape of exite filaments in most of the few trilobites for which limbs are known owing to postmortem deformation during compaction of the matrix. Serial sections through one of the three-dimensionally preserved specimens from Beecher's Trilobite Bed (Cisne, 1973b) (Fig. 7) show that exite filaments in *Triarthrus* are roughly circular in cross section (Fig. 6). By shape alone, the filaments are more like setae than they are like merostome gill blades. The new finding strongly supports Bergström's (1969, 1973), Cisne's (1975), and Whittington's (1975a) criticisms of the idea that the trilobite exite was primarily a

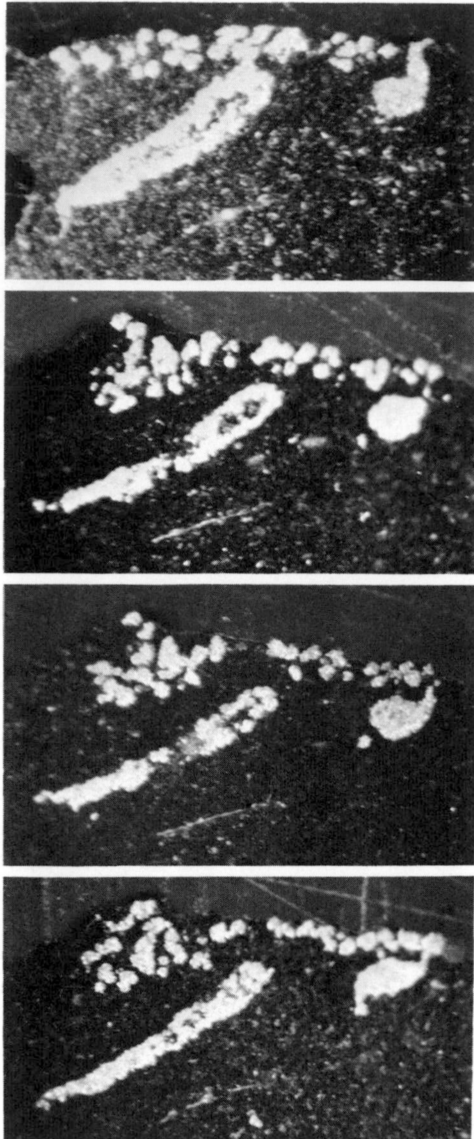

Fig. 7. Exite structure in the Middle Ordovician trilobite *Triarthrus eatoni:* four serial sections through the overlapping exite fringes of the third and fourth thoracic limbs on the left side of Museum of Comparative Zoology specimen MCZ 3638/4. Like the schematic cross section in Fig. 6, these sections are very nearly normal to the plane of the exite filaments. The pyritized filaments, about 0.03–0.05 mm in diameter, show up in these reflected light photomicrographs as the small, light-colored, round objects arranged in two horizontal, parallel

respiratory organ. Most likely it had manifold functions, including gas exchange. Bladelike exite filaments probably evolved independently in more specialized trilobites and in merostomes through modification of a more generalized structure as found in *Triarthrus*.

Chelicerata probably arose from an ancestor shared with Trilobita, not from Trilobita themselves. Discovery of the non-chelicerate nature of the trilobite-like Aglaspida (Briggs *et al.*, 1979) eliminates what had been a strong point favoring close relationship between them. Chelicerata probably represent an early offshoot from trilobite-like stock that diverged radically in becoming adapted for predaceous feeding.

3. TRILOBITA AND CRUSTACEA

As already discussed (Section II,C), certain evolutionary trends within Crustacea point back to an ancestor with a trilobite-like head structure and body tagmatization. The difference between gnathobasic mandibular mechanisms in trilobites and primitive crustaceans appears to be one of degree, not one of kind (Cisne, 1974, 1975). Otherwise, perhaps the most important similarities that indicate relationship between the two groups are those relating to a crustacean-type—specifically a cephalocarid-type (see Sanders, 1963)—trunk limb feeding mechanism in trilobites (Hessler and Newman, 1975; Cisne, 1974, 1975). With the exception of certain trilobitoids (see Tiegs and Manton, 1958), any indications of such a mechanism are unknown outside Trilobita and Crustacea (Manton, 1977). Though the mechanism may have been secondarily lost in Chelicerata, it was evidently never developed among Uniramia (Manton, 1977), and thus was not in any possible (and probably non-arthropodized) common ancestor of all the Arthropoda. The features that indicate this mechanism in trilobites are as follows:

(1) The four anatomically best known trilobites—*Triarthrus* (Cisne, 1975, 1981), *Olenoides* (Whittington, 1975a, 1980), the abherrant *Naraoia* (Whittington, 1977), and *Cryptolithus* (Cisne and Dempster, 1982—have a cephalocarid-type, uninvaginated food groove developed between paired postoral coxae. Food was evidently carried forward in the food groove while borne on endites, as in cephalocarids and phyllocarids.

(2) Possibly excepting *Cryptolithus,* these trilobites have relatively large,

---

rows. Each row corresponds to one of the two overlapping exites; the row above is probably that corresponding to the more anterior limb (see Cisne, 1975, 1981). The large, elongate, pyritized object inclined about 45° to the horizontal, and the large round object toward the upper right, are sections through the fourth and fifth thoracic telopods. The horizontal width of each picture corresponds to about 1.2 mm. (Sections and photomicrographs by Dr. Christine M. Hartman.)

blade-shaped coxae reminiscent of the cephalocarid protopod (Fig. 6). Variation in the space between adjacent coxae, if not in the whole interlimb space, could very well have created a feeding current system, specifically, currents that flowed medially between the limbs and into the midventral space and food groove (see Cannon and Manton, 1927; Sanders, 1963).

(3) Filaments on the featherlike exites in these trilobites (through not differently constructed filaments in the more derived phacopids: see Bergström, 1969) could perhaps have been involved in collection and concentration of particulate material suspended in the feeding current. The details of this process are unclear owing to the lack of any particularly close living analogs for trilobites as regards this point, and to ignorance concerning the precise pattern of limb movement in trilobites. For *Triarthrus,* the one trilobite for which the gut is fairly well known, the very small diameter of the mouth region and anus and the absence of any strong masticatory apparatus as seen in *Olenoides* and *Naraoia* suggest that the trilobite did in fact feed on the sort of particulate material seen to emanate from the gut in two specimens (Cisne, 1975, 1981). However, it is quite possible that trilobites fed in more than one way, and that more specialized forms not among the best known four secondarily lost the trunk limb mechanism, just as more specialized crustaceans have. Like trunk limb feeding branchiopods (see Cannon, 1933), phyllocarids (see Cannon, 1927), and mysids (see Cannon and Manton, 1927), which sometimes seize larger food items directly with the mandibles, trilobites may have seized larger food items directly in the food groove, as Whittington (1975a, 1977, 1980) suggested for *Olenoides* and *Naraoia*.

(4) *Triarthrus* had the same basic conformation of the trunk musculature that Hessler (1964) found to be characteristic of trunk limb feeding crustaceans: the dorsal longitudinal musculature is a sheet of parallel muscle fibers, the ventral longitudinal musculature is a pair of muscle bundles organized around a ladderlike endoskeleton, and the dorsal and ventral longitudinal musculatures are linked by a trusslike complex of dorosoventral muscles (Cisne, 1975, 1981). The exact trusslike pattern in which dorsoventral muscles are arranged is of no particular phylogenetic significance, as modern Crustacea show (see Hessler, 1964). The conformation of its trunk musculature merely reinforces the conclusion from other evidence that *Triarthrus* had a trunk limb food handling mechanism.

Trilobita and Crustacea probably descended from a trilobite-like common ancestor, a form that might be called a trilobitoid if known (see Section III,C,1). Now that the evolutionary conservatism of the trunk limb feeding mechanism seems established among primitive Crustacea (see Section II,B,2), it appears that the mechanism is a shared primitive character, an inheritance from a common ancestor, in Trilobita and Crustacea. That the ancestral crustacean's triramous limb (Fig. 3) is phyletically more primitive than trilobites' biramous limbs (Fig. 6) (see Section II,C) indicates that Crustacea could not have descended directly from the otherwise more primitive Trilobita, and probably that the trunk limb mechanism is represented in more primitive condition among the most primitive crustaceans. The mechanism probably underwent modification in trilobites, and perhaps even loss in some forms, just as it has among more derived crustaceans. Discovery of the seta-like, cylindrical exite filaments in *Triarthrus* (Figs. 6 and 7) shows that the difference in degree of phyletic specialization between trilobite and crustacean limbs is not nearly so great as it was supposed to have been when it was believed that trilobites categorically had specialized, bladelike exite filaments of the merostome sort. The difference seems to be understandable in terms of less radical specialization of trilobite limbs by anatomical reduction along lines analogous to those followed in specialization of crustacean limbs (see Sanders, 1957, 1963).

IV. THE ORIGIN OF THE CRUSTACEA

Crustacea appear to have arisen from trilobite-like ancestors during the first radiations of metazoan phyla and classes near the beginning of the Cambrian. They seem to have shared a common ancestor with Trilobita, Chelicerata, and some of the problematical arthropods now classed as trilobitoids—an ancestor that would seem to have been an arthropod. Following the classic pattern of an adaptive radiation (see Simpson, 1953), Crustacea and Chelicerata were initially parts of a more continuous spectrum that included trilobites and perhaps other extinct major groups (see Sections III,A and III,C,1), and they have grown progressively more distinct from one another as primitive, intermediate forms such as trilobites have become extinct, and as their own representatives have become more and more specialized. Very many problems concerning arthropod phylogeny remain to be solved, or at least solved more exactly. Among them are the relationships of Crustacea to Onychophora, Myriapoda, and Hexapoda (see Section III,B) and the evolution of arthropodization among the very most primitive arthropods and their ancestors.

Many evolutionary changes involved in the origin of Crustacea probably centered around the evolution of adaptations for increased dispersal and colonizing abilities through rather drastic reorganization of the ancestor's simpler developmental pattern and life history design. Diverse facts about primitive Crustacea fall into place when viewed in terms of this organizing theme, which is the main reason for giving any credence to the following scenario: The evolution of the ancestral crustacean's developmental pattern represented the evolution of adaptations for maintaining high intrinsic rates of natural increase (r_m in demography; see Andrewartha and Birch, 1954) in populations—a prime characteristic for species variously called colonizing, fugitive, or ecologically opportunistic (Lewontin, 1965). Rates of increase can be increased by two mechanisms:

(1) Increase in net reproductivity (R_0 in demography), the logarithm of which is directly proportional to the intrinsic rate of natural increase (see Andrewartha and Birch, 1954), through evolution of the nauplius larva.

(2) Decrease in the mean generation time (T in demography), which is inversely proportional to the intrinsic rate of natural increase (see Andrewartha and Birch, 1954), through progenesis (in the sense of Gould, 1977).

The extreme colonizing ability typical of branchiopods is consistent with this theory. Though this ability is in various ways enhanced through phylogenetically secondary ecological specializations [including another instance of progenesis in the origin of Cladocera (de Beer, 1958)], life history designs in more primitive branchiopods are perhaps least modified over the ancestral crustacean's. As discussed, the reproductive pattern in cephalocarids is interpreted as showing evidence of relatively minor phyletic specialization (see Section II,B,1).

The evolution of the nauplius probably involved reorganization of a simpler and more gradual trilobite-type developmental pattern (Garstang and Gurney, 1938). As Garstang and Gurney (1938) noted, the seemingly precocious development and differentiation of the second antenna marks the nauplius as a phyletically specialized larval type, just as the correspondingly differentiation of this segment marks the adult crustacean as being phyletically more specialized than an adult trilobite in terms of progressive tagmatization. The basis for this specialization is now recognized to be the naupliar net-sweep feeding mechanism (Gauld, 1959), a characteristically larval feeding mechanism distinct from the adult trunk limb mechanism of primitive forms. In line with the very gradual development expected for the ancestral crustacean (Sanders, 1963), one would expect the crustacean ancestor to have had still more gradual development, and to have had a single feeding mechanism developed in both larva and adult. Development in trilobites probably represents such a condition. The trilobite protaspis may represent the organizational grade of the larva, though probably not the exact larval type, that the nauplius supplanted. In *Triarthrus* (the trilobite for which the most, by far, is known of its early life history), protaspids and succeeding juvenile stages went through a relatively long pelagic dispersal phase, as early-hatching crustacean larvae often do, before becoming nektobenthic in later stages (Cisne, 1973a). The three postoral head limbs and whatever trunk limbs may have been functional in the protaspis probably participated in a trunk limb propulsion and feeding mechanism essentially like that in the adult. The high degree of serial similarity among all segments, including the preoral antennal segment, in the adult *Triarthrus* gives no indication that any segments were specialized in the manner of the crustacean second antennal segment at any earlier developmental stage (Cisne, 1975). Thus there is no reason to believe that the recently discovered phaselus larva (Fortey and Morris, 1978)—which the discoverers say is only probably a trilobite's—was nauplius-like as regards the second antenna, even if it is, in fact, a trilobite larva. Perhaps the admittedly nauplius-like remains represent the otherwise unknown phyllocarid nauplius (see Section II,B,2) that the leptostracan embryonic naupliar stage suggests (see Calman, 1909; Manton, 1934). But if the phaselus is a trilobite larva, it then points up the potential of the trilobite developmental pattern for hatching at a stage more nearly equivalent to the true nauplius in regard to the small number of segments bearing functional limbs.

The nauplius, with its specialized larval feeding mechanism, may have been the solution to "designing" a feeding larva that hatched at so early a stage that, with so few functional limbs, it could not support the adult trunk limb mechanism. Evolution of the net-sweep mechanism made it possible for the larva to hatch and feed itself at earlier stages and thus for it to foot a greater part of the energy bill for its own development. By reducing the parental energy investment per larva, all other factors remaining the same, evolution of the nauplius made it possible for parents to get a greater return on their total energy investment in terms of number of larvae. At one stroke, the nauplius increased colonizing ability in two ways: (1) by being adaptable as a dispersal stage in the life cycle (see Section II,B,1), and (2) by facilitating an increase in the populations' net reproductivity and intrinsic rate of natural increase (see Lewontin, 1965; Cisne, 1979).

Progenesis helps explain the small size and larva-like adult characteristics of the most primitive Crustacea (e.g., retention of the naupliar feeding mechanism in the adult; see Section II,B,1). Progenesis is one mechanism by which the mean length of the generation can be much reduced, and this in turn is generally the most effective means by which the intrinsic rate of natural increase can be greatly increased (Lewontin, 1965; Cisne, 1979). Progenesis has many times been an evolutionary route to ecological opportunism (Lewontin, 1965; Gould, 1977), and it seems to be implicated in the origins of a number of major groups (de Beer, 1958; Gould, 1977). According to this idea, the ancestral crustacean became reproductively mature at a precociously early developmental stage and small size relative to the crustacean ancestor's life history. This change is in accord with the size difference between the millimeters-long adult ancestral crustacean (see Section II,B,1) and the centimeters-long adult trilobitoids that might represent something close to the crustacean ancestor (e.g., *Emeraldella;* see Hessler and Newman, 1975).

Progenesis and evolution of the nauplius may very well have proceeded in concert during reorganization of the ancestor's developmental pattern. As a result of progenesis, the parent would presumably be reproducing on a lower total energy capital. It would have less energy to spend on larvae than the ancestral form. The nauplius can be viewed as a device for reducing the unit cost of the individual offspring through earlier hatching. This cost may have set limits for the earliest developmental stage at which reproduction could take place, and thus limits on the extent of progenesis. It is quite conceivable that the ancestral crustacean laid the minimum number of eggs and yet maintained high intrinsic rates of natural increase in populations by virtue of a very short mean length of the generation; for the generation time, and particularly the age at first reproduction, are overwhelmingly more important than net reproductivity in controlling the intrinsic rate of increase (Lewontin, 1965; Cisne, 1979). Cephalocarids bear only one or two eggs at a time (Sanders and Hessler, 1964). It is possible that this represents a survival of a primitive condition, and that delayed hatching represents a secondary accommodation to it in the evidently non-opportunistic modern cephalocarids (see Section II,B,1).

The nauplius-progenesis theory of crustacean origins finds some precedent in later crustacean evolution. Progenesis has been involved in the origins of several crustacean groups (see Section II,B,1). One may wonder to what extent Maxillopoda are united simply by the expression of progenesis in the maxillopodan condition. New larval types have appeared in several groups (e.g., the cladocera larva in the conchostracan *Cyclestheria,* the erichthus in stomatopods, various post-naupliar types among decapods). The major difference between these instances of developmental flexibility and that involving the origin of the nauplius is that they seem to have involved increase in the embryonization of development. Earlier hatching, like earlier first reproduction (see Lewontin, 1965), is probably more difficult to attain through evolution than later hatching; hence, it is probably a rarer evolutionary phenomenon.

ACKNOWLEDGMENTS

I thank L. G. Abele, R. R. Hessler, and anonymous reviewers for constructive criticism of the manuscript.

NOTE ADDED IN PROOF

Further studies on the newly discovered, aberrant, and apparently quite primitive crustacean Class Remipedia Jager, 1981 should be of much interest regarding crustacean origins and early evolution. Of particular importance for its bearing on questions of arthropod head structure and classification (see Section II, A) is the question of whether the limb-like pre-antennular process represents a pre-antennular segment, and if it does, the question of whether this condition is primitive in Crustacea or merely a developmental anomaly in Remipedia comparable to embryonic pre-antennal limbs in certain uniramians.

Additional important information on crustacean and arthropod origins will no doubt come from Müller's for the most part as yet unpublished work on a spectacularly preserved Late Cambrian arthropod fauna, which includes agnostid trilobites, cephalocarid and ostracode crustaceans, and other arthropods preserved complete with the setae on the limbs (K. J. Müller, personal communication; Müller, 1979).

REFERENCES

Anderson, D. T. (1973). "Embryology and Phylogeny in Annelids and Arthropods." Pergamon, Oxford.

Andrewartha, H. G., and Birch, L. C. (1954). "The Distribution and Abundance of Animals." Univ. of Chicago Press, Chicago, Illinois.

Baccetti, B. (1979). Ultrastructure of sperm and its bearing on arthropod phylogeny. *In* "Arthropod Phylogeny" (A. P. Gupta, ed.), pp. 609-644. Van Nostrand-Reinhold, Princeton, New Jersey.

Bergström, J. (1969). Remarks on the appendages of trilobites. *Lethaia* **2,** 395-414.

Bergström, J. (1973). Organization, life, and systematics of trilobites. *Fossils Strata* **2,** 1-69.

Boudreaux, H. B. (1979). Significance of intersegmental tendon system in arthropod phylogeny and a monophyletic classification of Arthropoda. *In* "Arthropod Phylogeny" (A. P. Gupta, ed.), pp. 551-586. Van Nostrand-Reinhold, Princeton, New Jersey.

Briggs, D.E.G. (1978). The morphology, mode of life, and affinities of *Canadaspis perfecta* (Crustacea: Phyllocarida), Middle Cambrian, Burgess Shale, British Columbia. *Philos. Trans. R. Soc. London, Ser. B* **281,** 439-487.

Briggs, D.E.G., Bruton, D. L., and Whittington, H. B. (1979). Appendages of the arthropod *Aglaspis spinifer* (Upper Cambrian, Wisconsin) and their significance. *Palaeontology* **22,** 167-180.

Calman, W. T. (1909). Crustacea. *In* "Treatise on Zoology" (E. R. Lankester, ed.), Vol. VII, Fasc. 3. Adam & Black, London.

Cannon, H. G. (1927). On the feeding mechanism of *Nebalia bipes. Trans. R. Soc. Edinburgh* **55,** 355-369.

Cannon, H. G. (1933). On the feeding mechanism of the Branchiopoda. *Philos. Trans. R. Soc. London, Ser. B* **222,** 267-352.

Cannon, H. G., and Manton, S. M. (1927). On the feeding mechanism of the mysid crustacean *Hemimysis Lamornae. Trans. R. Soc. Edinburgh* **55,** 219-254.

Cisne, J. L. (1973a). Life history of the Ordovician trilobite *Triarthrus eatoni. Ecology* **54,** 135-142.

Cisne, J. L. (1973b). Beecher's Trilobite Bed revisited: Ecology of an Ordovician deepwater fauna. *Postilla* **160.**

Cisne, J. L. (1974). Trilobites and the origins of arthropods. *Science* **186,** 13-18.

Cisne, J. L. (1975). Anatomy of *Triarthrus* and the relationships of Trilobita. *Fossils Strata* **4,** 45-63.

Cisne, J. L. (1979). Population dynamics. *In* "Encyclopedia of Paleontology" (R. W. Fairbridge and D. Jablonski, eds.), pp. 628-635. Dowden, Hutchinson & Ross, Stroudsburg, Pennsylvania.

Cisne, J. L. (1981). *Triarthrus eatoni* (Trilobita): Comparative anatomy of its exoskeletal, skeletomuscular, and digestive systems. *Paleontogr. Am.* **9** (53), 95-142.

Cisne, J. L., and Dempster, K. (1982). External anatomy of *Cryptolithus bellulus* from Beecher's Trilobite Bed. (In preparation.)

Clark, K. U. (1979). Visceral anatomy and arthropod phylogeny. *In* "Arthropod Phylogeny" (A. P. Gupta, ed.), pp. 467-549. Van Nostrand-Reinhold, Princeton, New Jersey.

Dahl, E. (1963). Main evolutionary lines among Recent Crustacea. *In* "Phylogeny and Evolution of Crustacea" (H. B. Whittington and W.D.I. Rolfe, eds.), Spec. Publ., pp. 1-15. Mus. Comp. Zool., Cambridge, Massachusetts.

de Beer, G. R. (1958). "Embryos and Ancestors." Oxford Univ. Press, London and New York.

Firstman, B. (1973). The relationship of the chelicerate arterial system to the evolution of the endosternite. *J. Arachnol.* **1,** 1-54.

Fisher, D. C. (1979). Evidence of subaerial activity of *Euproops danae* (Merostomata, Xiphosurida). *In* "Mazon Creek Fossils" (M. H. Nitecki, ed.), pp. 379-447. Academic Press, New York.

Fortey, R. A., and Morris, S. F. (1978). Discovery of nauplius-like trilobite larvae. *Palaeontology* **21,** 823-833.

Garstang, W., and Gurney, R. (1938). The descent of Crustacea from trilobites and their larval relations. *In* "Evolution: Essays Presented to Professor Goodrich on the Occasion of His Seventieth Birthday" (G. R. de Beer, ed.), pp. 271-286. Oxford Univ. Press, London and New York.

Gauld, D. T. (1959). Swimming and feeding in crustacean larvae: The nauplius larva. *Proc. Zool. Soc. London* **132,** 31-50.

Gould, S. J. (1977). "Ontogeny and Phylogeny." Belknap Press, Cambridge, Massachusetts.

Gupta, A. P., ed. (1979a). "Arthropod Phylogeny." Van Nostrand-Reinhold, Princeton, New Jersey.

Gupta, A. P. (1979b). Arthropod hemocytes and phylogeny. *In* "Arthropod Phylogeny" (A. P. Gupta, ed.), pp. 669-735. Van Nostrand-Reinhold, Princeton, New Jersey.

Hessler, A. Y., Hessler, R. R., and Sanders, H. L. (1970). Reproductive system of *Hutchinsoniella macracantha*. *Science* **168,** 1464.

Hessler, R. R. (1964). The Cephalocarida. Comparative skeletomusculature. *Mem. Conn. Acad. Arts Sci.* **16,** 1-97.

Hessler, R. R., and Newman, W. A. (1975). A trilobitomorph origin for the Crustacea. *Foossils Strata* **4,** 437-459.

Hessler, R. R., and Sanders, H. L. (1973). Two new species of *Sandersiella* (Cephalocarida), including one from the deep sea. *Crustaceana* **24,** 181-196.

Lankester, E. R. (1904). The structure and classification of the Arthropoda. *Q. J. Microsc. Sci.* [N.S.] **47,** 523-582.

Lewontin, R. C. (1965). Selection for colonizing ability. *In* "Genetics of Colonizing Species" (H. G. Baker and G. L. Stebbins, eds.), pp. 77-91. Academic Press, New York.

Manton, S. M. (1934). On the embryology of the crustacean, *Nebalia bipes*. *Philos. Trans. R. Soc. London, Ser. B* **223,** 168-238.

Manton, S. M. (1950). The evolution of arthropod locomotory mechanisms. Part 1. *J. Linn. Soc. London, Zool.* **41,** 529-570.

Manton, L. M. (1963). The evolution of arthropod locomotory mechanisms. Parts 2, 3, and 4. *J. Linn. Soc. London, Zool.* **42,** 93-117, 118-166, 299-368.
Manton, S. M. (1964a). The evolution of arthropod locomotory mechanisms. Parts 5 and 6. *J. Linn. Soc. London, Zool.* **43,** 153-187, 487-556.
Manton, S. M. (1964b). Mandibular mechanisms and the evolution of arthropods. *Philos. Trans. R. Soc. London, Ser. B* **247,** 1-183.
Manton, S. M. (1965). The evolution of arthropod locomotory mechanisms. Part 7. *J. Linn. Soc. London, Zool.* **44,** 383-461.
Manton, S. M. (1966). The evolution of arthropod locomotory mechanisms. Part 8. *J. Linn. Soc. London, Zool.* **45,** 251-484.
Manton, S. M. (1967). The evolution of arthropod locomotory mechanisms. Part 9. *J. Linn. Soc. London, Zool.* **46,** 103-141.
Manton, S. M. (1969). Introduction of classification of Arthropoda. *In* "Treatise on Invertebrate Paleontology" (R. C. Moore, ed.), Part R, Arthropoda 4, pp. R3-R14. Geol. Soc. Am., Boulder, Colorado, and the Univ. of Kansas Press, Lawrence.
Manton, S. M. (1972). The evolution of arthropod locomotory mechanisms. Part 10. *J. Linn. Soc. London, Zool.* **51,** 203-400.
Manton, S. M. (1973). The evolution of arthropod locomotory mechanisms. Part 11. *J. Linn. Soc. London* **53,** 257-375.
Manton, S. M. (1977). "The Arthropoda. Habits, Functional Morphology, and Evolution." Oxford Univ. Press (Clarendon), London and New York.
Mileikovsky, S. A. (1971). Types of larval development in marine bottom invertebrates, their distribution and ecological significance: A re-evaluation. *Mar. Biol.* **10,** 193-213.
Müller, F. (1864). "Fur Darwin." ["Facts and Arguments for Darwin" (transl. by W. S. Dallas). Murray, London, 1869.]
Müller, K. J. (1979). Phosphatocopine ostracodes with preserved appendages from the Upper Cambrian of Sweden. *Lethaia* **12,** 1-27.
Paulus, H. F. (1979). Eye structure and the monophyly of Arthropoda. *In* "Arthropod Phylogeny" (A. P. Gupta, ed.), pp. 299-383. Van Nostrand-Reinhold, Princeton, New Jersey.
Sanders, H. L. (1955). The Cephalocarida, a new subclass of Crustacea from Long Island Sound. *Proc. Natl. Acad. Sci. U.S.A.* **41,** 61-66.
Sanders, H. L. (1957). The Cephalocarida and crustacean phylogeny. *Syst. Zool.* **6,** 112-128.
Sanders, H. L. (1960). Benthic studies in Buzzards Bay. III. The structure of the soft-bottom community. *Limnol. Oceanogr.* **5,** 138-153.
Sanders, H. L. (1963). The Cephalocarida. Functional morphology, larval development, and comparative external anatomy. *Mem. Conn. Acad. Arts Sci.* **15,** 1-80.
Sanders, H. L., and Hessler, R. R. (1964). Larval development of *Lightiella incisa* Gooding (Cephalocarida). *Crustaceana* **7,** 81-97.
Schram, F. R. (1978). Arthropods: A convergent phenomenon. *Fieldiana, Geol.* **39,** 61-108.
Simpson, G. G. (1953). "The Major Features of Evolution." Columbia Univ. Press, New York.
Snodgrass, R. E. (1938). Evolution of the Annelida, Onychophora, and Arthropoda. *Smithson. Misc. Collect.* **97,** 1-59.
Størmer, L. (1933). Are trilobites related to arachnids? *Am. J. Sci.* [5] **26,** 147-157.
Størmer, L. (1939). Studies on trilobite morphology. Part I. The thoracic legs and their phylogenetic significance. *Nor. Geol. Tidsskr.* **19,** 143-273.
Størmer, L. (1942). Studies on trilobite morphology. Part II. The larval development, the segmentation and the sutures, and their bearing on trilobite classification. *Nor. Geol. Tidsskr.* **21,** 49-163.

Størmer, L. (1944). On the relationships of fossil and Recent Arachnomorpha. *Skr. Nor. Vidensk.-Akad.* [*Kl.*] *1: Mat.-Naturvidensk. Kl.* No. 5, pp. 1-158.

Størmer, L. (1951). Studies on trilobite morphology. Part III. The ventral cephalic structures with remarks on the zoological position of trilobites. *Nor. Geol. Tidsskr.* **29,** 108-158.

Stürmer, W., and Bergström, J. (1976). The arthropods *Mimetaster* and *Vachonisia* from the Devonian Hunsrück Shale. *Palaeontol. Z.* **50,** 78-111.

Tiegs, O. W., and Manton, S. M. (1958). The evolution of the Arthropoda. *Biol. Rev. Cambridge Philos. Soc.* **33,** 255-337.

Weygoldt, P. (1979). Significance of later embryonic stages and head development in arthropod phylogeny. *In* "Arthropod Phylogeny" (A. P. Gupta, ed.), pp. 107-135. Van Nostrand-Reinhold, Princeton, New Jersey.

Whittington, H. B. (1957). The ontogeny of trilobites. *Biol. Rev. Cambridge Philos. Soc.* **32,** 421-469.

Whittington, H. B. (1971). Redescription of *Marrella splendens* (Trilobitoidea) from the Burgess Shale, Middle Cambrian, British Columbia. *Geol. Surv. Can., Bull.* **209,** 1-24.

Whittington, H. B. (1975a). Trilobites with appendages from the Middle Cambrian, Burgess Shale, British Columbia. *Fossils Strata* **4,** 97-136.

Whittington, H. B. (1975b). The enigmatic animal *Opabinia regalis,* Middle Cambrian, Burgess Shale, British Columbia. *Philos. Trans. R. Soc. London, Ser. B* **284,** 1-43.

Whittington, H. B. (1977). The Middle Cambrian trilobite *Naraoia,* Burgess Shale, British Columbia. *Philos. Trans. R. Soc. London, Ser. B* **280,** 409-433.

Whittington, H. B. (1979). Early arthropods, their appendages and relationships. *In* "The Origin of Major Invertebrate Groups" (M. R. House, ed.), pp. 253-268. Academic Press, New York.

Whittington, H. B. (1980). Exoskeleton, moult stage, appendage morphology, and habits of the Middle Cambrian trilobite *Olenoides serratus. Palaeontology* **23,** 171-204.

Yager, J. (1981). Remipedia, a new class of Crustacea from a marine cave in the Bahamas. *J. Crust. Biol.* **1,** 328-333.

4

The Fossil Record and Evolution of Crustacea

FREDERICK R. SCHRAM

In the modern world the celibacy of the medieval learned class has been replaced by the celibacy of the intellectual which is divorced from the concrete contemplation of the complete facts.

A. N. Whitehead

THE BIOLOGY OF CRUSTACEA, VOL. 1

ISBN 0-12-106401-8

He [Lamarck] had little patience with those who did not realize a theory could be founded upon a great number of facts and still be false.

R. W. Burkhardt

There is something about writing on arthropod phylogeny that brings out the worst in people.

J. W. Hedgepeth

I. THE NATURE AND LIMITS OF THE CRUSTACEAN FOSSIL RECORD

Past speculations on crustacean phylogeny have been largely based on analysis of the comparative anatomy and/or functional morphology of living forms (e.g., Calman, 1909; Dahl, 1963; Siewing, 1956, 1963). But an analysis of the fossil record and of the form and function of modern Crustacea have much to contribute toward a comprehensive and coherent understanding of how the Crustacea may have evolved.

The fossil record of crustaceans, as now understood, can provide useful insights into many aspects of crustacean phylogeny. Some groups have no recognized fossil record, e.g., cephalocarids, mystacocarids, branchiurans, and some of the minor cirriped groups. Other taxa possess records which are difficult to assess because of the nature of the fossils themselves as opposed to what can be learned from living forms (e.g., diplostracans) or because of a general scarcity of material (e.g., copepods or anostracans). Still other groups possess generally fine records with much material of good to excellent quality (e.g., ostracodes, thoracican cirripeds, and malacostracans).

The types of preservation in the fossil record are variable and consequently affect any information or conclusions that can be drawn from it. A crucial element is the degree of preservation of the appendages, since many functional considerations and phylogenetic speculations on crustaceans depend on sound knowledge of appendage structure. For example, the record of the calmanostracan Branchiopoda (Tasch, 1969) is composed almost entirely of carapaces or tail parts; thus, little reference can be made to the appendage morphology of modern forms. Malacostracan fossils contain a great deal of appendage information (see Schram and Horner, 1978; or Schram, 1979a) which can be directly related to knowledge gained from modern forms. Certain unusual fossil assemblages, in which entire faunas have been preserved virtually intact (soft-bodied worms and jellyfish, entire arthropods, as well as more traditional shell fossils), are termed *Konservat-Lagerstätten* (Seilacker, 1970). Such assemblages have been especially helpful in elucidating malacostracan history. Faunas, like those found in the Mississippian Bear Bulch Limestone (Schram, 1976a) or the Jurassic Solenhofen Limestone (Münster, 1840; Oppel, 1862), have been preserved as whole suites of crustacean species; in some cases, this is almost as good as having the actual animals. Frequently, however, crustacean fossils are only parts of bodies or appendages (e.g., Herrick and Schram, 1978), which can make even identification of suborder, let alone genus and species, difficult. The extensive fossil record of ostracodes, on the other hand, is based entirely on carapaces, which presents some serious limitations to interpretation. However, some of the fossil ostracode carapace anatomy can be related to structures seen in the living exemplars, and even such discrete shell structures as the pit conuli (Benson, 1976, 1977) have been exploited in discerning fossil and living ostracode lineages.

These problems of preservation result from the nature of the crustacean body plan, composed as it is of discrete body and appendage segments. These segments are made of a base substance of chitin variously reinforced with sclerotin and mineral salts. Small amounts of reinforcing material make for poor fossils, and large amounts make for good ones. These differences result in such anomalies of the fossil record as numerous *Callianassa* claws but few intact bodies. Only quick burial under anoxic conditions can produce intact fossil crustaceans of high quality, and such conditions did not always prevail at the initial stages of crustacean fossilization. Warping and distortion are also problems in crustacean fossils which are not typical of purely calcareous fossils like mollusks or branchiopods. One needs to be aware of the vagaries of preservation and to realize the limits of interpretation placed on any particular fossil. Fossil data should never be pushed too far; however, fossils should not be ignored completely in phylogenetic analysis.

The crustacean record extends from the Lower Cambrian to the Recent. As a group, crustacean fossils are found world wide and are preserved in all kinds of lithologies (but particularly in limestones and shales). All sorts of habitat types are recognized to have preserved crustacean fossils. Shallow-water marine habitats are most common, though brackish, freshwater, and deep-water crustacean fossils are also known.

Attention has been devoted to individual taxa to produce cladistic trees (e.g., Förster, 1967, 1973; Glaessner, 1960), but habitat groupings and biotic associations have also been helpful in elucidating the faunistic history, e.g., Late Paleozoic malacostracans (Schram, 1981a) and the important *Dakoticancer* assemblage of the North American midcontinent Cretaceous (Bishop, 1981). Insight into the evolution of crustaceans as part of entire faunas in space–time is being developed from such studies.

In short, while the crustacean record as a whole might be considered uneven, it is no more so than that of any other major phylum and in some respects, it is better than many. Since Crustacea are almost entirely aquatic, the completeness of their record stands in strong contrast to a phylum like

Uniramia, which is almost completely terrestrial in its habits, and whose fossil record leaves much to be desired. In the Crustacea, as in all arthropodous phyla, a great deal of the biology of many "visceral" systems are reflected in their exoskeletons. Thus in this regard, they are less frustrating to study than a group such as Mollusca or Bryozoa, where much of the biology is "soft" and not directly reflected in the shell or theca.

A final element which must be dealt with in order to understand crustacean history is the geologic context of the fossils. The positions of ancient continents, past climatic conditions, and sediment types are useful and frequently essential in getting at "the whole story" when assessing crustacean evolution. Unless attention is paid to such major phenomena as plate tectonic movements, valuable data can be obscured and crucial conclusions never drawn. For example, Schram (1977) examined the Late Paleozoic biogeography of hoplocaridans and eumalacostracans and concluded that these groups had been endemic to the equatorial continent Laurentia (North America and Europe). They did not disperse to other parts of the world until the Permian formation of the supercontinent Pangaea. It was this Permo-Triassic dispersal and concomitant radiation of new stocks that essentially gave us the basic pattern of the current malacostracan world fauna. G. A. Bishop (personal communication) has made a similar analysis of climatic conditions and sediment types in connection with Cretaceous crab assemblages of North America to yield some interesting insights into the location of shallow-water faunas of that time.

In light of the above, there is a sound basis for the use of crustacean fossils to develop a phylogeny for the group. In some cases, only general statements can be made and in others, very specific insights can be gained. Like attempts to employ data exclusively from living forms, analyses based entirely on fossil material are of limited use. A perspective that includes past history as well as modern anatomical variety is necessary in order to develop a phyletic understanding that is both adequate and complete. What follows, then, is a statement on crustacean phylogeny which utilizes both modern and fossil data. Some of it may be rather provocative for some readers. However, pursuit of a clearer understanding will surely entail disagreement and debate in order to reach a concensus on Crustacean evolution.

II. ORIGINS OF THE CRUSTACEA

The fossil record tells us little about the origin of Crustacea, which is not surprising, since it tells us little about the origin of any phylum. However, scant information has not deterred anyone from engaging in phyletic speculation: lack of data only seems to encourage speculation. Recent studies of the functional morphology of locomotion by Manton (1973, 1977) and of early developmental stages by Anderson (1973) have made very strong arguments for the polyphyletic nature of the arthropods. Manton and Anderson recognized at least three phyla (Uniramia, Crustacea, and Chelicerata), though neither considered trilobites or pycnogonids in real detail. Cisne (1974, 1975), examining trilobites, and Hessler and Newman (1975), reconstructing the ancestral crustacean, both concluded that there were only two phyla, Uniramia and Trilobita-Chelicerata-Crustacea. Schram and Hedgpeth (1978) considered the Pycnogonida as a separate subphylum from Chelicerata, and Schram (1978) recognized four phyla: Uniramia, Crustacea, Trilobitomorpha, and Cheliceriformes. Schram admitted, however, that there may still be some merit to the old concept of Arachnomorpha (Størmer, 1944), which united Trilobitomorpha and Cheliceriformes into a single group, although Whittington (1979) feels that there is no basis in fact for this group. Manton (1977) and Manton and Anderson (1979) suggested the possibility of trilobites forming a fourth phylum.

As far as this volume is concerned, we need only deal with the dispute over the origin of the Crustacea. There are two opposing views based in large part on evidence from the fossil record. Cisne (1975) and Hessler and Newman (1975) argued for a trilobite or trilobitoid origin. Schram (1978) defended an origin of Crustacea separate from any arthropodous group. The protrilobite arguments are several and diffuse, and will be evaluated here one by one.

Cisne maintained that the feeding process in the Ordovician trilobite *Triarthrus* is like that of branchiopod, leptostracan, and cephalocarid crustaceans. This conclusion is based on the presence of detritus in the gut and midventral food groove of *Triarthrus* in the specimens Cisne studied. He admitted difficulty in explaining how the trilobite limb could have operated like that of a cephalocarid limb, but suggested that the vibrating coxa of the trilobite might have brought detritus to the midventral food groove: Cisne avoided any speculation as to how the laterally directed components of the trilobite limb might have assisted in such action. It was this same lateral orientation of the biramous stenopodous trilobite limb that caused Schram to emphasize the differences in functional morphology that exist in the ventrally-oriented polyramous (with endites, endopod, exopod, and epipods) foliaceous limb of the crustaceans. Cannon and Manton (1927) and Størmer (1939) hypothesized functional patterns for the trilobite limb as a feeding device, and their suggested arrangements are quite different from what exists in any of the crustaceans (Schram, 1978). Hessler and Newman (1975) said that "their" crustacean (derived from a cephalocarid plan) had a "trilobitoid" feeding mechanism, though the term Trilobitoidea is now known to be a spurious category (Whittington, 1979). They anticipated the

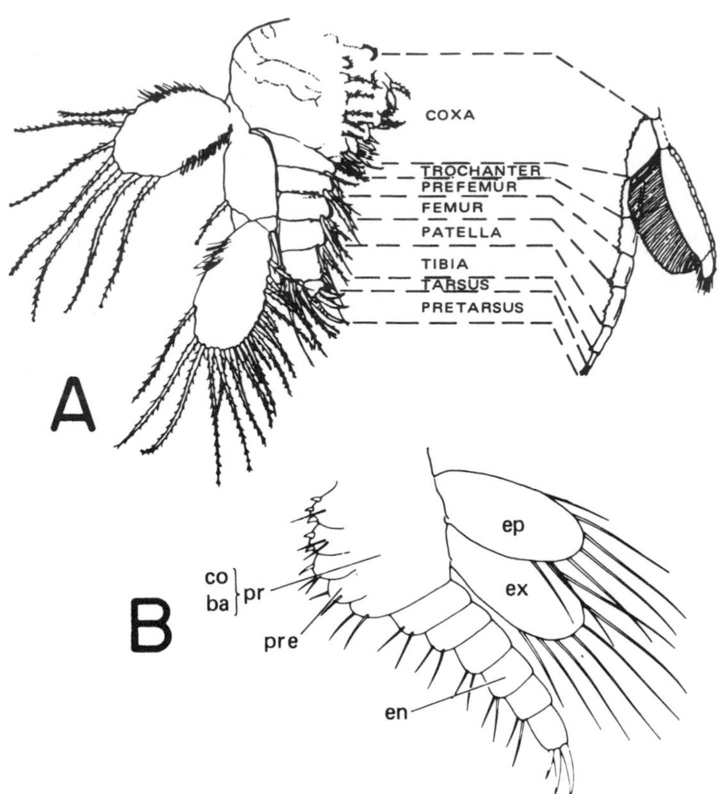

Fig. 1. (A) Comparison of limbs of *Hutchinsoniella* and a trilobite (from Sanders, 1957). (B) Hypothetical limb of the "urcrustacean" (from Hessler and Newman, 1975). Abbreviations: ep, epipodite; ex, exopod; en, endopod; pr e, endites; pr, protopod; co, coxa; ba, basis.

argument that "trilobitoids" had various different body plans by maintaining that the processing of food into a midventral groove is the only process that "satisfies all known morphologies" (p. 448). However, Hughs (1975) suggested an entirely different interpretation of feeding in at least one Middle Cambrian "trilobitoid," *Burgessia bella* (p. 433 and Fig. 5 p. 430). Hughs concluded that the food was grasped under the body and perhaps sucked into the mouth after some external digestion. The degree of the variety in the anatomy and inferred functions of the limbs of trilobitomorphs (e.g., Whittington, 1975; Schram, 1978, Fig. 10) discourages blanket statements on supposed similarities of trilobite and crustacean feeding types, since trilobites obviously possessed many different modes of feeding.

The limbs themselves have been cited as proof that trilobites and cephalocarids (and thus all crustaceans) have similar origins. A curious bit of juggling of limb components has gone on in order to achieve such comparisons (Fig. 1). Sanders (1957, Fig. 4, p. 118) reoriented the trilobite limb 90° in order to achieve comparison with cephalocarids. Hessler and Newman (1975, Fig. 7, p. 451) concluded that the primitive cephalocarid limb can be "ascribed to the urcrustacean," but they proceeded to describe a urcrustacean that was not like a cephalocaridan, but that had a limb whose branches were shifted laterally 45°. Cisne (1975) maintained that the shape of the coxa and the ambulatory nature of the telopod of trilobites were supposed to be like that of cephalocarids; but he admitted that limb structure "does not appear to provide solid evidence" (p. 56) by which to link trilobites with any other major group. Hessler and Newman advanced long arguments for homology of trilobite limb parts to cephalocarid limb parts, but neither they nor Cisne dealt with the fact that known trilobitomorph limbs were basically biramous, stenopodous, and ventrolaterally directed, while the primitive crustacean limb appears to have been polyramous, foliaceous, and directed ventrally. Manton (1977) believed the trilobite coxa could function in a fashion other than strict promotor-remotor mode: its weak articulation and extrinsic muscle supply imply movements possibly in a rotary mode. Schram (1978) maintained that the limb structures of these two groups were quite incompatible without extensive hypothetical phylogenies which would unite them in some more distant ancestor.

Cisne's elegant work on *Triarthrus* revealed details of soft internal anatomy, including muscles, although as pointed out (Cisne, 1975), there were limits to the X-ray technique he used. Nevertheless, his work will remain one of the high points of trilobite studies. Cisne maintained that the box truss arrangement of *Triarthrus* body muscles inserting on ventral ladderlike bars was too much like that of the cephalocarids not to imply common origin. Schram (1978) pointed out that muscle patterns can be convergent depending on function. In addition, this box truss arrangement is seen in many articulate phyla (Fig. 2), including polychaete annelids, pauropod, diplopod, and chilopod uniramians, crustaceans, and trilobites. Campbell (1975), in an analysis of the mechanics of enrollment in the trilobite *Cryptolithus,* considered the possibility of the box truss arrangement in achieving enrollment. He concluded that in *Cryptolithus* it would have been impractical to use the box truss, because the contraction of such muscles would have produced extreme compression of the heart and gut. Thus, we might conclude that, although *Triarthrus* employed the box truss, certainly not all trilobites did. In other words, the box truss arrangement of muscles is convergent and related to habits, and phylogenetic statements based on it are risky.

Midgut diverticula were used by Cisne and by Hessler and Newman to prove relationships between trilobites and crustaceans. Cisne claimed that

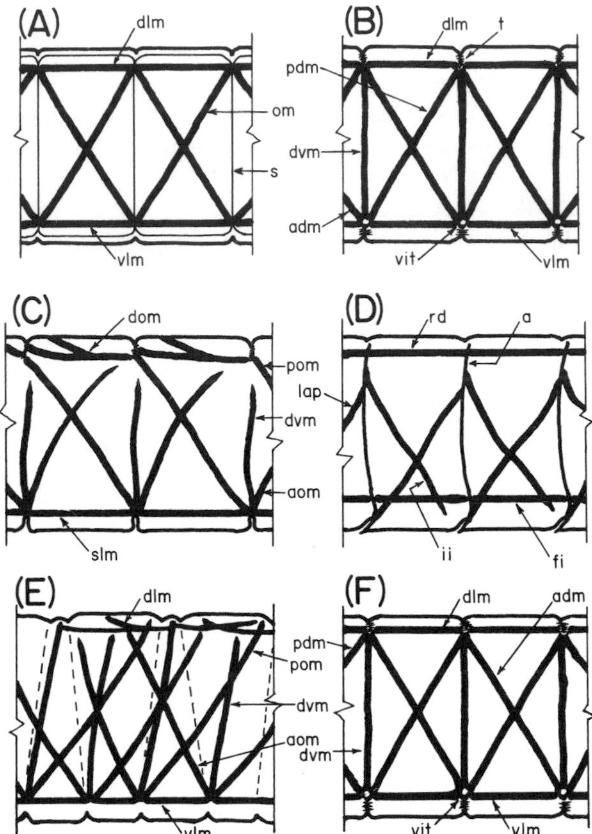

Fig. 2. Diagrams of box truss arrangement of body muscles in various articulate groups: (A) a polychaete annelid, *Nereis;* (B) a trilobite, *Triarthrus;* (C) a scolopendromorph chilopod, *Cryptops;* (D) a diplopod, *Cylindroiulus;* (E) a pauropod, *Pauropus* (dotted lines indicate segment boundaries); (F) a cephlocarid crustacean, *Hutchinsoniella.* Abbreviations: a, apodeme; adm, anteriorly descending dorsoventral muscle; aom, anterior deep oblique muscle; dlm, dorsal longitudinal muscle; dom, dorsal oblique muscle; dvm, dorsoventral muscle; fi, flexor inferus; ii, involens inferus; lap, levator apophysis posticae; om, oblique muscle; pdm, posteriorly descending dorsoventral muscle; pom, posterior deep oblique muscle; rd, retractor dorsalis; s, septum; slm, sternal longitudinal muscle; t, tendon; vit, ventral intersegmental tendon; vlm, ventral longitudinal muscle.

the cephalic digestive glands in *Triarthrus* were like those of the cephalocarids and chelicerates. Hessler and Newman maintain that "the presence of midgut digestive diverticula was a feature that relates crustaceans and trilobitomorphs" (p. 454). However, midgut hepatic diverticula are a feature of all arthropodous phyla; the diverticula are endodermal derivatives and, given their importance to digestion, there are only so many places they can go depending on what other viscera may be fitted into an arthropodous body. Along these same lines, great phyletic significance is attached to the posterior orientation of the mouth in trilobites and primitive crustaceans by Cisne and by Hessler and Newman. Schram would agree that there may be functional significance to mouth orientation in any particular organism, but this is of questionable phyletic import given what is known of the plasticity of mouth orientation in various crustaceans with respect to particular feeding habits.

The various arguments based on larval forms frequently employ logic that is difficult to follow. Cisne compared trilobite adults to the crustacean nauplius and post-nauplius larvae. He pointed out that, in crustacean larvae, the second antennae are postoral and the mandibles and both pair of maxillae are similar in structure to the trunk appendages. Cisne (1975, p. 59) minimized the similarity of the four primary segments of the *Limulus* blastodisc to the four cephalic segments of the trilobite head. Though he felt the merostome division of four was independently evolved, he preferred to believe that the trilobite division of four, with its entirely different pre- and post-oral segment pattern and appendage conditions, was closer to the crustacean larval stages than to merostomes. Hessler and Newman, in their comparison of trilobite adults and crustacean nauplii, made an extensive documentation of how the nauplius appendage arrangement is different from that of the adult crustacean: they claimed, however, that there is no reason to suspect the trilobite protaspid larval appendage arrangement was any different than that of a trilobite adult, and they further claimed that nauplii have the primary cephalic appendages of a trilobite (?? a nauplius has two appendage sets, second antenna, and mandible, behind the mouth, as opposed to the trilobite having an antenna and three sets of legs on the head). Therefore, they concluded that nauplii must be like protaspids. Hessler and Newman also make a labored comparison of crustacean adults and protaspid larvae. The development of the protaspid segments leads to the four cephalic appendages of the adult trilobite. They compare this to "the fact that the adult crustacean also possesses four postoral cephalic somites" (p. 449) (?? the basic adult crustacean cephalon has three postoral segments with the mandibles and two pairs of maxillae). Hessler and Newman went on to document the variations which occur in the number of mesodermal segments (Anderson, 1965; Solland, 1923), as opposed to those which can occur in the appendage and ganglia bearing segments of the crustacean nauplius. They decided that these extra segments are probably primary, i.e., not teloblastic. They used this assumption to conclude that nauplii have the same number of segments in the cephalon as do trilobites. However, Hessler and Newman remark on Manton's (1928) observation that extra mesodermal bands in *Hemimysis* early development are teloblastic, and also point out that Anderson (1967) discovered that in some brachiopods the anterior thoracic segments are not teloblastic. Admittedly, the reasoning here is a tangled web, but Hessler and Newman in the end conclude "that the number of somites appearing before activation of the mesodermal teloblast is not of great phylogenetic significance" (p. 450). Thus, these larval arguments are much ado about nothing.

Fortey and Morris (1978) described what they termed the trilobite preprotaspis "phaselus" larva from Ordovician beds in Spitsbergen. The phaseli are shell-like, 200–250 μm long, ovoid in outline, and hemispherical in shape. They are composed of apatite, thought to be replacement. Fortey and Morris interpret these as nauplius-like trilobite larvae, since the phaseli have "different symmetry from any described anaprotaspis" (p. 827). The suggestion of trilobite affinities for these fossils is based on the presence of raised polygonal ridges on the surface, a doublure along the margin, and a comparable number of protaspids in the same beds as the phaseli. However, several points must be raised in argument against these reasons for phaseli as trilobites. Fortey and Morris point out that the polygonal features have some preservational questions associated with them (p. 826): the doublure is not a unique feature of trilobites, and a comparison by the authors of phaselus-protaspis growth show that, on a log-log plot of the data, the phaseli from one of the samples lie more than one standard deviation from the regression line developed from the protaspid data. Fortey and Morris also point out that the phaselus material occurs in a planktonic association in adjacent beds of radiolarians, chitinozoans, and the problematic *Janospira* (similarly preserved). Thus, size at the end of a protaspid sequence could be merely a phenomenon of the plankton nature of the assemblage. There are other enigmatic elements known in these beds, and so I would differ with the contention that the phaseli are probably trilobites and maintain that the data indicates they are only possibly so.

Homology of the phaselus with the nauplius of crustaceans is quite beyond the realm of even "speculative extension" (p. 827). Fortey and Morris maintain that similarities of size and form of phaseli to nauplii argue for a nauplius "homology." However, the general argument of size is questioned above, and it places too much emphasis on similarities of such nondescript shapes as ovoid hemispheres. The authors go quite off the speculative deep end, however, when they compare phaselus appendage number with that of nauplii. They argue that the stage preceding any protaspid has only three appendage pairs: however, this is based on the assumed homology of supposed phaselus and nauplius appendages. Schram (1978) has already questioned the comparison of trilobite adult cephalic appendages with those of crustaceans. Size argues only for phaseli being plankton: the simple form of parts of organisms means little, and the assumption of three appendages is the most blatant of speculations and is less than justifiable. The arguments, like those of Cisne and of Hessler and Newman, are propelled more by the desire to arrive at a preconceived end rather than an impartial consideration of fact.

Schram (1978) maintained that one should only compare larvae to larvae and adults to adults. Protaspids have four segments which appear to develop into four segments of the adult trilobite head. The nauplius pattern is quite distinct, with one pre-oral and two post-oral segments which will eventually comprise only part of the adult head. In terms of the adult morphology of *Triarthrus,* trilobites appear to have no embryonically derived pre-oral segments (Schram, 1978, Table 2) and are thus more like cheliceriformes than crustaceans or uniramians.

Hessler and Newman also attached significance to tri-lobed bodies. The axial and lateral lobes of trilobites are compared to the body and ventrolaterally directed pleura of cephalocarids. That the arrangement of these parts is different in the two animals "is of minor importance" (p. 448). They also compared the caudal furca in trilobites and the cephalocarid-derived urcrustacean. But not all trilobites have furca: *Olenoides* does (Whittington, 1975), but *Triarthrus* does not (Cisne, 1975). To add to the confusion, Bowman (1971) concluded that cephalocarids have an anal segment and uropods rather than a telson with furca, though Schminke (1976) took a more orthodox stand.

Neither Cisne nor Hessler and Newman attempted to address the evidence from crustacean comparative embryology (Anderson, 1973) for the separation of the phylum from all other arthropodous groups.

From the foregoing, it is apparent that the evidence for uniting trilobitomorphs and crustaceans is at best poorly drawn and certainly does not prove the case. Manton (1977, p. 488) stated "a recent attempt to group arthropods into a diphyletic scheme . . . rests on spurious arguments." Schram (1978) approached the problem with more pragmatic and parsimonious logic. All one can say is that the Crustacea were derived from a form that was homonomously multilegged; possessed polyramous, foliaceous ventrally directed appendages; ultimately manipulated food with the base of one appendage (jaw); was a filter feeder; and had a crustaceoid blastomere pattern, leading to anamorphic development. Whether this form can be ultimately related to any other arthropodous form is a moot point without recourse to elaborate, unprovable, and unnecessary "paper phylogenies." The question of the possible form of ancestral crustaceans will be taken up in a later section after consideration of primitive groups.

III. PRIMITIVE GROUPS

A. Cephalocarida

Much of modern carcinological phylogenetic theory is based on the generally accepted premise that the Cephalocarida are the most primitive living crustaceans. As such, the cephalocarids form a focal point from which other crustacean groups can be derived (Sanders, 1957, 1963a,b; Hessler and Newman, 1975). However, it is usually forgotten that Sanders, himself (in Whittington and Rolfe, 1963, p. 177), pointed out the danger of presenting the cephalocarids as the primitive condition; at the same time, Dahl (1963, pp. 177–178) pointed out some abberant conditions in *Hutchinsoniella,* especially of the central nervous system, which had to be interpreted as adaptations to a possible habitat. This last point is frequently overlooked or minimized; e.g., Hessler and Newman (1975) dismiss such features as lack of eyes, abdominal limbs, and possibly a carapace as "apparently secondary" (p. 442). Indeed, a curious dissonance is sometimes apparent in the literature on cephalocarids. Sanders and Hessler (1963, p. 82) and Sanders (1963b) pointed out how the analysis of *Hutchinsoniella* functional morphology reveals specializations which limit its niche to organic-rich flocculent layers and make it "unlikely" that the animal feeds anywhere else, whereas Hessler and Sanders (1972, p. 196) claimed it is "unlikely" that such a primitive animal can be a specialist. The difficulty lies in the trap which Glaessner warned against (in Whittington and Rolfe, 1963, p. 179), that modern forms are the descendants of "ancestors" and a great deal has happened to them since the "ancestor." Savory (1971, p. 13) put it another way: "There are no longer primitive animals. It is probable that every animal . . . is a mass of specialization built into a scaffolding which alone is a vestige of the primitive condition."

It is not my intention to completely overthrow the central position of cephalocarids in crustacean phyletic theory. But some new insights might be gained. Gould (1977) has most recently reconsidered the role of paedomorphosis in macroevolution; he has advanced the idea that progenetic forms (those in which functional gonads prematurely appear in some juvenile or larval stage) will be associated with *r*-selection regimes and that neotenic forms (those in which juvenile characters are held over into the adult stage) will be associated with *K*-selection regimes. The lack of eyes, of abdominal and sometimes posterior thoracic appendages, and possibly of a carapace; the great similarity of all appendages; and the very small size would mark cephalocarids as progenetic paedomorphs. Such would not be an unusual interpretation; e.g., Hessler (1971) and Hessler and Newman (1975) suggest the same for mystacocarids, and Gurney (1942) postulated this in the case of copepods.

r-Selection regimes would be those with large and frequent environment changes, catastrophic mortalities, a superabundance of resources, or with lack of crowding in the niche. Any or several of these could apply to the flocculent zones that cephalocarids seem to inhabit. Indeed, a progenetic *r*-selection strategy may be the only life cycle that could have served such an ancient group for so long under such conditions. The attributes of *r*-selection strategies are early maturation, high fecundity, rapid development, short life span, limited parental care, and diversion of the better part of available resources to reproduction. Some of these, i.e., high fecundity and lack of parental care, would appear at first glance not to be applicable to cephalocarids as *r*-strategists. But Gould (1977) points out that progenetic *r*-selection will likely produce the "tiniest animals" and, when combined with the allometry of egg production and body size (Gould, 1966), might produce conditions as seen in cephalocarids. "Constrained by their design to produce relatively few eggs, they opt for what is usually taken as a mark of *K*-selection: brood protection of relatively few offspring, since they cannot afford to liberate such a small number freely into the plankton." (Gould, 1977, pp. 293-294). What is known of cephalocarid reproduction (Sanders and Hessler, 1963; Hessler *et al.,* 1970) is in accordance with such an interpretation. Cephalocarids apparently reproduce throughout most of the year, are constantly preparing eggs, and brood one or two eggs at a time.

Hessler *et al.* (1970) tried to discount the possibility of self-fertilization, since cephalocarids are unique among crustaceans in being hermaphrodites with a common genital duct for both systems. Can an *r*-selection organism afford the luxury of relying only on cross-fertilization? This may be an unrealistic constraint. Population densities of cephalocarids under some conditions (Sanders, 1960) may accommodate cross-fertilization. Many cephalocarid localities (Hessler and Sanders, 1972; Wakabara and Mizoguchi, 1976) have yielded only a few individuals, and self-fertilization may be at least a reserve option for such a progenetic organism. Indeed, the great morphological conservativeness of all cephalocarids (Hessler and Sanders, 1972) may be a reflection of a high incidence of self-fertilization.

If Gould's observations are applicable to cephalocarids, and all indications seem to say they are, then speculations on crustacean phylogeny using cephalocarids must be tempered with the knowledge that although the group seems to be an ancient one (Hessler and Sanders, 1972), it is specialized in its morphology and life cycle to a particular environmental regime. The reasoning and conclusions here, however, stand at odds with those of Cisne (Chapter 3 of this volume).

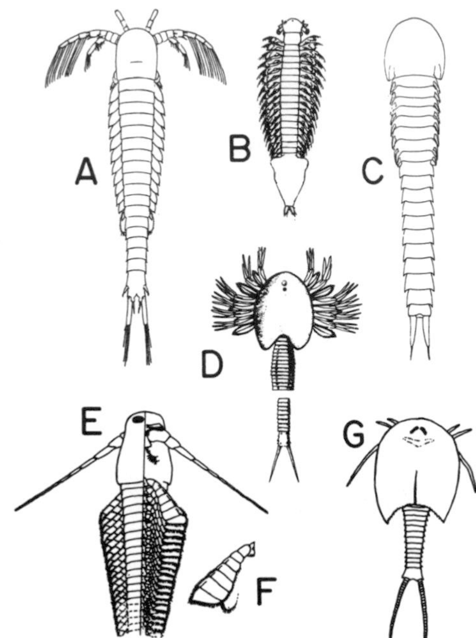

Fig. 3. Various types of "primitive" crustaceans: (A) Lipostraca, *Lepidocaris rhyniensis,* Devonian (from Scourfield, 1926); (B) Anostraca, *Polyartemia,* Recent (from Calman, 1909); (C) Cephalocarida, *Hutchinsoniella,* Recent (from Sanders, 1955); (D) Kazacharthra, *Jeanrogerium,* Jurassic (from Tasch, 1969); (E) Enantiopoda, *Tesnusocaris,* Pennsylvanian (from Brooks, 1955); (F) enlargement of limb of *Tesnusocaris* (from Brooks, 1955); (G) Notostraca, *Triops,* Recent.

There is no known fossil record for cephalocarids. Brooks (1955) placed the Pennsylvanian fossil *Tesnusocaris goldichi* in the cephalocarids, and Birshtein (1960) made this the basis of a higher classification of orders for cephalocarids. Hessler (1969b) rightly rejects *Tesnusocaris* as a cephalocarid. What is known of the appendages of *Tesnusocaris* invites comparison to the branchiopods or to the new class of Crustacea, Remipedia (Yaeger, 1981). However, the animal is strange (Fig. 3E,F): the forms of both sets of its antennae are unusual, reminiscent of but not identical to those of the remipedes. The details of the fourth and fifth cephalic appendages are unclear. Brooks felt that the third pair of cephalic appendages "modified as mandibles . . . characterized it a crustacean" (p. 852). The trunk appendages are unusual with segmented exopods and flaplike endopods, again not unlike Remipedia. Brooks' reconstruction appears to be influenced by distortions of preservation in the one known specimen. *Tesnusocaris goldichi* appear to have a midventral food groove, but the mouth and "maxillae" seem to be too far forward from the anterior termination of the groove to make functional sense. However, in light of the distinct separation of function in Remipedia between cephalic feeding and trunk locomotion, it may not be entirely unreasonable. Until more and better material is found for *Tesnusocaris,* the creature should probably be maintained as an uncertain order of the branchiopods, the Enantiopoda; or, pending a better understanding of the Remipedia, it may be eventually aligned with that class.

B. Branchiopoda

The branchiopods have figured prominently in comparisons with cephalocarids. Although appendages are quite similar in basic format to those of cephalocarids, the variety of existing body forms gives the group a rather diverse character.

The subclass Sarsostraca contains forms without a carapace (Fig. 3b). An important species that has been compared extensively to cephalocarids is the Devonian lipostracan, *Lepidocaris rhyniensis* Scourfield (1926, 1940). The elegant preservation of this tiny creature in chert nodules allowed Sanders (1957, 1963a) to make extensive use of *Lepidocaris* (see Fig. 7A) limbs as anatomical intermediates between *Hutchinsoniella* and various entomostracan groups. Even in size (3 mm) the two forms are similar. However, the anatomy of *Lepidocaris* is more akin to that of anostracan branchiopods. Indeed it has been used to help visualize how anostracans and other branchiopods evolved from primitive forms. However, *Lepidocaris* had thoracic appendages that were not homonomous, and the first maxilla was specialized as a clasper in the male. These characters mark Lipostraca as a distinct line. *Lepidocaris* also apparently inhabited a rather special environment. Scourfield (1926) concluded that the Rhynie Chert represented a hot water silica-rich habitat.

Actual fossil anostracan material is rare (Tasch, 1969) and generally of poor quality. Some of it may prove to be insect material, such as *Rochdalia* of the British Carboniferous (Rolfe, 1967), which was once believed to be an anostracan and is now considered to be an insect nymph. However, good fossil anostracan material has been described by several authors. Chirocephalids were named from the Lower Cretaceous of Transbaykal (Trusova, 1971) and the Eocene of the Isle of Wight (Woodward, 1879), and silicified specimens possibly representing a new family were descirbed from the Miocene of California (Palmer, 1957).

The subclass Sarsostraca contains forms without a carapace (Fig. 3B). An (Tasch, 1969). Fossils assigned to the modern notostracan genus *Triops* are known from the Carboniferous onward, and those compared to the modern genus *Lepidurus* are known from the Triassic onward. This fossil notostracan material generally consists of only parts of animals, but these parts closely resemble those of living forms.

An interesting series of genera comes from the Lower Jurassic of Kazakhstan and is considered to be a separate order, Kazacharthra (Novozhilov, 1957). Although Sharov (cited in Novozhilov, 1959, p. 266) would have placed these forms in the Notostraca, Novozhilov felt that the fewer number of trunk appendages (six) and the nature of the tail warranted a separate order. Indeed, from the reconstruction of the one well known species (Fig. 3D), *Jeanrogerium sornayi*, and in light of Bowman's (1971) comments about the differences between telsons and anal segments, there may be some basis to the belief that at least some members of this group had an anal segment and uropods rather than the true telson noted by Bowman in *Lepidurus*. The habitat of kazacharthrans appears to have been freshwater since they are found in association with plants and insect remains.

Finally, a Lower Devonian deep water species from the Hunsrück Shale, *Vachonisia rogeri* Lehmann (1955) that was once believed to be a branchiopod in its own order, Acercostraca, has recently been shown to be related to trilobitoids (Stürmer and Bergström, 1976).

The calmanostracans, taken as a whole, display a rather small radiation on a basic body plan. The fossil record of these freshwater forms is an ancient one, dating from the Devonian (Tasch, 1977), but it may represent only a small part of those that actually existed. Tasch (1969) pointed out the apparently great conservativeness of the group. His comments about cladistic relationships among calmanostracans must be viewed as quite speculative, however, until better knowledge of appendage structure in fossil forms comes to light.

The only preserved parts of fossil Diplostraca are the bivalved carapaces characteristic of the group. As a result, fossils are classified on the basis of carapace outline and sculpturing. Although there are a fair number of diplostracan fossils, especially of conchostracans, the fossil record can not be compared with the modern forms because of lack of appendages. The Conchostraca extend from the Devonian to the Recent. Most of the fossils occur in brackish or freshwater deposits such as lagoons and lake beds. Tasch (1969) is of the opinion that those few fossils that are found in association with marine fossils were probably secondarily introduced from nearby fresh and brackish water habitats. Cladocera, while common in modern ephemeral freshwater pools, are largely known only from subfossil finds in lake deposits. Some cysts of *Daphnia* from Miocene lake beds were mentioned by Tasch (1969), and Heyden (1861) described *Daphnia* from Oligocene lake deposits of Germany. Smirnov (1970) has reported Cladocera from the Permian.

C. The Ancestral Crustacean

Two points should be made about all the "primitive" forms described in this section. First, they are generally adapted to life in what might be termed *r*-selection environments. The living branchiopods, for the most part, exploit ephemeral freshwater habitats; *Lepidocaris* lived in hot, silica-saturated pools; and cephalocarids occupy marine flocculent layers. Second, a case has been made (Section II,A) for cephalocarids as progenetic paedomorphs, and this might also be extended to the various branchiopod groups as well. This should have some important implications for crustacean phylogeny.

Hessler and Newman (1975, Fig. 6, p. 445) suggested a reconstruction of the "urcrustacean." They developed two versions, one with (Newman) and one without (Hessler) a carapace. I have some disagreement in details but might accept the basic form of their beast. Their urcrustacean is essentially derived from the cephalocarid body plan. Any approach to the problem should be tempered by the fact that cephalocarids seem to be rather specialized in their own right. One should more properly approach the issue by treating all the primitive groups equally in order to deduce any "ancestors." I have come to be opposed in principle to the drawing of such "ancestors" since they are little more than "paper animals" (a theme I shall return to repeatedly in this chapter); however, I offer an "ancestral type" in contrast to Hessler and Newman for the sole purpose of presenting an example of the basic subjectivity of this sort of thing (Fig. 4), even though such paper ani-

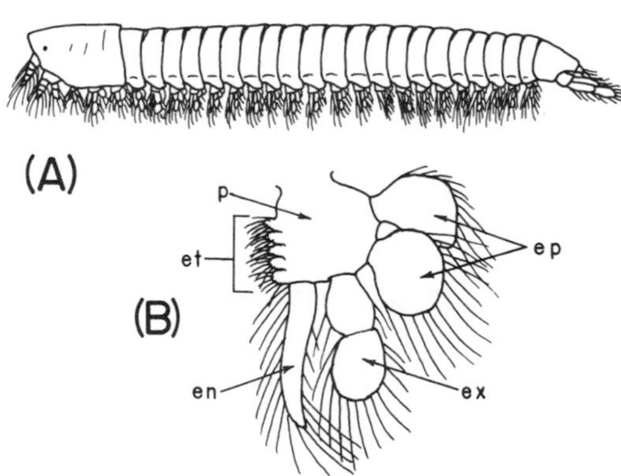

Fig. 4. (A) One possible visualization of a hypothetical ancestral crustacean, with homonomous trunk segments, homonomous postmandibular appendages, simple eyes, and no carapace. (B) Possible hypothetical foliaceous postmandibular appendage from which a variety of more advanced types of limbs could have been derived. Abbreviations: ep, epipods; ex, exopod; en, endopod; et, endopodites; p, protopod.

mals have little meaning in phylogenetic studies. Thus, I would tend to disagree with Hessler's approach and conclusions (Chapter 5 of this volume).

I would tend to opt for something (Fig. 4A) without a carapace. The history of the carapace in the entire phylum seems to indicate that carapaces were evolved repeatedly and variably with circumstance. The eyes in my creature are simple ocelli. This seems to offer the optimal neutral condition from which to evolve several anatomical conditions, whether it was to elaborate the simple ocelli, lose the eyes altogether, or to develop compound eyes of several types and forms. The appendage is a general type (Fig. 4B) from which any number of crustacean limb types might be derived, including that of a cephalocarid. Such an arrangement seconds the conclusion of Manton (1977, p. 148), who states that ". . . one suspects a simpler limb gave rise to both Branchiopoda and *Hutchinsoniella*." The recent discovery of the class, Remipedia (Yaeger, 1981), may overturn all current speculations about crustacean ancestors, since this new class seems in many respects more primitive than anything known to date; and, in light of my statements above, it may require a biramous limb on an ancestor.

IV. OSTRACODA

This is the oldest known of the crustacean groups. The Archaeocopida appeared in the Early Cambrian and the Leperditicopida in the Late Cambrian. It is also the most difficult group to relate to the other classes since the ostracodes have a most distinctive and divergent body plan.

The main adaptive thrust of the group is in the elaboration in form and function of the antennae for locomotion (Howe *et al.*, 1961), reproduction (Kesling, 1961), and feeding in connection with the mandibles (Cannon, 1931, 1940; Dahl, 1956a).

There are several peculiar and unique anatomical features. The ostracode body is essentially a head with an atrophied trunk enclosed in a calcareous bivalved shell (carapace). Only the extinct archaeocopids show evidence of a shell with more chitin than calcite. The mandible has a large palp which is typically biramous. There is only one pair of maxillae. The first trunk limb is variously developed: Calman (1909) agreed with Müller (1894) that this limb is more thoracic in character than cephalic, and that as a result, the ostracodes lack the second maxillae (but see Maddox, Chapter 5 of this volume). The development of these creatures is of no help in determining their relationships. The eggs hatch as a distinctive nauplius with a bivalved carapace. The development takes place gradually through a series of six to nine instars. All these unique anatomical conditions, combined with the very ancient status of the group, indicates that the ostracodes are a very distinct and early offshoot with little connection to other crustacean lines (see, e.g., Dahl, 1963; Schram, 1978).

Though the fossil record of the ostracodes is an extensive one, little has been done to determine evolutionary relationships within the class using these fossils. Efforts to determine a phylogeny are of necessity restricted to the study of modern forms (Hartmann, 1963). Recent considerations of shell form in light of strict mechanical principles (Benson, 1975) indicates that close attention to shell functional morphology is useful in determining habitat and of critical importance in basic taxonomy. Some efforts are now being made to evaluate reticular, ridge, and pore patterns of the shell (Benson, 1972, 1974; Liebau, 1969; Pokorný, 1969a,b) in order to arrive at a way to measure the degree of relationship between species. The best results, though of limited use, are the "theta-rho difference" analyses of Benson (1976, 1977) on pore conuli used in measuring taxonomic closeness in certain lines of trachyleberid Podocopa. Such functional analyses as these are breathing new life into fossil ostracode studies; even so, the use of ostracodes in phylogenetic evaluations is still frustrating because of the constraints of using carapace features only.

Müller (1979) described the unusually preserved appendages of Cambrian phosphatocopine ostracodes, a suborder of Archaeocopida. However, there is a serious difficulty in considering these fossils as ostracodes. The appendages, as reconstructed by Müller (Fig. 34, p. 22), are actually more like maxillopodan than ostracodan types, especially when one considers cirriped and naupliar copepod forms. Müller also comments on the fact that only the anterior end of his phosphatocopines was adequately preserved and that the posterior end of the body was usually lacking. He avoids reconstructing the posterior end of *Vestrogothia spinata* (Fig. 35, p. 23), but it may well have been a segmental and appendaged tagma. These very interesting fossils certainly deserve further study; however, they may have more to tell us about maxillopodan rather than ostracodan evolution.

V. MAXILLOPODA

Dahl (1956b, 1963) allied the Mystacocarida, Copepoda, Branchiura, and Cirripedia into a subclass Maxillopoda. This roughly coincided with Beklemishev's (1969) superorder Copepodoidea for Copepoda, Cirripedia, Branchiura, and Ascothoracica. Siewing (1960) argued for also placing Ostracoda into the Maxillopoda (although as we have seen above, the ostracodes seem to stand completely off by themselves). Subsequent authors (Ax, 1960; Hessler and Newman, 1975; Schram, 1978; Siewing, 1963) have followed Dahl, though Kaestner (1970) preferred to treat all these groups as separate subclasses.

The Maxillopoda are distinguished by the various ways in which they employ the maxillae and anterior thoracopods in locomotion and in food getting. Maxillopoda have especially exploited parasitic life styles. Of the four subgroups, the Branchiura are entirely parasitic, and the Copepoda and Cirripedia exhibit repeated adaptations to the parasitic habit.

A. Mystacocarida

No fossil record is known for this group, and it is unlikely that any ever will be, in view of the interstitial sandy beach habitat preferred by these animals. Hessler (1971) outlined the gaps in our knowledge of mystacocarid biology even though, as Hall and Hessler (1971) pointed out, the Mystacocarida are among the most studied groups of interstitial organisms. Hessler (1971, p. 87) and Kaestner (1970, p. 145) raised the possibility of neoteny (paedomorphosis) as a factor in the evolution of this group, whereas Gould (1977, p. 336) urged a progenetic interpretation to their evolution in contrast to earlier claims that mystacocarids are primitive. In any case, mystacocarids would appear to be a rather specialized interstitial side branch of the maxillopodan stock.

Seven species of *Derocheilocaris* are recognized (Hessler, 1972) in an essentially circum-Atlantic distribution. Since the Atlantic Ocean has only come into being with the Cretaceous, it would now appear that mystacocarid evolution is tied in with the Cenozoic evolution of the Atlantic. This illustrates the importance of paying attention to the geologic setting of fossil and Recent forms (as mentioned in Section I). Though little can be said of the anatomical evolution of mystacocarids, paleobiogeographic information can suggest a time frame for their history.

B. Copepoda

The copepods contain the primary pelagic radiation of the maxillopodans, and also display several independent parasitic evolutions. Gurney (1942, p. 20) pointed out the importance of paedomorphosis in the development of this group. The fossil record of copepods is sparse; only a few free-living Miocene fossils are known (Palmer, 1969). Cressy and Patterson (1973) discussed some beautifully preserved dichelesthioid siphonostomes from the gill chambers of some Lower Cretaceous bony fish. These Aptian parasites are important in that they are intermediate in form between siphonostomes afflicting invertebrates and those infesting fish. Although a chapter itself might be devoted to copepod evolution, it can best be done within a framework of the comparative anatomy and embryology of modern forms. Space restrictions here do not allow such a detailed consideration of these important maxillopodans.

C. Branchiura

This entirely parasitic group was formerly assigned to the copepods. They closely resemble, in some aspects, the caligoid copepods. The branchiurans can be considered a small side branch of the maxillopodans. They have no fossil record.

D. Cirripedia

The barnacles have one of the best fossil records of the Crustacea. The earliest form is from the Middle Cambrian Burgess Shale (Collins and Rudkin, in Newman, 1979). Another early form is the Upper Silurian *Cyprilepas holmi* (Fig. 5E), found attached to the appendages of Estonian eurypterids (Wills, 1963). The only other Paleozoic forms are two lepadimorph species of *Praelepas* in the Carboniferous, *P. jaworskii* Chernyshev (1931) (Fig. 5D) and *P. damrowi* Schram (1975) (Fig. 5A), and a probable acrothoracican in the Permo-Carboniferous *Trypetesa caveata* Tomlinson, 1963. The greatest number of thoracicans is found in the Mesozoic and Cenozoic. While it appears that cirripeds are an old group extending well back in the Paleozoic, the great radiation of calcareous forms may not have begun until Triassic time.

It is probably that a morphotypic maxillopodan can be most easily derived among the cirripeds, rather than the copepods. W. A. Newman (personal communication) feels that because of primitive ascothoracican forms, such as *Synagoga sandersi* (Newman, 1974), one could derive other maxillo-

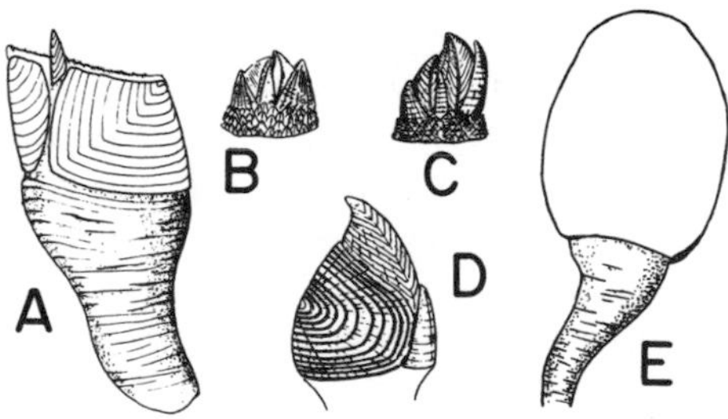

Fig. 5. Types of extinct barnacles: (A) *Praelepas damrowi,* Middle Pennsylvanian, U.S.A. (from Schram, 1975); (B and C) Brachylepadomorpha, Jurassic-Miocene (from Newman *et al.*, 1969), (B) *Brachylepas,* (C) *Pycnolepas;* (D) *Praelepas jaworskii,* Middle Carboniferous, U.S.S.R. (from Newman *et al.*, 1969); (E) *Cyprilepas holmi,* Upper Silurian.

podan types from some "urascothoracican" by progenesis. Furthermore, Newman points out that it is difficult to distinguish the organic shells of the supposed archaeocopid ostracodes from those of ascothoracicans. The morphology of Müller's phosphatocopids, discussed above, would tend to reinforce Newman's contention in that though phosphatocopids have been considered ostracodes, their appendages are in fact maxillopodan.

The cirripeds as a whole, despite a distinct adult form, are most closely allied with the copepod-branchiuran branch of the maxillopodans (Newman *et al.*, 1969). The harpacticoid copepod *Longipedia* has a cirriped-like nauplius with prominent anterolateral horns. The cirrus-like appendages and carapace of the branchiurans are similar to those of cirripeds, and the larval cirriped and branchiuran eyes are so alike as to possibly indicate some relationship. The cirripeds, however, have evolved a distinct bivalved carapace, natatory thoracopods, and prehensile first antennae that have formed the basis for their unique and most effective radiation.

The exact origin of the cirripeds is not clear. The free-moving Ascothoracica, though specialized in their parasitic habits, retain what is probably the most primitive of body plans, with a bivalved carapace and six thoracic and five abdominal somites. It is not difficult to envision the rest of the cirripeds evolving from a similar morphotype, though with a filter feeding rather than a parasitic lifestyle (Newman *et al.*, 1969).

The Silurian thoracican, *Cyprilepas,* forms a transitional type between a free-living form and the advanced sessile barnacles. The body of *Cyprilepas* was divided into a capitulum and peduncle with a bivalved carapace, which contained little or no calcite. The absence of growth lines on the carapace would seem to indicate that *Cyprilepas* molted the entire exoskeleton (Fig. 5E).

The Carboniferous genus *Praelepas* is more advanced than *Cyprilepas* (Fig. 5A,D). The five chitinous plates of *Praelepas* dramatically confirm the predictions of Broch (1922) of the appearance or "facies" of a common ancestor of the advanced lepadimorph families. Newman *et al.* (1969) consider *Praelepas* the closest thing to such an ancestral stock, and they compare it to the living oxynaspid thoracicans. Based on *Praelepas* and the general lack of Paleozoic thoracicans, they also concluded that calcareous plates on barnacles did not evolve until the Mesozoic.

Though the Scalpellidae have the oldest record of the calcareous thoracicans, the Lepadidae are actually more primitive and closer to the Paleozoic families (Newman *et al.*, 1969). The scalpellids, however, seem to be closest to the stock from which the nonpedunculate barnacles arose, i.e., the modern dominant Balanomorpha, the Verrucomorpha, and the extinct Brachylepadomorpha (Fig. 5B,C). Newman and Ross (1976) consider the possibility that Balanomorpha are polyphyletic.

The Rhizocephala, a purely parasitic group, have no fossil record. The particular mode of infection by rhizocephalans is prefigured in certain lepadomorph parasites of sharks and polychaete worms. Newman *et al.* (1969) indicated that the group may be polyphyletic, with members derived several times from within the thoracicans. But more recently W. A. Newman (personal communication) has offered some evidence of a single origin of Rhizocephala from Ascothoracica, based on detailed study of kentrogon attachment.

The acrothoracicans are recognized indirectly as fossils from their borings in calcitic substrates. Tomlinson (1963) has described the only definitely identifiable acrothoracican, *Typetesa caveata,* in Late Paleozoic myalinid clams. Other borings attributed to acrothoracicans have been named but should only be considered as ichnofossils of uncertain affiliation to cirripeds (e.g., Sainte-Seine, 1951, 1954; Codez and Sainte-Seine, 1957).

Ascrothoracicans are suspected to be in the fossil record. Madsen and Wolff (1965) attributed scars in a Cretaceous echinoid to an ascrothoracican similar to *Ulophysema*. Voight (1959) described some cysts on a Cretaceous octocoral and named them *Endosacculus*. He suggested ascrothoracican affinities for these.

VI. MALACOSTRACA

Malacostraca is the dominant and most successful of crustacean classes. It also possesses the most extensive and complete fossil record for Crustacea. The Phyllocarida are known from the Middle Cambrian, the Hoplocarida from the Late Devonian, and the Eumalacostraca from the Early Mississippian (Lower Carboniferous) and possibly from the Middle Devonian.

A. Phyllocarida

1. LEPTOSTRACA

The subclass Phyllocarida has always presented problems for phylogenists. Until recently, the only order that was adequately known was the living Leptostraca. The Paleozoic orders were generally deficient in information on appendages. It had been considered that phyllocarids differed from other Malacostraca in the possession of seven abdominal somites and a telson with well developed furca; however, Bowman (1971) and Sharov (1966, p. 215) seriously question both these points. In their opinion, the Leptostraca do not have a telson with furca, i.e., a nonsegmental somite with well-developed spines, but possess a true eighth abdominal somite with terminal anus and uropods. Thus, in discarding the controversial caudal furca as a separating characteristic of Leptostraca, they increase the gap with

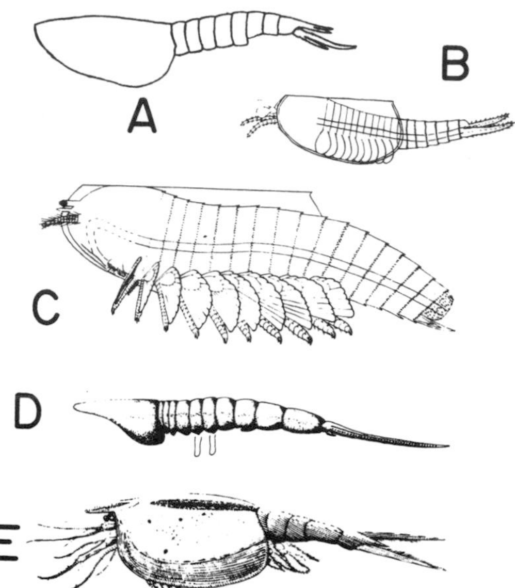

Fig. 6. Examples of extinct orders of Phyllocarida: (A) Hymenostraca, *Hymenocaris,* Lower Cambrian to Lower Ordovician; (B and C) Canadaspidida, Middle Cambrian, (B) *Perspicaris* (from Briggs, 1977), (C) *Canadaspis,* left valve of carapace removed to fully reveal appendage structure and location, and path of gut drawn in (from Briggs, 1978); (D) Hoplostraca, *Sairocaris* Carboniferous (from Schram, 1979b); (E) Archaeostraca, *Nahecaris,* Lower Devonian (from Rolfe, 1969).

the other malacostracans by interpreting eight segments in the leptostracan abdomen.

The possession of an anal segment rather than a telson helps to draw the Leptostraca even closer to the primitive groups (Section III). Bowman believes that the primitive groups, for the most part, have anal segments with terminal anuses. Hessler and Newman (1975) wanted to unite the primitive groups with the Malacostraca in a taxon called Thoracopoda (on a par with Maxillopoda) because of the similarity of the leptostracan's polyramous foliaceous thoracopods to those of cephalocarids and branchiopods.

Leptostraca have no fossil record, with the possible exception of a poorly known carapace from the Permian of Germany (Rolfe, 1969).

2. CANADASPIDIDA

Briggs (1978), in a restudy of the Middle Cambrian species, *Canadaspis perfecta,* revealed an order of animals which may have solid affinities with the Leptostraca (Fig. 6C). Briggs also placed two other Middle Cambrian species, *Perspicaris dictynna* and *P. recondita* (Fig. 6B), in this order (Briggs,

1977). The maxillary and thoracopodal appendages are homonomous and of a peculiar foliaceous type. The lack of a rostrum and abdominal appendages and the peculiar nature of the distal end of the abdomen set the canadaspids off by themselves. Briggs' reconstructions clearly show the anus opening terminally on what he calls a telson, with *Perspicaris* having what he termed furca. Briggs (1978, p. 484) suggested that similarities between canadaspidans and leptostracans were so close that one might be tempted to elevate these two groups to a separate subclass level and perhaps align them with the primitive orders.

3. HYMENOSTRACA

Hymenostraca is a small order containing Middle Cambrian to Lower Ordovician material. Rolfe (1969) placed the hymenostracans as a separate order because of their unhinged carapace and peculiar telson (Fig. 6A). Rolfe interpreted the telson of *Hymenocaris* as a short unit with three sets of spines. Bowman (1971) suggests it might better be interpreted as a short eighth anal somite with a pair of biramous uropods and a deeply divided true telson. The fossils themselves do not reveal the anus, which may have been terminal or on the ventral proximal base of that eighth abdominal unit. If Bowman's interpretation proves correct, then the hymenostracans would seem to be closely allied with the canadaspidans and leptostracans.

4. ARCHAEOSTRACA

Archaeostraca is the largest order now classified in the Phyllocarida. It extends from the Lower Ordovician to the Upper Triassic. Rolfe (1969) placed 22 genera in 6 families with certain archaeostracan affinities and listed almost two dozen other genera of uncertain affinity. The peak radiation of the group occurred in the Devonian. Most of the uncertain genera occur in the Cambrian, and some of this material has already been reassigned to other known groups, e.g., *Canadaspis* and *Sairocaris*.

The order is well characterized with a bivalved carapace, articulated rostrum, and distinctive styliform telson and furca (Fig. 6E). Little is known of the appendages of various archaeostracans (Rolf, 1969, p. R304). This paucity of leg data has always been a frustrating fact in classifying and assessing the phylogeny of these animals. Siewing (1956) suggested that *Nahecaris* was an ancestral type for Eumalacostraca, but Rolfe (1969) disagreed.

The Bowman view is of great relevance here. The traditional view of the very distinct styliform tail fan of archaeostracans has been a telson with furca. Rolfe (1963) and Copeland and Rolfe (1978) present the only knowledge we have of the archaeostracan anus in *Shugurocaris* and *Dithyrocaris,* respectively. The anus is on the proximal ventral side of the telson. Bowman suggests that the proximal telson head should be considered an anal somite and the distal styliform terminus an unarticulated telson; the furca then are the uropods of the anal segment. In my opinion, however, the arrangement of the archaeostracan tail is more like that seen in hoplocaridans and eumalacostracans, i.e., a true telson with a ventral and proximal anus, rather than the unarticulated telson that is more like the anal flaps recorded by Bowman in such groups as the mystacocarids, lipostracans, and cladocerans. It would appear to me that archaeostracans are not closely allied to the Leptostraca, Canadaspidida, and Hymenostraca.

5. HOPLOSTRACA

The Hoplostraca is a small order of Carboniferous forms that lacks a rostrum, has an unhinged carapace covering only the thorax, an exceedingly long abdomen in relation to the cephalothorax, and tiny caudal furca (Fig. 6D). Schram (1973, 1979a) called attention to the distinctive anatomical form of these beasts in relation to that of the Hoplocarida. Although the sairocarids were too specialized to have given rise to the hoplocaridans, they are an example of the diversity possible in the phyllocarid body form. Olsen *et al.* (1978) have some questionable sairocarids from Upper Triassic deposits of Virginia and North Carolina.

What has been said about the archaeostracan telson versus an anal somite (Section VI, A, 4) also applies here. Bowman would probably interpret the hoplostracan tail as an eighth anal somite with unarticulated telson and uropods, but I feel that one might equally argue for a true telson with articulated spines. In any case, hoplostracans and archaeostracans appear to be more closely related to each other on the basis of tail fan form than either is to the canadaspidans, leptostracans, and possibly hymenostracans. Whether this difference involves a seven-segmented abdomen with telson as opposed to an eight-segmented abdomen without telson remains to be seen. It does involve a proximal ventral anus in the former two orders as opposed to a terminal anus in at least two of the latter three orders.

B. Hoplocarida

1. PHYLOGENETIC CONSIDERATIONS

A modest debate has erupted over the suggestion by Schram (1969a,b) that Hoplocarida were independently derived from the Phyllocarida, separate from the "caridoid" eumalacostraca, i.e., syncarids, peracarids, and eucarids. The arguments are essentially as follows. The fossil hoplocaridans show no convergence with the caridoid archetype that Calman (1909) said was ancestral to all Eumalacostraca (including Hoplocarida). Rather, a distinct array of autapomorphic characters can be delineated which uniquely define the Hoplocarida. These are: three flagella on the first antenna; thoracopods with a three-segment protopod, flaplike outer branch, and four-segment inner branch; abdomen enlarged, containing most of the viscera; dendrobranchiate pleopod gills; "quasi-caridoid" musculature in abdomen; and possibly the fusion pattern in the abdominal somites. The possession of gills on all appendages and the appearance of uropods last in development are autplesiomorphic characters of the hoplocarids. Although other plesiomorphic characters are shared with Eumalacostraca and Phyllocarida, and both these groups have autapomorphic characters in their own right, the Hoplocarida do not apparently share any synapormorphies with these two groups. Schram (1969a, Fig. 4) considered the possibility of deriving caridoid and hoploid types from some common primitive eumalacostracan, but such a scheme is considered needlessly complicated as contrasted with that of a more parsimonious independent development of the two groups from within the phyllocarids, or some basal stock in the sense of Dahl (1976).

A problematic apomorphy concerns the manner in which six abdominal segments may have been derived from a primitive seven. It had been suggested by Siewing (1963) that hoplocarids may have fused segments in the anterior part of the primitive series of seven abdominal segments rather than in the posterior part as caridoids apparently did. This was based on the studies of Komai and Tung (1931) on the circulatory system of *Squilla,* which showed that in this genus, the first abdominal segment had a double blood supply from the heart. Burnett (1973) discovered that such an arrangement did not exist in *Hemisquilla*. Siewing's suggestion contradicted the developmental studies of Shiino (1942) on *Squilla* which indicated a possible fusion of segments in the posterior of the abdomen. Certainly the dispute here has not been disproved either way and could be answered with complete studies of gross anatomy and development in all four living families of Stomatopoda. In support of an independent origin, Reaka (1975) noted that the stomatopods had distinctive molt suture patterns that suggested a fusion of segments in the posterior thorax-anterior abdomen series. A median suture of the thorax exists on the sixth and seventh thoracomeres. The anterior thorax has characteristically lateral sutures along the sides of the carapace. The eighth thoracomere does not have a suture and is shed with the abdomen.

Burnett and Hessler (1973) took issue with Schram's hypothesis. They claim that Schram's characterization of the hoplocaridan abdominal gills as epipodal is "without adequate foundation" (p. 389). Yet, they chose to ignore the well-preserved gills on the pleopodal protopods of the aeschronectid genera, *Kallidecthes* and *Aratidecthes* (Schram, 1969b). In addition, there is now evidence for such gills on the palaeostomatopod *Bairdops*

(Schram, 1979a). However, I would agree with Hessler (Chapter 5 of this volume) that the presence of gills on all appendages appears to be plesiomorphic, and not relevant to phyletic relationships of the Hoplocarida.

Hessler (1964), and Burnett and Hessler (1973) maintained that stomatopods have a caridoid-like musculature in the abdomen. Careful reading of Hessler (1964) reveals, however, that the details of musculature in the abdomen are not the same between hoplocarids and eumalacostracans, and are best considered two distinct patterns which might have been derived from a common type. In addition, mantis shrimp simply do not have the classic caridoid tail-first escape reaction, but use an essentially different type of escape reaction (Manning, in Whittington and Rolfe, 1963, p. 18). Interpretation of form must always be tempered with an understanding of function.

Burnett and Hessler questioned as enigmas the meaning of the apomorphic characters listed above for hoplocarids. Schram, however, did not suggest an explanation of these character states: there may be none. These character states were probably stochastic derivatives, randomly arrived at in the dim recesses of antiquity; but once achieved and stabilized they marked the line of descent as quite separate and distinct from any other.

Again the antiquity, completeness, and conservativeness of the hoploid line is the most telling argument for independence of Hoplocarida and Eumalacostraca: we cannot lightly shunt aside, as Burnett and Hessler (1973, p. 390) would have us do. Indeed, hoplocaridans are at least as old as the eumalacostracans (Rodendorf, 1972; Schram, 1977). The phyletic scheme of Burnett and Hessler (1973, Table 3) is essentially a reworking of the phylogenetic alternative presented by Schram and alluded to above (1969a, Fig. 4). It was true in 1969—and is still true today—that one may erect any "paper phylogeny" one may wish to, but one should logically adopt the more parsimonious of alternatives. A separate origin of Hoplocarida is such an alternative. Schram (1978) and Schram and Horner (1978) essentially recognized this by elevating Hoplocarida to subclass rank coequal with Phyllocarida and Eumalacostraca.

2. AESCHRONECTIDA

This entirely natant order was apparently a Carboniferous radiation (Fig. 7C). Many of them were quite specialized in appendage structure. The kallidecthids had large natant second maxillae and the aenigmacarids had stenopodous pleopods. It appears that the order was an important component of the pelagic and epibenthic filter feeding crustaceans of the Late Paleozoic. As mentioned above (Section VI, B, 1), this animal has played an important role in illustrating the constancy of the hoplocaridan body plan.

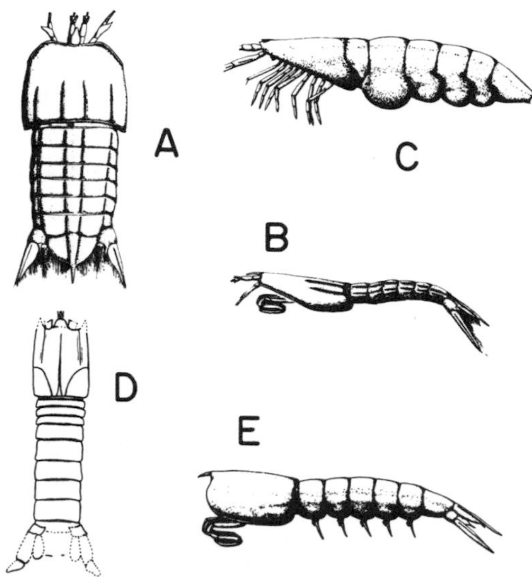

Fig. 7. Examples of the orders of Hoplocarida: (A, B, and E) Palaeostomatopoda, Carboniferous, (A and B) dorsal and lateral views of *Perimecturus* (from Schram, 1979b), (E) *Bairdops* (from Schram, 1979b); (C) Aeschronectida, *Kallidecthes,* Upper Carboniferous (from Schram, 1979b); (D) Stomatopoda, *Paleosquilla,* Cretaceous (from Schram, 1968).

3. PALAEOSTOMATOPODA

This was an early experiment on a rapacious carnivore body plan (Fig. 7A,B,C). A battery of subchelate thoracopods and the morphological trends through time toward increasing spinescence and decoration indicate that this order possessed rapacious and probable agonistic behaviors (Schram, 1979c). However, these do not necessarily imply direct ancestry to the stomatopods (Schram, 1969b). Palaeostomatopods declined toward the end of the Mississippian after the origin of tryannophontid archaeostomatopodeans.

4. STOMATOPODA

The tryannophontids are probably present in Dinantian Lower Carboniferus deposits of Britain (Schram, 1979a). By the Upper Carboniferous, after coexisting with palaeostomatopods for some time (Schram and Horner, 1978), the tyrannophontids became the sole rapacious hoplocarids. A hiatus exists in the record between these Paleozoic archaeostomatopodeans and the essentially modern unipeltatan forms of the Jurassic (Holthuis and Manning, 1969). With the exception of the extinct Mesozoic Sculdidae, all the recognized unipeltatan fossils can be assigned to living families (Fig. 7D), (e.g., Berry, 1939; Rathbun, 1926; Remy and Avnimeleck, 1955; Schram, 1968). Unipeltatans are rare as fossils, though they have a modern radiation of over 350 species.

C. Eumalacostraca

1. SYNCARIDA

Syncarida contain the eumalacostracans with schizopodous thoracopods and no carapace. This superorder has a good Paleozoic record and was actually known through fossils (Packard, 1885) some years before the first living form was described from Tasmania (Calman, 1896). The extinct palaeocaridaceans and the living family Anaspididae have a generalized form that, except for the lack of a carapace, conforms perfectly to Calman's (1909) generalized caridoid morphotype. All the other families in the two living orders are adapted to interstitial or ground water life styles and display various degrees of paedomorphosis in their development. Indeed paedomorphosis has been a hallmark of syncarid evolution with probably three independent paedomorphic lines.

Palaecaridacea is the principal Paleozoic order and contains the most primitive and generalized of the syncarids. Five families are currently recognized, exhibiting various adaptations of the basic body plan. The palaeocarids are the most primitive (Fig. 8A,C). The Lower Carboniferous genus (Fig. 8A) *Minicaris* Schram, 1979a has a small body size, eight equally developed thoracomeres, and unspecialized appendages which make it the most primitive of known syncarids. The Middle to Upper Carboniferous squillitids (Schram and Schram, 1974; Schram, 1979a) are very close to an anaspidacean condition, including annulate pleopods, except that they still have a free first thoracomere (Fig. 8D). The order was widely distributed in the brackish and near shore marine environments of the Paleozoic continent Laurentia (Schram, 1977).

Anaspidacean fossils include *Anaspidites antiquus* from the Triassic of Australia and *Clarkecaris brasilicus* from the Brazilian Permian; both occur in brackish to freshwater deposits. The living stygocarids are ground water forms found in Australia, Chile, and Argentina. The lack of eyes, reduced size and number of pleopods, and the reduced nature of the thoracopodal exopods are paedomorphic attributes. Siewing (1963) advanced the idea that the stygocarids had a lacinia mobilis on the mandible. This was based on a misinterpretation of material presented by Noodt (1963). Gordon (1964) and Noodt (1965) reject this notion. They both conclude that the "penicillae" or "Sägeborsten" that Siewing referred to were developed from

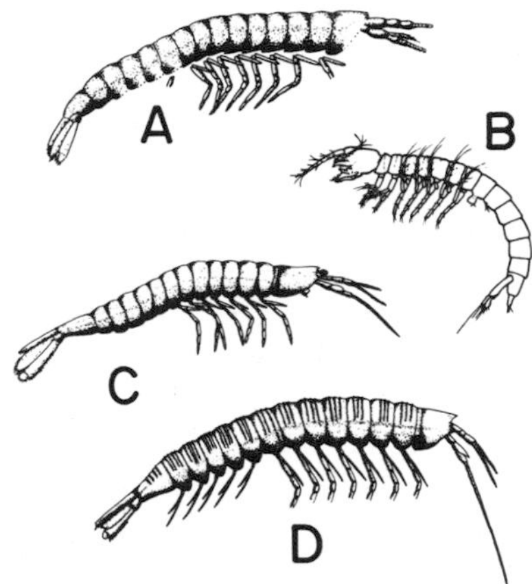

Fig. 8. Examples of Syncarida: (A) *Minicaris,* the earliest and most primitive known palaeocaridacean, Lower Carboniferous; (B) *Allobathynella,* typical bathynellacean displaying many paedomorphic features such as vermiform body, reduced thoracopods, and absent pleopods (from Brooks, 1969); (C) *Palaeocaris,* typical palaeocaridacean, flaplike pleopods are known from this genus but merely not preserved on the species illustrated (from Schram, 1979b); (D) *Praeanaspides,* palaeocaridacean closest to the origin of Anaspidacea (from Schram, 1979b).

primitive cusp series *c* on the mandible (Manton, 1928) rather than cusp *b* from which the lacinia mobilis evolved.

Manton (1930) showed that the feeding habits of the living *Anaspides* are of a rather random, inefficient grazing-foraging type. The animals process food as they stumble across it. The anatomy of the anaspids is much like that of the palaeocaridaceans. This primitive, somewhat inefficient feeding behavior may explain why the Paleozoic syncarids went into a decline and why their descendents today are in a refuge pattern of distribution (Noodt, 1965; Schram, 1977).

Bathynellacea are interstitial ground water forms (Fig. 8B). They are the only syncarid group which can be said to be successful. They are found on all continents (Noodt, 1965, 1974a; Schminke, 1973, 1974; Schminke and Wells, 1974). Serban (1970), in a review of the species *Bathynella natans,* elevated the Bathynellacea to its own superorder, Podophallocarida, with no detailed explanation or justification. Bowman (1971) suggested that the terminal end of the abdomen was different in bathynellids than in other syncarids; however, Schminke (1976) disputed this. Bathynellacea should remain as an order of the Syncarida. The bathynellids are only an extreme expression of a repeated pattern of paedomorphic forms in the Syncarida. Their paedomorphic features include reduction or in some cases elimination of the pleopods and also the thoracopods in some species, a pleotelson, their tiny size, a large and sometimes prehensile mandibular palp, and setiferous antennae.

2. EOCARIDA

Brooks (1962) erected this superorder to accommodate what was thought to be certain primitive Paleozoic caridoid forms. All eocarids were thought to have a single segment protopod and caudal furca. The pygocephalomorphs (Schram, 1974a) were seen to have a coxa and basis along with oöstegites and were reassigned as a specialized suborder of mysidaceans. The Eocaridacea have not been completely abolished but it is now known that at least some of the protopods of species in that order have two segments, and in almost all cases caudal furca have not proven to be very diagnostic (Schram, 1979a).

The eocarids are retained to accommodate certain caridoid malacostracan types which are not clearly mysidacean or euphausiacean. These include five families for the time being (Fig. 9). Some former eocaridans have been assigned to other groups. *Crangopsis* is now an aeschronectid hoplocaridan (Schram, 1979a). The paleopalaemonids are recognized as the earliest decapods (Schram *et al.*, 1978), and the holotype of *Eocaris* has been re-examined and may be an aeschronectid (Schram, 1977). *Eocarida* is thus a group that is now understood to include generalized schizopodous caridoids.

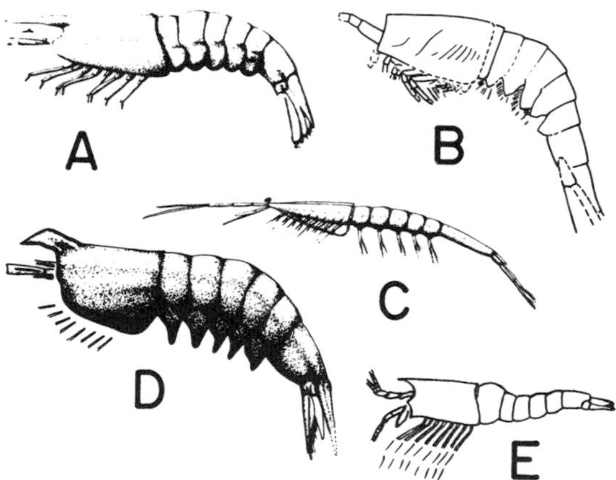

Fig. 9. Examples of the families of Eocaridacea: (A) Belotelsonidae, *Belotelson*, Carboniferous (from Schram, 1979b); (B) Eocarididae, *Eocaris*, Upper Devonian (from Brooks, 1962); (C) Waterstonellidae, *Waterstonella*, Lower Carboniferous (from Schram, 1979b); (D) Anthracophausiidae, *Anthracophausia*, Carboniferous (from Schram, 1979b); (E) Essoidiidae, *Essoidia*, Upper Carboniferous.

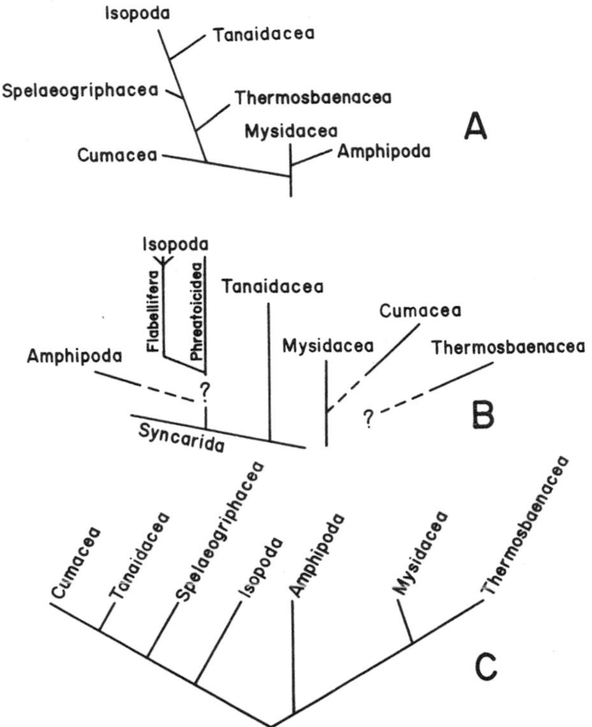

Fig. 10. Various phylogenetic schemes for the orders of Peracarida: (A) traditional arrangement of most authors with Mysidacea a stem group from which two lines arose, one leading to Isopoda and the other to Amphipoda; (B) derived from Glaessner (1957) with essentially carapace (right) and noncarapace (left) forms separate; (C) derived from Watling (1981) with the three separate (mysidoid, gammaroid, and mancoid) types.

3. PERACARIDA

The generally accepted scheme of peracarid relationships, based on the comparative anatomy of modern forms, is of a single family tree (Fig. 10A), with Mysidacea as an expression of a basal stock, one main branch leading to the isopods and another main branch leading to the amphipods (Siewing, 1956, 1963; Fryer, 1964; Hessler, 1969b). Glaessner (1957) proposed a different view, based largely on fossil evidence. In his scheme (Fig. 10B) the carapaceless eumalacostracans were a separate line from those with a carapace. Glaessner's views have been largely rejected or ignored by carcinologists. Watling (1981) examines fossil evidence, comparative anatomy, and comparative embryology in a cladistic context and believes that there are probably independent lines of peracarid types (Fig. 10C). They are:

gammaroids (Amphipoda), mysidoids (Mysidacea), and mancoids (everything else). Only the mysidoid line was primitively carapaced. The mancoid sequence proposed by Watling, a reverse of the traditional view, places the isopods as the most primitive and the cumaceans as the most specialized. Watling views gammaroids as a separate line with distant connections with the mysidaceans. In this regard, Dahl (1977) suggested in passing that certain features of amphipod form were inconsistent with supposed mysid origins. In some respects the Watling hypothesis is more in line with what is known in the fossil record. Mysidoid and mancoid types appear simultaneously in the Lower Carboniferous (Schram, 1979a), with tanaids, spelaeogriphaceans, and isopods appearing before cumaceans. The Watling hypothesis, intermediate between Siewing's and Glaessner's, at the very least will demand a serious and detailed reconsideration of peracarid relationships and evolution.

The Mysidacea underwent an extensive radiation in the Carboniferous. The specialized suborder Pygocephalomorpha (Fig. 11E,F,G) was one of the

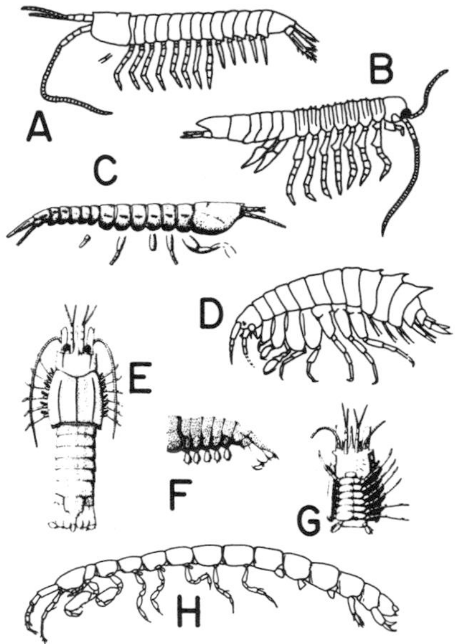

Fig. 11. Example of various peracaridan types: (A) Spelaeogriphacea; *Acadiocaris,* Middle Carboniferous; (B) Isopoda, *Hesslerella,* Middle Pennsylvanian; (C) Tanaidacea, *Anthracocaris,* Lower Carboniferous (from Schram, 1979b); (D) gammarid Amphipoda, *Praegmelina*, Middle Miocene (from Hessler, 1969); (E, F, and G) pygocephalomorph Mysidacea, *Tealliocaris,* Lower Carboniferous (from Schram, 1979b); (E) dorsal body; (F) lateral abdomen; (G) ventral thorax; (H) ingolfiellid Amphipoda, Recent, displaying paedomorphic form (from Hessler, 1969).

dominant groups of that time. The pygocephalomorphs possessed certain convergent similarities to palinuran eucarids (Schram, 1974c) and inhabited all types of habitats (Schram, 1981a). Of the more generalized mysidaceans, Lophogastrida, have no fossil record, and Mysida have a few poorly-preserved Mesozoic forms. Schram (1974a, 1979a) has suggested that some of the eocaridaceans might be mysidaceans.

Amphipoda is the only peracarid order with a fossil record that does not go back to the Paleozoic. The earliest known amphipods are Upper Eocene in age. The distribution of the fossils is the Baltic area in the Eocene, Alsace in the Oligocene, and the Caucasus and Caspian region in the Miocene (Fig. 11D). All of these fossils are gammarids. J. L. Barnard (personal communication) feels that the origin of freshwater gammaridans lies in the Mesozoic when Pangaea existed, and he has some cogent biogeographic arguments for this position. Therefore, amphipod origins go back at least to the Middle to Late Mesozoic. Barnard also remarks that amphipods in general seem to be poorly adapted to tropical conditions; most of their evolution seems to have taken place in a northern or southern temperate context (Barnard, 1976). In this regard, I then wonder if their origin postdated the formation of Pangeae in the Permian; i.e., amphipod origins, whatever they be, may have been quite separate in space-time from most of the other peracarid orders in the Devono-Carboniferous tropical continent Laurentia (Schram, 1977). Amphipods such as caprellids and ingolfiellids exhibit paedomorphic features (Fig. 11H).

Cumacea are virtually unknown as fossils. Rolfe (1969, p. R313) records an Upper Permian carapace as a cumacean. Bachmeyer (1960) described *Palaeocuma* from the Jurassic of France.

Spelaeogriphacea are known from two species: *Acadiocaris novascotica* (Fig. 11A) from the Carboniferous of Canada (Schram, 1974b), and the living *Spelaeogriphus lepidops* from a pool in Bat Cave at Table Mountain in Cape Town, South Africa (Gordon, 1957). The reduced nature of the carapace, thoracopods, and pleopods in *Spelaeogriphus* as compared to *Acadiocaris* suggests paedomorphosis as a factor in spelaeogriphacean evolution.

Tanaidacea are among the oldest of peracarid fossils and are known from a series of remarkably preserved Paleozoic species (Glaessner and Malzahn, 1962; Schram, 1974b, 1979a). Most of these are monokonophorans. The Paleozoic family Anthracocarididae is the most primitive of the tanaids (Schram, 1979b); its telson and sixth pleomere are separate (Fig. 11C).

Isopoda have a moderately good fossil record. The phreatoicideans (Fig. 11B) are known from the Middle Pennsylvanian (Schram, 1970) and are the only isopods in the Paleozoic (Hessler, 1969b). Schram (1974b, Table 3) compared anatomical states in the isopod suborders and concluded that the phreatoicideans are the most primitive of isopod groups, with the dorsoventrally flattened suborders being derived. Schram (1977) pointed out that the early phreatoicids were marine, whereas by the Permo-Triassic they had become adapted to freshwater. Since that time they have become restricted to a freshwater Gondwana refugium pattern of distribution. The rest of the isopod fossils (Hessler, 1969b), although of generally good quality, are not particularly helpful phylogenetically. The Flabellifera have an extensive fossil record, from the Triassic to Recent. Valvifera and Oniscoidea occur from the Eocene.

4. PANCARIDA

Thermosbaenacea apparently have no fossil record, though the waterstonellid eocaridans may have some affinities to them. Siewing (1958) treated these as a separate superorder Pancarida. Stella (1959) and Taramelli (1954) tried to link thermosbaenaceans with the syncarids. However, a number of recent authorities (Barker, 1962; Gordon, 1958; Hessler, 1969b; Monod, 1927, 1940; Fryer, 1964) rejected these views in favor of a peracarid affinity. But such an arrangement does have the disadvantage of making the peracaridans a one character group, the character being a lacinia mobilis on the mandible. Thermosbaenacea brood eggs dorsally under the carapace rather than in an oöstegite brood pouch. Small body size, brooding of few eggs, reduction in appendages, and the presence of a pleotelson (in *Thermosbaena*) suggest progenetic paedomorphosis in the evolution of the order. Knight (1978) has clearly shown the presence of a lacinia mobilis in the juvenile stages of the euphausiaceans. Thus the lacinia mobilis cannot be considered a unique feature of the peracarids.

5. EUCARIDA

Eucarida are sometimes thought to be the most successful of malacostracans mainly because of the great radiation of brachyuran crabs (though in actual species numbers and diversity of types the peracarids may exceed eucarids). The eucarids are divided into three orders: Euphausiacea and Amphionidacea, which lack any fossil record, and Decapoda, which has an extensive record that begins in the Late Devonian. The eocarid family Anthracophausiidae bears a striking resemblance in some regards to euphausiaceans, but the exact relationships of euphausiaceans and amphionidaceans to decapods is not clear.

The earliest decapod fossil is *Palaeopalaemon newberryi* of the Late Devonian of North America (Schram *et al.,* 1978). This animal has characters of both astacidean and palinuran pleocyemates (Fig. 12). The occurrence of *Palaeopalaemon* in the Devonian indicates that all the basic eumalacostracan superorders appeared very early in the history of the subclass. Such circumstances suggest a "phyletic grass" approach to eumalacostracan

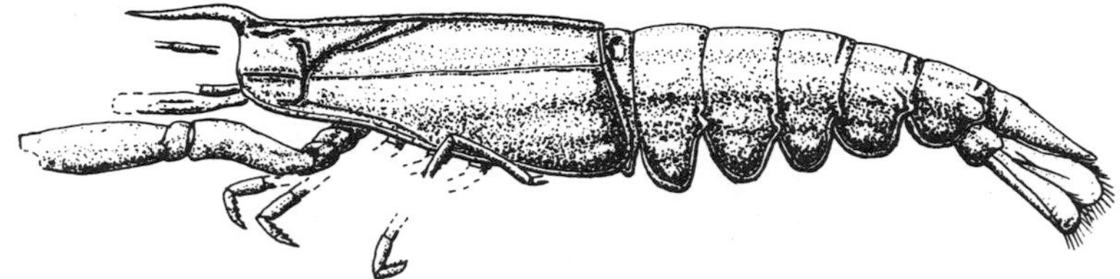

Fig. 12. *Palaeopalaemon newberryi,* the earliest decapod, Lower Devonian; pereiopods incompletely known but with hypertrophied first pereiopod, subchelate second through fourth pereiopods, apparently achelate fifth pereiopod (from Schram *et al.,* 1978).

evolution rather than the "phyletic tree" concept of an ancestor-descendent branching slowly and unfolding through time.

The current classification of decapods is a compromise arrangement reflecting an incomplete understanding of how the group evolved. The suborders Dendrobranchiata and Pleocyemata were erected by Burkenroad (1963) as the preliminary first step in a proposed revisionary classification of decapods. The complete classification was never published. The first attempts at a modern classification of decapods was that of Beurlen and Glaessner (1930) and Beurlen (1930). Their efforts stemmed from a desire to incorporate data from the fossil record and replace the old and entirely inadequate Borradaile (1907) classification of Natantia-Reptantia. This latter system was based solely on modern forms and still persists today in textbooks. The neocarcinologists balked at the paleocarcinological efforts, and the classification of Glaessner (1969) is largely zoological, tempered by some fossil input. Figure 13 presents a view of decapod relationships based on current knowledge of recent and fossil forms. The knowledge (Schram *et al.,* 1978) that there is an extensive Paleozoic history of decapods to be discovered should eventually result in enough material to complete our understanding of the early decapod radiation. This will help us settle on a more effective classification for the decapods as our ideas of their evolution become clearer.

Dendrobranchiata (Late Triassic-Recent) is a largely Mesozoic radiation. Our fossil knowledge of this natant suborder is derived from the excellent preservations in lithographic limestones in the Jurassic and Cretaceous of Germany and the Middle East.

Pleocyemata is divided into several infraorders. The natant Uncinidea (Jurassic) and Caridae (Jurassic-Recent) have fossil records with few poorly preserved specimens. The reptant infraorders have good fossil records, mainly due to their well-sclerotized and mineralized exoskeletons. The Astacidea (Late Permian-Recent) underwent a succession of radiations in the Mesozoic: the erymids flourished during the Triassic and Jurassic, and the

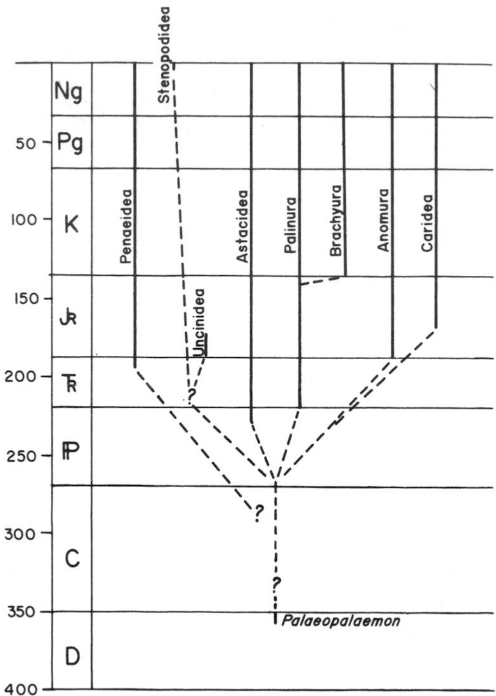

Fig. 13. Possible phyletic scheme for decapods based on current understanding of group; *Palaeopalaemon* the closest known form to an ancestral stock, and with penaeids, stenopodid-uncinid, and carideans as three independent natant stocks.

nephropids during Late Jurassic-Cretaceous time. Current astacideans are now largely exploiters of deep-water marine and freshwater habitats. The Palinura (Triassic-Recent) also exhibit waves of radiations: glypheoids first came on in the Triassic and only one living form, *Neoglyphea,* survives; the eryonoids reached their height in the Jurassic; and the palinuroids built through the Jurassic and Cretaceous to a Cenozoic peak, and are still important today. Anomura (Jurassic-Recent) have remained at fairly steady levels since their origin. All the anomuran superfamilies start in the Jurassic except the hippoids of the Oligocene. Brachyura appear to the Jurassic, but the radiation of this infraorder does not really get underway until the Cretaceous. The brachyurans have undergone three clearly marked stages in their history. The first was the initial Jurassic deployment of the Dromiacea. The second stage was a radiation of Cretaceous groups: the dromiaceans persisted and Oxystomata arose and radiated into the Cenozoic. The last stage was the grand Eocene radiation of the Brachyrhyncha, Cancridea, and Oxyrhyncha that has persisted to the present.

VII. PATTERNS IN CRUSTACEAN HISTORY

Having reviewed the "nuts and bolts" of the origin, history, and evolution of the Crustacea, certain patterns are evident and deserve some detailed comment and discussion.

First, there is the great diversity of Crustacea. Insects have had a grand radiation, but are all based on the same *Bauplan*. Crustacea exhibit several distinctly different morphotypes. Implicit in this is a great capacity for experimentation. As an example of the diversity that could theoretically be generated, Savory (1971) described 15 basic character states for arachnids, which, if they varied by merely presence or absence of each state, could result in nature of 32,768 different combinations, i.e., 32,768 possible orders that might have been formed. If we consider just the Malacostraca, with 20 possible segments and appendages thereon, and consider, for example, whether a particular set of appendages might or might not be specialized from some generalized type, we would have 1,048,567 possible basic variants of Malacostraca alone, and this would not include orders based on variant specializations that might be found on any one segment. True, if such existed, we should probably lump rather than split such variety in recognizing orders. But the potential for diversity is truly staggering to contemplate.

Crustacea seem to have been able to generate distinct body formats randomly and quickly as circumstances allowed, producing many heteronomous combinations. The optimal arrangements established themselves to repeat the process at another level at a later time. For example, hoplocaridans evolved into a series of morphotypes in the Late Paleozoic. But only one rapacious-carnivore line survived to form the basis of a Late Mesozoic–Cenozoic radiation of mantis shrimp. Another example is the eumalacostracans which produced several basic, yet distinct, morphotypes in the Late Devonian (palaeocaridacean syncarids, various basic peracarid forms, decapods, and eocarid types).

Second, the Crustacea are a very repetitive group with the ability to come on with wave after wave of the same basic form. The Brachyura are a fine example of this, with a history of proliferating genera. Though the average extinction rate may be high for these genera, the development of new taxa is equally high, with three basic bursts of radiation for the true crabs.

Third, there are striking patterns of morphological convergence (Schram, 1974c). Certain body forms, despite the potential for diversity mentioned above, were repeatedly and independently evolved in unrelated groups, e.g., reduction of the abdomen in Permian notocarid pygocephalomorphs and post Jurassic brachyuran decapods. The number of possible heteronomous combinations in crustaceans is probably astronomic and the basic turnover of taxa is high. Yet it appears that, crustaceans being crustaceans, there are certain constraints on what can and what cannot be an optimal functioning arrangement of parts. So we end up with gross similarities of form and function in a habitat type, e.g., similarities between such obviously unrelated forms as mystacocarids, bathynellaeceans, and ingolfiellan amphipods in interstitial habitats.

Finally, the crustacean record is now known well enough to indicate that biogeographic patterns are going to be important in ultimately revealing the intricacies of crustacean history. For example, as already mentioned above, Schram (1977) analyzed the Late Paleozoic and Triassic biogeography of malacostracans and advanced arguments for the idea that the present day broad patterns of distribution stem from events of that previous time. Certainly, the relict distributions of anaspidaceans, spelaeogriphaceans, and phreatoicid isopods can be seen as the result of vicariance of formerly contiguous faunas which were dispelled with the Cretaceous breakup of Pangaea.

The fossil record of Crustacea has a wealth of information in it. We can no longer afford to dismiss this data bank as "insufficient" or as "disputed," as some neocarcinologists have done in the past (e.g., Siewing, 1963, p. 85). The same pejorative adjectives can be applied to purely neocarcinological analyses.

VIII. THEORETICAL MATTERS

We can now address several crucial theoretical matters: Some are immediately applicable to the crustacean fossil record and evolution, and some are of broader application.

The first issue is the widespread occurrence of paedomorphosis as an agent of evolution in Crustacea. Of the 50 orders of crustaceans (see Appendix and Bowman and Abele, Chapter 1 of this volume), at least 23 of these show clear obvious attributes of paedomorphosis. These are the Brachypoda, the orders of Branchiopoda, Mystacocaridida, Arguloida, the orders of Copepoda, and some families of Anaspidacea, Bathynellaecea, Thermosbaenacea, and Spelaeogriphaecea. One could probably make good cases for several other orders as well, e.g., the ostracode and cirripede groups. But the point is well made: the Crustacea are rife with paedomorphosis—most, if not all of it, progenetic.

Why have crustaceans repeatedly exploited progenesis as an agent of evolution? Gould (1977, p. 338) points out that progenesis achieves "the redirection of selection towards the timing of maturation . . ." releasing ". . . the rigid selection usually imposed upon morphology." It was pointed out in Section VII that crustaceans exhibit a great deal of diversity and ability to develop great numbers of taxa as variants on a basic body plan. It would be just such a progenetic strategy that could allow these diverse morphological variations to arise and move to a point where they might radiate. This would be facilitated by the minimization or removal of the usual constraint of the selection of anatomy. The intense experimentive nature of crustacean evolution thus finds a vehicle in paedomorphosis.

The crustacean *Bauplan* had the theoretical capacity to produce millions of orders. That it produced as many as it did (at least 50) is a testimony to the phylum's innate capacity to diverge. Although progenesis offers an explanation for the great experimental versatility of the group, it places some serious constraints on phylogenetic speculations within the phylum. Certain forms that have been viewed as primitive must now be seriously reassessed, e.g., cephalocarids. Gould (1977, p. 281) most effectively points out that we must not confuse generalized structure with temporal priority.

We carcinologists have frequently delineated facies that result in the visualizations of archetypes—paper animals that supposedly acted as ancestors (e.g., Hessler, Chapter 5 of this volume). We all make them and use them. They are convenient crutches to support our phylogenetic speculations. But we must constantly remind ourselves we engage in typology when we do this, and seriously consider facies limitations and their purely pragmatic nature. Not to do this is to court disaster. Malacologists fell so much under the spell of the "archetypal mollusc," that when the segmented *Neopilina* was discovered, and a whole host of Paleozoic monoplacophorans were thereby thrown into perspective, many of the neomalacologists stuck with the archetype and attempted to explain away *Neopilina*. A few prominent neomalacologists even said *Neopilina* was "irrelevant." The paleomalacologists, who were not as deeply devoted to the "honorable ancestor," not only reassessed the monoplacophorans, but were effectively able to interpret and include newly-discovered Lower Paleozoic forms in a coherent phylogeny of Mollusca (Yochelson *et al.*, 1973; Runnegar and Pojeta, 1974; Runnegar and Jell, 1976; Pojeta and Runnegar, 1976). This was done with little reference to the "archetypal mollusc" that still appears in textbooks.

We have our "sacred cows" in carcinology, and like the holy beasts of India, they are allowed to roam through our papers and discussions at will. For example, the caridoid facies of Calman (1909) have taken on this role. All phylogenetic speculations on eumalacostracans must be referred back to the caridoid archetype. This still persists despite the opposition of several eminent authorities. Tiegs and Manton (1958, p. 295) questioned whether the caridoid facies might not have been convergently developed in several groups. Dahl (1963, p. 4) presented an analysis that cast doubt on the caridoid facies as a primitive pattern for malacostracans, and he later (Dahl,

1976, p. 165) questioned the necessity for a caridoid facies at all. And Watling (1981) builds a perfectly feasible phylogenetic scheme for peracarids without recourse to a caridoid type. Unless we guard against the excesses of such facies theories, we run the risk of having our ideas frozen into concepts based on the limited understanding and the particular prejudices of the time in history in which they were formulated.

And yet, we must strive to arrive at some cohesive theory of the evolution of the groups we study. The fossil record suggests an alternative framework to that of facies theories—I term it "stochastic mosaicism." It appears that in the initial phases of the origin of a major group, there is a rapid and random development of several variant morphotypes. The constituent features of anatomy go together in a mosaic pattern. The exact combination of characters that produces the full scale radiation is most likely governed by chance as much as any actual selection for a particular character state. Savory (1971) called this aleatory evolution.

As an example, consider the Devonian deployment of malacostracans, and within "caridoid-schizopod" types, consider two sets of character alternatives: (a) carapace fused or not fused to the thoracomeres, and (b) having or not having a brood pouch. There are four combinations of these four mosaic characters, three of which we recognize in living and fossil groups: (1) fused carapace and no brood pouch (Euphausiacea), (2) unfused carapace with brood pouch (Mysidacea), (3) fused carapace with a brood pouch, and (4) unfused carapace without a brood pouch (waterstonellid Eocaridacea). Number three has not yet been recognized in the fossil or Recent record. At least three of these four possible combinations were evolved. There is no *a priori* reason why any one set should be considered superior to any of the others. Indeed, I do not think they were. The combinations were randomly evolved more or less simultaneously, and there may have been a large stochastic element in which combinations actually survived to the present. Why one combination may survive when another becomes extinct is then an irrelevant question for the most part. Raup (1978) arrives at a similar position in his consideration of extinction as a probabilistic phenomenon.

Now take the four character states above and add a third couplet for the mosaic, viz, with or without schizopodous thoracopods. There are now eight possible sets of combinations and the new alternatives include three more we can recognize: (5) fused carapace, no brood pouch, and uniramous thoracopods (Decapoda), (6) fused carapace, brood pouch, and uniramous thoracopods (Amphionidacea), (7) unfused carapace, no brood pouch, and uniramous thoracopods (belotelsonid Eocaridacea), and (8) unfused carapace, brood pouch, and uniramous thoracopods (no recognized group). The game can be expanded indefinitely. Add another set of characters—

without any carapace, or with a short carapace only over the anterior most thoracomeres. Eight more alternative combinations are made of which four more can be recognized as one is brought into the realm of syncarids, and carapaceless and short carapace peracarids. Among the carapaceless forms there are: (9) no brood pouch and schizopods (Syncarida), (10) brood pouch and schizopods, (11) no brood pouch and uniramous thoracopods, and (12) broad pouch and uniramous thoracopods (Amphipoda and Isopoda). Among the short carapaced forms there are: (13) no brood pouch and schizopods (Thermosbaenacea), (14) brood pouch and schizopods (Cumacea, Tanaidacea, and Spelaeogriphacea), (15) no brood pouch and uniramous thoracopods, and (16) brood pouch and uniramous thoracopods. As above, not all lines have been recognized in the fossil or Recent record and may or may not have ever come to exist. But distinct *Baupläne* can be recognized and has suggested how Eumalacostraca can be classified more reasonably (Schram, 1981b) than is done with the traditional Calman method.

This sort of analysis also suggests that the identification of lines of evolution is as important, perhaps more so, than manufacturing cladograms. Trying to determine what is plesiomorphic or apomorphic, let alone if it is syn- or not, may be irrelevant in a stochastic universe. Raup *et al.* (1973) and Schopf *et al.* (1975) have suggested stochastic models for duplicating phylogenetic patterns seen in nature. If the evolving biological universe is stochastic, then only approximate relationships may be knowable between lines of evolution, and these with degrees of uncertainty (an evolutionary uncertainty principle). Raup and Gould (1974) challenged us to arrive at a nomothetic paleontology. I freely admit that the consequences of such may be unacceptable (hopefully, only for a short time) to those of us accustomed to approaching our science ideographically. In the words of Slobodkin (1968, p. 189) who found ". . . it repulsive [stochastic events] in that its acceptance would invite cutting off an apparently legitimate empirical question from further investigation and in that sense, a kind of intellectual despair."

Stochastic mosaicism suggests a logical and philosophical approach to phylogeny of pragmatic parsimony, and I am suggesting that such a stochastic and pragmatic approach is the only one in which one can effectively and coherently handle a group of innate properties such as those found in the Crustacea (indeed, in any animal group). In this respect the method suggested by Dahl (1976) and Manton (1977) for constructing crustacean and arthropod phylogenies is likely to prove more fruitful than earlier facies schemes. The determination of integrated functional systems should afford a viable pragmatic parsimonious approach to interpreting a stochastic universe.

Such concepts, as mentioned above, have already afforded new insight into the interpretation of crustacean history. Schram (1974c) considered the convergences in form and function that occur between different caridoid types in different times. He concluded that not only did similarity of habitat coincide with similarity of form, but that open niches available to a taxon are always filled; i.e., diversity is an intrinsic property of biological systems. Diversity develops to the limits allowed by the exigencies of the *Bauplan* and by parameters external to the organism proper such as ecologic, geographic, and even geologic factors. This same property was noted to operate in community evolution, as well (Olson, 1975; Schram, 1979b; Walker and La Porte, 1970). In the case of crustaceans, community types are seen to have appeared at different times and places with the same types of animals in them (Schram, 1976b, 1981a).

It then follows (Schram, 1978, 1981b) that when attempting to develop phylogenetic schemes, one cannot be confined to consideration of anatomy alone. All aspects of a system are needed for an effective understanding of the evolution of a group. Such a holistic approach stands in strong contrast to past ideographic, reductionist philosophies, which assumed that by dissecting out and considering parts separately, one could know the whole. However, the uncertainty principle alluded to above operates, since we can either recognize lines of evolution or parts of systems and be unsure of the relationships; or, we can postulate hypothetical relationships and be unsure that these paper creations ever existed or that they are in fact connecting the lines or parts we seek to understand.

IX. SOOTH

Paleontology in the late nineteenth century was at the cutting edge of evolutionary theory. Ideas such as orthogenesis and racial senility originated in the study of fossils. With the development of genetics at the turn of the twentieth century and the genetic theories of selection in the ensuing decades, paleontology became the "country cousin" of the other evolutionary sciences. Now, as mechanistic reductionist approaches to evolution have reached a plateau in their applicability—as do all paradigms in the course of their history—paleontology is coming into its own again by offering new evolutionary insights. The increase in knowledge of the crustacean fossil record in the past two decades has been paralleled in other phyla. Such growth demands a period of consolidation and generalization which we have now entered.

There is a sustained effort in ecological as well as evolutionary theory to arrive at a set of nomothetic laws. A necessary first step in that process is to perceive the history and evolution of animal groups in light of recently expanded knowledge of the fossil record. I am not offering paleontology as an ultimate authority on the evolution of Crustacea, or any other group for that matter. But the record does offer us the opportunity to re-examine and test long-established ideas of phylogeny. If they are found wanting in any way, we cannot hesitate to disregard falsified concepts and seek new ones. To do otherwise is to invite a stolid and stodgy orthodoxy, which will surely sound the knell for crustacean phylogenetics as a vibrant and productive exercise.

APPENDIX

What follows is a taxonomic arrangement of the higher catagories of Crustacea with their stratigraphic ranges. Certain aspects of crustacean classification are still in flux and are elaborated on in the text. For example, we may wish to rethink the status given to the cephalocarids; at least some of the archaeocopids may be assigned to other groups, such as maxillopodans; and some of the extinct branchiopod groups may bear further consideration with better materials. Other taxonomic concepts, however, should not be abandoned.

(1) Maxillopoda seems to be a valid category; it reflects a real relationship between several similar subclasses. Maxillopoda should not be juxtaposed against a taxon "Thoracopoda": there seems to be little justification in uniting cephalocarids and branchiopods with malacostracans. In fact, one of the original authors now feels (W. A. Newman, personal communication) that the term "thoracopod" is quite meaningless taxonomically. (2) Hoplocarida are fundamentally different in their organization from the Eumalacostraca and deserve separate subclass status. (3) The order arrangement within the Eumalacostraca has been reformulated here in accordance with the principles discussed above. The sequence of subtaxa under a major heading can be construed as indicating approximate phylogenetic sequence.

Class Cephalocarida Sanders, 1955—Recent
- Order Brachypoda Birshtein, 1960—Recent

Class Branchiopoda Latreille, 1817—Devonian-Recent
- Subclass Sarsostraca Tasch, 1969—M. Devonian-Recent
  - Order Lipostraca Scourfield, 1926—M. Devonian
  - Order Anostraca Sars, 1967—(L. Devonian?) L. Cretaceous-Recent
  - Order Enantiopoda Birshtein, 1960—L. Pennsylvanian
- Subclass Calmanostraca Tasch, 1969—Carb.-Recent
  - Order Notostraca Sars, 1867—U. Carboniferous-Recent
  - Order Kazacharthra Novozhilov, 1957—L. Jurassic

Subclass Diplostraca Gerstaecker, 1866—Devonian-Recent
Order Conchostraca Sars, 1867—Devonian-Recent
Order Cladocera Latreille, 1829—Permian-Recent
Class Ostracoda Latreille, 1802 emend. 1804—Cambrian-Recent
Order Archaeocopida Sylvester-Bradley, 1961—Cambrian-L. Ordovician
Order Leperditicopida Scott, 1961—(U. Cambrian?) Ordovician-Devonian
Order Palaeocopida Henningsmoen, 1953—Ordovician-Permian
Order Podocopina Müller, 1894—Ordovician-Recent
Order Myodocopida Sars, 1866—Ordovician-Recent
Class Maxillopoda Dahl, 1956—Silurian-Recent
Subclass Cirripedia Burmeister, 1834—U. Silurian-Recent
Order Ascothoracica Lacaze-Duthiers, 1880—(Cretaceous?) Recent
Order Rhizocephala Müller, 1862—Recent
Order Acrothoracica Gruvel, 1905—U. Carboniferous-Recent
Order Thoracica Darwin, 1854—M. Cambrian-Recent
Suborder Lepadomorpha Pilsbry, 1916—M. Cambrian-Rec
Suborder Verrucomorpha Pilsbry, 1916—M. Cretaceous-Recent
Suborder Brachylepadomorpha Withers, 1923—U. Jurassic-U. Miocene
Suborder Balanomorpha Pilsbry, 1916—U. Cretaceous-Recent
Subclass Copepoda Milne-Edwards, 1840—Cretaceous-Recent
Order Calanoida Sars, 1903—Recent
Order Harpacticoida Sars, 1903—Miocene-Recent
Order Cyclopoida Sars, 1903—Miocene-Recent
Order Misophrioida Gurney, 1933—Recent
Order Monstrilloida Sars, 1903—Recent
Order Siphonostomatoida Thorell, 1859, emend. Sars, 1918—Cretaceous-Recent
Order Poecilostomatoida Thorell, 1859, emend. Sars, 1918—Recent
Order Notodelphyoida Sars, 1903—Recent
Subclass Branchiura Thorell, 1864—Recent
Order Arguloida Wilson, 1932—Recent
Subclass Mystacocarida Pennak and Zinn, 1943—Recent
Order Mystacocaridida Pennak and Zinn, 1943—Recent
Class Malacostraca Latreille, 1806—L. Cambrian-Recent
Subclass Phyllocarida Packard, 1879—L. Cambrian-Recent
Order Canadaspidida Novozhilov, 1960—M. Cambrian
Order Hymenostraca Rolfe, 1969—L. Cambrian-L. Ordovician
Order Leptostraca Claus, 1880—(U. Permian?) Recent
Order Hoplostraca Schram, 1973—Carboniferous-U. Triassic ?
Order Archaeostraca Claus, 1888—L. Ordovician-U. Triassic
Subclass Hoplocarida Calman, 1904—U. Devonian-Recent
Order Aeschronectida Schram, 1969—Carboniferous
Order Palaeostomatopoda Brooks, 1962—U. Devonian-L. Carboniferous
Order Stomatopoda Latreille, 1817—Carboniferous-Recent
Suborder Archaeostomatopodea Schram, 1969—Carboniferous
Suborder Unipeltata Latreille, 1825—Jurassic-Recent
Subclass Eumalacostraca Grobben, 1892—M. Devonian-Recent
Order Syncarida Packard, 1885—Carboniferous-Recent
Suborder: Palaeocaridacea Brooks, 1962—Carboniferous-Permian
Suborder Anaspidacea Calman, 1904—Permian-Recent
Suborder Bathynellacea Chappuis, 1915—Recent

Order Acaridea Schram, 1981
  Suborder Isopoda Latreille, 1817—U. Carboniferous-Recent
    Infraorder Phreatoicidea Stebbing, 1893—U. Carboniferous-Recent
    Infraorder Asellota Latreille, 1803—Recent
    Infraorder Flabellifera Sars, 1882—Triassic-Recent
    Infraorder Gnathiidea Leach, 1814—Recent
    Infraorder Anthuridea Leach, 1814—Recent
    Infraorder Valvifera Sars, 1882—Oligocene-Recent
    Infraorder Oniscoidea Latreiile, 1803—Eocene-Recent
    Infraorder Epicaridea Latreille, 1831—(U. Jurassic?) Recent
  Suborder Amphipoda Latreille, 1816—U. Eocene-Recent
    Infraorder Gammaridea Latreille, 1803—U. Eocene-Recent
    Infraorder Caprellidea Leach, 1814—Recent
    Infraorder Hyperiidea Latreille, 1831—Recent
Order Hemicaridea Schram, 1981
  Suborder Spelaeogriphacea Gordon, 1957—L. Carboniferous-Recent
  Suborder Tanaidacea Dana, 1853—L. Carboniferous-Recent
    Infraorder Monokonophora Lang, 1956—L. Carboniferous-Recent
    Infraorder Dikonophora Lang, 1956—Recent
  Suborder Cumacea Kröyer, 1846—U. Permian-Recent
Order Thermosbaenacea Monod, 1927—Recent
Order Euphausiacea Dana, 1852—(U. Carboniferous?)Recent
Order Decapoda Latreille, 1803—U. Devonian-Recent
  Suborder Dendrobranchiata Bate, 1888—Triassic-Recent
  Suborder Pleocyemata Burkenroad, 1963—U. Devonian-Recent
Order Amphionidacea Williamson, 1973—Recent
Order Mysidacea Boas, 1883—L. Carboniferous—Recent
  Suborder Lophogastrida Boas, 1883—Recent
  Suborder Pygocephalomorpha Beurlen, 1930—L. Carboniferous-Permian
  Suborder Mysida Boas, 1883—Triassic-Recent
Order Waterstonellidea Schram, 1981—L. Carboniferous
Order Belotelsonidea Schram, 1981—Carboniferous
Order Eocaridacea Brooks, 1962—Carboniferous (contains families of otherwise uncertain affinities).

ACKNOWLEDGMENTS

The initial draft of this chapter was produced with the use of the library and facilities of Professor Joel W. Hedgpeth at the Pacific Marine Station, Dillon Beach, California. The following people discussed certain issues and read all or parts of the manuscript: Drs. W. C. Cummings and H. Bertsch, San Diego Natural History Museum; R. M. Feldmann, Kent State University; S. J. Gould, Museum of Comparative Zoology; J. W. Hedgpeth, Santa Rosa, California; E. C. Olson, University of California at Los Angeles; D. M. Raup, Field Museum of Natural History; and J. M. Schram. In addition, the content of material in some sections was benefitted by discussion with Drs. J. L. Barnard, National Museum of Natural History; and L. Watling, University of Maine. This paper was drafted in 1978, with only minor alterations since.

My thoughts have profited immensely by informal debate over the years with Drs. J. L. Cisne, Cornell University, and R. R. Hessler, Scripps Institution of Oceanography. Our positions on arthropod phylogeny have not always coincided, but by agreeing to disagree we mutually ensure mental acuity and due deliberation.

REFERENCES

Anderson, D. T. (1965). Embryonic and larval development and segment formation in *Ibla quadrivalvis* Cuv. (Cirripedia). *Aust. J. Zool.* **13,** 1-15.

Anderson, D. T. (1967). Larval development and segment formation in the branchiopod crustaceans *Limnadia quadrivalvis* (Choncostraca) and *Artemia salina* (Anostraca). *Aust. J. Zool.* **15,** 46-91.

Anderson, D. T. (1973). "Embryology and Phylogeny in Annelids and Arthropods." Pergamon, Oxford.

Ax, P. (1960). "Die Entdeckung neuer Organisationtypen im Tierreich." Z. Ziemsen Verlag, Lutherstadt, Wittenburg.

Bachmeyer, F. (1960). Eine fossile Cumaceenart (Crustacea, Malacostraca) aus dem Callovien von La Voulte-sur-Rhône (Archèche). *Eclogae Geol. Helv.* **53,** 422-426.

Barker, D. (1962). A study of *Thermosbaena mirabilis* (Malacostraca, Peracarida) and its reproduction. *Q. J. Microsc. Soc.* [N.S.] **103,** 261-286.

Barnard, J. L. (1976). Amphipoda (Crustacea) from the Indo-Pacific Tropics: A review. *Micronesica* **12,** 169-181.

Beklemishev, B. N. (1969). "Principles of Comparative Anatomy of Invertebrates." Univ. of Chicago Press, Chicago, Illinois.

Benson, R. H. (1972). The *Bradleya* problem, with a description of two new psychrospheric ostracode genera, *Agrenocythere* and *Poseidonamicus*. *Smithson. Contrib. Paleobiol.* **12,** 1-150.

Benson, R. H. (1974). The role of ornamentation in the design and function of the ostracode carapace. *Geosci. Man* **6,** 47-57.

Benson, R. H. (1975). Morphologic stability in Ostracoda. *Bull. Am. Paleontol.* **65,** 13-45.

Benson, R. H. (1976). The evolution of the ostracode *Costa* analyzed by "theta-rho difference." *Abh. Verh. Naturwiss. Ver. Hamburg* **18/19,** 127-139.

Benson, R. H. (1977). Evolution of *Oblitacythereis* from *Paleocosta* during the Cenozoic in the Mediterranean and Atlantic. *Smithson. Contrib. Paleobiol.* **33,** 1-47.

Berry, C. T. (1939). A summary of the fossil Crustacea of the order Stomatopoda, and a description of a new species from Angola. *Am. Midl. Nat.* **21,** 461-471.

Beurlen, K. (1930). Vergleischende Stammesgeschichte Grundlagen, Methoden, Probleme unter besonderer Berücksichtigung der höheren Krebse. *Fortschr. Geol. Palaeontol.* **8,** 317-586.

Beurlen, K., and Glaessner, M. F. (1930). Systematik der Crustacea Decapoda auf stammesgeschichtlicher Grundlage. *Zool. Jahrb.* **60,** 49-84.

Birshtein, Ya. A. (1960). Podklass Cephalocarida. *In* "Osnovy Paleontologii" (Yu. A. Orlov, ed.), pp. 421-422. Moskva.

Bishop, G. A. (1981). Occurrence and fossilization of the *Dakoticancer* assemblage from the Upper Cretaceous Pierre Shale of South Dakota. *In* "Community Paleoecology" (J. Gray, A. Boucot, and W. Berry, eds.). Dowden, Hutchinson, & Ross, Stroudsburg, Pennsylvania (in press).

Bocquet-Vedrine, J. (1972). Suppression de l'ordre des Apodes (Crustacés Cirripèdes) et rattachement de son unique representant, *Protolepas bivincta,* à la famille des Crinoniscidae (Crustacés Isopodes, Cryptonisciens). *C. R. Hebd. Seances Acad. Sci., Part D* **275,** 2145-2148.

Borradaile, L. A. (1907). On the classification of the Decapoda. *Ann. Mag. Nat. Hist.* [7] **19,** 457–486.
Bowman, T. E. (1971). The case of the nonubiquitous telson and the fraudulent furca. *Crustaceana* **21,** 165–175.
Briggs, D. E. G. (1977). Bivalved arthropods from the Cambrian Burgess Shale of British Columbia. *Palaeontology* **20,** 595–621.
Briggs, D. E. G. (1978). The morphology, mode of life, and affinities of *Canadaspis perfecta,* Middle Cambrian, Burgess Shale, British Columbia. *Philos. Trans. R. Soc. London, Ser. B* **281,** 439–487.
Broch, H. (1922). Papers from Dr. Th. Mortensen's Pacific Expedition 1914–16. X. Studies on Pacific cirripeds. *Vidensk. Medd. Dan. Naturh. Foren.* **73,** 1–358.
Brooks, H. K. (1955). A crustacean from the Tesnus Formation of Texas. *J. Paleontol.* **29,** 852–856.
Brooks, H. K. (1962). The Paleozoic Eumalacostraca of North America. *Bull. Am. Paleontol.* **44,** 163–338.
Burkenroad, M. D. (1963). The evolution of the Eucarida in relation to the fossil record. *Tulane Stud. Geol.* **2,** 3–16.
Burnett, B. R. (1973). Notes on the lateral arteries of two stomatopods. *Crustaceana* **23,** 303–305.
Burnett, B. R., and Hessler, R. R. (1973). Thoracic epipodites in the Stomatopods: A phylogenetic consideration. *J. Zool.* **169,** 381–392.
Calman, W. T. (1896). On the genus *Anaspides* and its affinities with certain fossil Crustacea. *Trans. R. Soc. Edinburgh* **38,** 787–802.
Calman, W. T. (1909). Crustacea. *In* "A Treatise on Zoology" (R. Lankester, ed.), Vol. VII, pp. 1–346. Adam & Black, London.
Campbell, K. S. W. (1975). The functional morphology of *Cryptolithus. Fossils Strata* **4,** 65–86.
Cannon, H. G. (1931). On the anatomy of a marine ostracod, *Cypridina (Doloria) laevis. 'Discovery' Rep.* **2,** 435–482.
Cannon, H. G. (1940). On the anatomy of *Gigantocypris mulleri. 'Discovery' Rep.* **19,** 185–244.
Cannon, H. G., and Manton, S. M. (1927). On the feeding mechanism of a mysid crustacean, *Hemimysis lamornae. Trans. R. Soc. Edinburgh* **55,** 219–252.
Chernyshev, B. I. (1931). Cirripedien aus dem Bassin des Donez und von Kusnetzk. *Zool. Anz.* **92,** 26–28.
Cisne, J. L. (1974). Trilobites and the origin of arthropods. *Science* **186,** 13–18.
Cisne, J. L. (1975). The anatomy of *Triarthrus* and the relationships of the Trilobita. *Fossils Strata* **4,** 45–64.
Codez, J., and Sainte-Sein, R. (1957). Révision des Cirripèdes Acrothoraciques fossiles. *Bull. Soc. Geol. Fr.* [6] **7,** 699–719.
Copeland, M. J., and Rolfe, W. D. I. (1978). Occurrence of a large phyllocarid crustacean of Late Devonian-Early Carboniferous age from Yukon territory. *Geol. Surv. Pap. (Geol. Surv. Can.)* **78-1B,** 1–5.
Cressy, R., and Patterson, C. (1973). Fossil parasitic copepods from a Lower Cretaceous fish. *Science* **180,** 1283–1285.
Dahl, E. (1956a). On the differentiation of the topography of the crustacean head. *Acta Zool. (Stockholm)* **37,** 123–192.
Dahl, E. (1956b). Some crustacean relationships. *In* "Bertil Hanström: Zoological Papers in Honour of His Sixty-Fifth Birthday, November 20, 1956" (K. G. Winstrand, ed.), pp. 138–147. Lund Zool. Inst., Lund, Sweden.
Dahl, E. (1963). Main evolutionary lines among recent Crustacea. *In* "Phylogeny and evolution

of Crustacea" (H. B. Whittington and W. D. I. Rolfe, eds.), Spec. Publ., pp. 1-15. Mus. Comp. Zool., Cambridge, Massachusetts.

Dahl, E. (1976). Structural plans as functional models exemplified by the Crustacea Malacostraca. *Zool. Scr.* **5,** 163-166.

Dahl, E. (1977). The amphipod functional model and its bearing upon systematics and phylogeny. *Zool. Scr.* **6,** 221-228.

Förster, R. (1967). Die reptanten Dekapoden der Trias. *Neues Jahrb. Geol. Palaeontol., Abh.* **128,** 136-194.

Förster, R. (1973). Untersuchungen an oberjurassischen Palinuridae. *Mitt. Bayer Staatssamml. Palaeontol. Hist. Geol.* **13,** 31-46.

Fortey, R. A., and Morris, S. F. (1978). Discovery of nauplius-like trilobite larvae. *Palaeontology* **21,** 823-833.

Fryer, G. (1964). Studies on the functional morphology and feeding mechanism of *Monodella argentarii Stella* (Crustacea: Thermosbaenacea). *Trans. R. Soc. Edinburgh* **66,** 49-90.

Glaessner, M. F. (1957). Evolutionary trends in Crustacea. *Evolution* **11,** 178-184.

Glaessner, M. F. (1960). The fossil decapod Crustacea of New Zealand and the evolution of the order Decapoda. *N. Z. Geol. Surv., Paleontol. Bull.* **31,** 1-63.

Glaessner, M. F. (1969). Decapoda. *In* "Treatise on Invertebrate Paleontology" (R. C. Moore, ed.), Part R, Arthropoda 4, Vol. II, pp. R399-R533. Geol. Soc. Am., Boulder, Colorado, and the Univ. of Kansas Press, Lawrence.

Glaessner, M. F., and Malzahn, E. (1962). Neue Crustaceen aus dem niederrheinischen Zechstein. *Fortschr. Geol. Rheinl. Westfalen* **6,** 245-264.

Gordon, I. (1957). On *Spelaeogriphus,* a new cavernicolous crustacean from South Africa. *Bull. Br. Mus. (Nat. Hist.), Zool.* **5,** 31-47.

Gordon, I. (1958). A thermophilous shrimp from Tunisia. *Nature (London)* **182,** 1186.

Gordon, I. (1964). On the mandible of the Stygocarididae and some other Eumalacostraca with special reference to the lacina mobilis. *Crustaceana* **7,** 150-157.

Gould, S. J. (1966). Allometry and size in ontogeny and phylogeny. *Biol. Rev. Cambridge Philos. Soc.* **41,** 587-640.

Gould, S. J. (1977). "Ontogeny and Phylogeny." Belknap Press, Cambridge, Massachusetts.

Gurney, R. (1942). "Larvae of Decapod Crustacea." Ray Society, London.

Hall, J., and Hessler, R. R. (1971). Aspects in the population dynamics of *Derocheilocaris typica* (Mystacocarida, Crustacea). *Vie Milieu* **22,** 305-326.

Hartmann, G. (1963). Zum Problem polyphyletischer Merkmalsentstehung bei Ostracoden. *Zool. Anz.* **171,** 148-164.

Herrick, E. M., and Schram, F. R. (1978). Malacostracan crustacean fauna from the Sundance Formation (Jurassic) of Wyoming. *Am. Mus. Novit.* **2652,** 1-12.

Hessler, A., Hessler, R. R., and Sanders, H. L. (1970). Reproductive system of *Hutchinsoniella macracantha. Science* **168,** 1464.

Hessler, R. R. (1964). The Cephalocarida comparative skeletomusculature. *Mem. Conn. Acad. Arts Sci.* **16,** 1-97.

Hessler, R. R. (1969a). A new species of Mystacocarida from Maine. *Vie Milieu* **20,** 105-116.

Hessler, R. R. (1969b). Cephalocarida *and* Peracarida. *In* "Treatise on Invertebrate Paleontology" (R. C. Moore, ed.), Part R, Arthropoda 4, Vol. I, pp. R120-R128, R360-R393. Geol. Soc. Am., Boulder, Colorado, and the Univ. of Kansas Press, Lawrence.

Hessler, R. R. (1971). Biology of the Mystacocarida: A prospectus. *Smithson. Contrib. Zool.* **76,** 87-90.

Hessler, R. R. (1972). New species of Mystacocarida from Africa. *Crustaceana* **22,** 259-273.

Hessler, R. R., and Newman, W. A. (1975). A trilobitomorph origin for the Crustacea. *Fossils Strata* **4,** 437-459.

Hessler, R. R., and Sanders, H. L. (1972). Two new species of *Sandersiella* (Cephalocarida), including one from the deep sea. *Crustaceana* **13,** 181-196.

Heyden, C. (1861). Gliederthiere aus der Braunkohle des Niederrhein's der Wetterau und der Rohn. *Palaeontographica* **10,** 62-63.

Holthuis, L. B., and Manning, R. B. (1969). Stomatopoda. *In* "Treatise on Invertebrate Paleontology" (R. C. Moore, ed.), Part R. Arthropoda 4, Vol. II, pp. R535-R552. Geol. Soc. Am., Boulder, Colorado, and the Univ. of Kansas Press, Lawrence.

Howe, H. V., Kesling, R. V., and Scott, H. W. (1961). Morphology of living Ostracoda. *In* "Treatise on Invertebrate Paleontology" (R. C. Moore, ed.), Part Q, Arthropoda 3, pp. Q3-Q17. Geol. Soc. Am., Boulder, Colorado, and the Univ. of Kansas Press, Lawrence.

Hughs, C. P. (1975). Redescription of *Burgessia bella* from the Middle Cambrian Burgess Shale. *Fossils Strata* **4,** 415-435.

Johnson, R. G. (1970). Variations in diversity within benthic marine communities. *Am. Nat.* **104,** 285-300.

Kaestner, A. (1970). "Invertebrate Zoology," Vol. II. Wiley (Interscience), New York.

Kesling, R. V. (1961). Reproduction of Ostracoda. *In* "Treatise on Invertebrate Paleontology" (R. C. Moore, ed.), Part Q, Arthropoda 3, pp. Q17-Q19. Geol. Soc. Am., Boulder, Colorado, and the Univ. of Kansas Press, Lawrence.

Knight, M. D. (1978). Larval development of *Euphausia fallax* with a comparison of larval morphology within the *E. gibboides* species group. *Bull. Mar. Sci.* **28,** 255-281.

Komai, T., and Tung, Y. M. (1931). On some points of the internal structure of *Squilla oratoria. Mem. Coll. Sci., Kyoto Imp. Univ.* **6,** 1-15.

Lehmann, W. M. (1955). *Vachonisia rogeri* ein Branchiopod aus dem unterdevonischen Hunsrückschiefer. *Palaeontol. Z.* **29,** 126-130.

Liebau, A. (1969). Homologisierende Korrelationen von Trachyleberididen-Ornamenten. *Neues Jahrb. Geol. Palaeontol., Monatsh.* pp. 390-402.

Madsen, N., and Wolff, T. (1965). Evidence of the occurrence of Ascothoracica (parasitic-cirripedes) in the Upper Cretaceous. *Medd. Dan. Geol. Foren.* **15,** 556-558.

Maguire, B. (1965). *Monodella texana* n. sp., an extension of the range of the crustacean order Thermosbaenacea into the western hemisphere. *Crustaceana* **9,** 149-154.

Manton, S. M. (1928). On the embryology of the mysid crustacean, *Hemimysis lamornae. Philos. Trans. R. Soc. London, Ser. B* **216,** 363-463.

Manton, S. M. (1930). On the habits and feeding mechanisms of *Anaspides* and *Paranaspides. Proc. Zool. Soc. London* pp. 791-800.

Manton, S. M. (1973). Arthropod phylogeny—a modern synthesis. *J. Zool.* **171,** 111-130.

Manton, S. M. (1977). "The Arthropoda." Oxford Univ. Press, London and New York.

Manton, S. M., and Anderson, D. T. (1979). Polyphyly and the evolution of arthropods. *In* "The Origin of Major Groups" (M. R. House, ed.), pp. 269-321. Academic Press, New York.

Monod, T. (1927). *Thermosbaena mirabilis,* remarques sur sa morphologie et sa position systematique. *Faune Colon. Fr.* **1,** 29-51.

Monod, T. (1940). Thermosbaenacea. *Bronn's Klassen* **5,** Abt. 1, Buch 4, 1-24.

Müller, G. W. (1894). Die Ostracoden. *Fauna Flora Golfes Neapel* **21,** 1-404.

Müller, K. J. (1979). Phosphatocopine ostracodes with preserved appendages from the Upper Cambrian of Sweden. *Lethaia* **12,** 1-27.

Münster, G. (1840). "Beitrage zur Petrefactenkunde." Beyreuth.

Newman, W. A. (1979). A new scalpellid (Cirripedia); a Mesozoic relic living near an abyssal hydrothermal spring. *Trans. San Diego Soc. Nat. Hist.* **19,** 153-167.

Newman, W. A., and Ross, A. (1976). Revision of the balanomorph barnacles; including a catalog of the species. *Mem. San Diego Soc. Nat. Hist.* **9,** 1-108.

Newman, W. A., Zullo, V. A., and Withers, T. H. (1969). Cirripedia. *In* "Treatise on Invertebrate

Paleontology" (R. C. Moore, ed.), Part R, Arthropoda 4, Vol. I, pp. R206-R295. Geol. Soc. Am., Boulder, Colorado, and the Univ. of Kansas Press, Lawrence.

Noodt, W. (1963). Subterrane Crustaceen der zentralen Neotropis. Zur Frage mariner Relikte im Bereich des Rio Paraquay—Paraná—Amazonas—Systems. *Zool. Anz.* **171,** 114-147.

Noodt, W. (1965). Natürliches System und Biogeographic der Syncarida. *Gewäesser Abwasser* **37/38,** 77-186.

Noodt, W. (1974a). Bathynellacea (Crustacea, Syncarida) auch in Nord-Amerika! *Naturwissenschaften* **61,** 132.

Noodt, W. (1974b). Anpassung an interstitielle Bedingungen: Ein Factor in der Evolution höherer Taxa der Crustacea? *Faun.—Oekal. Mitt.* **4,** 445-452.

Novozhilov, N. (1957). Un nouvel ordre d'arthropodes particuliers: Kazacharthra du Lias des monts Ketmen. *Bull. Soc. Geol. Fr.* [6] **7,** 171-184.

Novozhilov, N. (1959). Position systematique des Kazacharthra d'après de nouveaux materiaux des monts Ketmen et Sajkan, S. E. et N. E. *Bull. Soc. Geol. Fr.* [7] **1,** 265-269.

Olsen, P. E., Remington, C. L., Cornet, B., and Thompson, K. S. (1978). Cyclic change in Late Triassic lacustrine communities. *Science* **201,** 729-733.

Olson, E. C. (1975). Permo-Carboniferous paleoecology and morphotypic series. *Am. Zool.* **15,** 371-389.

Oppel, A. (1862). Ueber jurassiche Crustacean. *Mus. Bayer. Staates, Palaeontol. Mitt.* **1,** 1-120.

Packard, A. S. (1885). The Syncarida, a group of Carboniferous Crustacea. *Am. Nat.* **19,** 700-703.

Palmer, A. R. (1957). Miocene arthropods from the Mojave Desert, California. *Geol. Surv. Prof. Pap. (U.S.)* **294-G,** 1-280.

Palmer, A. R. (1969). Copepoda. *In* "Treatise on Invertebrate Paleontology" (R. C. Moore, ed.), Part R, Arthropoda 4, Vol. I, pp. R200-R203. Geol. Soc. Am., Boulder, Colorado, and the Univ. of Kansas Press, Lawrence.

Pojeta, J., and Runnegar, B. (1976). The paleontology of rostroconch mollusks and the early history of the Phylum Mollusca. *Geol. Surv. Prof. Pap. (U.S.)* **968,** 1-88.

Pokorný, V. (1969a). The genus *Radimella* in the Galapagos Islands. *Acta Univ. Carol. Geol.* **4,** 293-334.

Pokorný, V. (1969b). *Radimella,* a new genus of the Hemicytherinae. *Acta Univ. Carolinae, Geol.* **4,** 359-373.

Rathbun, M. J. (1926). The fossil stalk-eyed Crustacea of the Pacific Slope of North America. *Bull.—U.S. Nat. Mus.* **138,** 1-155.

Raup, D. M. (1978). Approaches to the extinction problem. *J. Paleontol.* **52,** 517-523.

Raup, D. M., and Gould, S. J. (1974). Stochastic simulation and evolution of morphology—toward a nomothetic paleontology. *Syst. Zool.* **23,** 305-322.

Raup, D. M., Gould, S. J., Schopf, T. J. M., and Simberloff, D. S. (1973). Stochastic models of phylogeny and the evolution of diversity. *J. Geol.* **81,** 525-542.

Reaka, M. L. (1975). Molting in stomatopod crustaceans. I. Stages of the molt cycle, setagenesis, and morphology. *J. Morphol.* **146,** 55-80.

Remy, J. M., and Avnimeleck, M. (1955). *Eryon yehoachi* et *Cenomanocarcinus* cf. *vanstraeleni,* crustacés décapodes du Crétacé supérieur de l'état d'Israël. *Bull. Soc. Geol. Fr.* [6] **5,** 311-314.

Rodendorf, B. B. (1972). Devonskie eopteridi—ne nasekomie, a raboobraznie Eumalocostraca. *Entomol. Obozr.* **51,** 96-97.

Rolfe, W. D. I. (1963). Morphology of the telson in *Ceratiocaris? cornwallensis* from Czechoslovakia. *J. Paleontol.* **37,** 486-488.

Rolfe, W. D. I. (1967). *Rochdalia,* a Carboniferous insect nymph. *Palaeontology* **10,** 307-313.

Rolfe, W. D. I. (1969). Phyllocarida. *In* "Treatise on Invertebrate Paleontology" (R. C. Moore, ed.), Part R, Arthropoda 4, Vol. I, pp. R296-R331. Geol. Soc. Am., Boulder, Colorado, and the Univ. of Kansas Press, Lawrence.

Runnegar, B., and Jell, P. A. (1976). Australian Middle Cambrian molluscs and their bearing on early molluscan evolution. *Alcheringa* **1,** 109-138.

Runnegar, B., and Pojeta, J. (1974). Molluscan phylogeny: The paleontological viewpoint. *Science* **186,** 311-317.

Sainte-Seine, R. (1951). Un cirripède acrothoracique de Crétacé, *Rogerella lecointre*. *C. R. Hebd. Seances Acad. Sci.* **233,** 1051-1053.

Sainte-Seine, R. (1954). Existence de cirripèdes acrothoraciques des le Lias, *Zapfella pattei*. *Bull. Soc. Geol. Fr.* [6] **4,** 447-451.

Sanders, H. L. (1957). The Cephalocarida and crustacean evolution. *Syst. Zool.* **6,** 112-129.

Sanders, H. L. (1960). Benthic studies in Buzzards Bay III. The structure of the soft bottom community. *Limnol. Oceanogr.* **5,** 138-153.

Sanders, H. L. (1963a). The Cephalocarida functional morphology, larval development, comparative external anatomy. *Mem. Conn. Acad. Arts Sci.* **15,** 1-80.

Sanders, H. L. (1963b). Significance of the Cephalocarida. *In* "Phylogeny and Evolution of Crustacea" (H. B. Whittington and W. D. I. Rolfe, eds.), pp. 163-175. Mus. Comp. Zool., Cambridge, Massachusetts.

Sanders, H. L., and Hessler, R. R. (1963). The larval development of *Lightiella incisa*. *Crustaceana* **7,** 81-97.

Savory, T. (1971). "Evolution in the Arachnida." Merrow, Watford Herts., England.

Schminke, H. K. (1973). Evolution, System und Verbreitungsgeschichte der Familie Parabathynellidae. *Abh. Math.-Naturwiss. Kl., Akad. Wiss. Lit., Mainz, Mikrofauna Meersboden* **24,** 1-192.

Schminke, H. K. (1974). Mesozoic intercontinental relationships as evidenced by bathynellid Crustacea. *Syst. Zool.* **23,** 157-164.

Schminke, H. K. (1976). The ubiquitous telson and the deceptive furca. *Crustaceana* **30,** 292-300.

Schminke, H. K., and Wells, J. B. J. (1974). *Nannobathynella africana* and the zoogeography of the family Bathnellidae. *Arch. Hydrobiol.* **73,** 122-129.

Schopf, T. J. M., Raup, D. M., Gould, S. J., and Simberloff, D. S. (1975). Genomic versus morphological rates of evolution: Influence of morphologic complexity. *Paleobiology* **1,** 63-70.

Schram, F. R. (1968). *Paleosquilla,* a stomatopod crustacean from the Cretaceous of Colombia. *J. Paleontol.* **42,** 1297-1301.

Schram, F. R. (1969a). Polyphyly in the Eumalacostraca? *Crustaceana* **16,** 243-250.

Schram, F. R. (1969b). Some Middle Pennsylvanian Hoplocarida and their phylogenetic significance. *Fieldiana, Geol.* **12,** 235-289.

Schram, F. R. (1970). Isopod from the Pennsylvanian of Illinois. *Science* **169,** 854-855.

Schram, F. R. (1973). On some phyllocarids and the origin of the Hoplocarida. *Fieldiana, Geol.* **26,** 77-94.

Schram, F. R. (1974a). Mazon Creek caridoid Crustacea. *Fieldiana, Geol.* **30,** 9-65.

Schram, F. R. (1974b). Late Paleozoic Peracarida of North America. *Fieldiana, Geol.* **33,** 95-124.

Schram, F. R. (1974c). Convergences between Late Paleozoic and modern caridoid Malacostraca. *Syst. Zool.* **23,** 323-332.

Schram, F. R. (1975). Lepadomorph barnacle from Mazon Creek. *J. Paleontol.* **49,** 928-930.

Schram, F. R. (1976a). Some notes on Pennsylvanian crustaceans of the Illinois basin. *Fieldiana, Geol.* **35,** 21-28.

Schram, F. R. (1976b). Crustaceans of the Pennsylvanian Linton vertebrate beds of Ohio. *Palaeontology* **19,** 411-412.

Schram, F. R. (1977). Paleozoogeography of Late Paleozoic and Triassic Malacostraca. *Syst. Zool.* **26,** 367-379.

Schram, F. R. (1978). Arthropods: A convergent phenomenon. *Fieldiana, Geol.* **39,** 61-108.

Schram, F. R. (1979a). British Carboniferous Malacostraca. *Fieldiana, Geol.* **40,** 1-129.

Schram, F. R. (1979b). The Mazon Creek biotas in the context of a Carboniferous faunal continuum. *In* "Mazon Creek Fossils" (M. H. Nitecki, ed.), pp. 159-190. Academic Press, New York.

Schram, F. R. (1979c). The genus *Archaeocaris,* and a review of the order Palaeostomatopoda. *Trans. San Diego Soc. Nat. Hist.* **19,** 57-66.

Schram, F. R. (1981a). Late Paleozoic crustacean communities. *J. Paleontol.* **55.**

Schram, F. R. (1981b). On the classification of Eumalacostraca. *J. Crustacean Biol.* **1.**

Schram, F. R., and Hedgpeth, J. W. (1978). Locomotory mechanisms in Antarctic pyconogonids. *J. Linn. Soc. London, Zool.* **63,** 145-169.

Schram, F. R., and Horner, J. (1978). Crustacea of the Mississippian Bear Gulch Limestone of central Montana. *J. Paleontol.* **52,** 394-406.

Schram, F. R., Feldman, R. M., and Copeland, M. J. (1978). The Late Devonian Palaeopalaemonidae and the earliest decapod crustaceans. *J. Paleontol.* **52,**

Schram, J. M., and Schram, F. R. (1974). *Squillites spinosus* from the Mississippian Health Shale of central Montana. *J. Paleontol.* **48,** 95-104.

Scourfield, D. J. (1926). On a new type of crustacean from the Old Red Sandstone (Rhynie Chert Bed, Aberdeenshire)—*Lepidocaris rhyniensis. Philos. Trans. R. Soc., London, Ser. B* **214,** 153-187.

Scourfield, D. J. (1940). Two new and nearly complete specimens of young stages of the Devonian fossil crustacean *Lepidocaris rhyniensis. Proc. Linn. Soc. London* **152,** 290-298.

Seilacker, A. (1970). Begriff und Bedeutung der Fossil-Lagerstätten. *Neues Jahrb. Geol. Palaeontol. Monatsh.* pp. 34-39.

Serban, E. (1970). A propos deu genre *Bathynella. Trav. Inst. Speol. "Emile Racovitza"* **11,** 265-273.

Sharov, A. G. (1966). "Basic Arthropodan Stock." Pergamon, Oxford.

Shiino, S. M. (1942). Studies on the embryology of *Squilla oratoria. Mem. Coll. Sci., Kyoto Imp. Univ.* **17,** 77-174.

Siewing, R. (1956). Untersuchungen zur Morphologie der Malacostraca. *Zool. Jahrb., Abt. Anat. Ontog. Tiere* **75,** 39-176.

Siewing, R. (1958). Anatomie und Histologie von *Thermosbaena mirabilis.* Ein Beitrag zur Phylogenie der Reihe Parcarida. *Abh. Math.-Naturwiss. Kl., Akad. Wiss. Lit., Mainz* pp. 197-270.

Siewing, R. (1960). Neuere Ergebnisse der Verwandtschaftsforschung bei den Crustaceen. *Wiss. Z. Univ. Rostock, Math.-Naturwiss. Reihe* **9,** 343-358.

Siewing, R. (1963). Studies in malacostracan morphology: Results and problems. *In* "Phylogeny and Evolution of Crustacea" (H. B. Whittington and W. D. I. Rolfe, eds.), pp. 85-103. Mus. Comp. Zool., Cambridge, Massachusetts.

Slobodkin, L. B. (1968). Toward a predictive theory of evolution. *In* "Population Biology and Evolution" (R. C. Lewontin, ed.), pp. 187-205. Syracuse Univ. Press, Syracuse, New York.

Smirnov, N. N. (1970). Cladocera (Crustacea) iz permskikh otlozheniy Vostochnogo Kazakhstana. *Paleontol. Zh.* **3,** 95-100.

Solland, E. (1923). Recherche sur l'embryogenic des Crustaces decapods de la sous-damille des Palaemoninae. *Bull. Biol. (Woods Hole, Mass.)* **5,** 1-234.

Stella, E. (1959). Ulteriore osservazioni sulla riproduzione e lo sviluppo di *Monodella argentarii. Riv. Biol.* **51,** 121-144.

Størmer, L. (1939). Studies on trilobite morphology. Part I. The thoracic appendages and their phylogenetic significance. *Nor. Geol. Tidsskr.* **19,** 143-273.

Størmer, L. (1944). On the relationships of the fossil and recent Arachnomorpha. *Skr. Nor. Vidensk.-Acad. [Kl.] 1: Mat.-Nuaturvldensk. Kl.* **5,** 1-158.

Stürmer, W., and Bergström, J. (1976). The arthropods *Mimetaster* and *Vachonisia* from the Devonian Hunsrück Shale. *Palaeontol. Z.* **58,** 78-111.

Taramelli, E. (1954). La posizione sistematica dei Thermosbenacei quale risulta dallo studio anatomico di *Monodella argentarii. Monit. Zool. Ital.* **62,** 9-24.

Tasch, P. (1969). Branchiopoda. *In* "Treatise on Invertebrate Paleontology" (R. C. Moore, ed.), Part R, Arthropoda 4, Vol. I, pp. R128-R191. Geol. Soc. Am., Boulder, Colorado, and the Univ. of Kansas Press, Lawrence.

Tasch, P. (1977). Ancient Antarctic freshwater ecosystems. *In* "Adaptations Within Antarctic Ecosystems," pp. 1077-1089. Smithson. Inst., Washington, D.C.

Tiegs, O. W., and Manton, S. M. (1958). The evolution of the Arthropoda. *Biol. Rev. Cambridge Philos. Soc.* **33,** 255-337.

Tomlinson, J. T. (1963). Acrothoracican barnacles in Paleozoic myalinids. *J. Paleontol.* **37,** 164-166.

Trusova, Ye. K. (1971). O pervoy nakhodke v mezozoye predstavitoley otryada Anostraca. *Paleontol. Zh.* pp. 68-73.

Voight, E. (1959). *Endosacculus moltkiae,* ein vermutlicher fossiler Ascothoracide als Gystenbildner bei der Oktocoralle *Moltkia minuta. Palaeontol. 2* **33,** 211-223.

Wakabara, Y., and Mizoguchi, S. M. (1976). Record of *Sandersiella bathyalis* from Brazil. *Crustaceana* **30,** 220-221.

Walker, K. R., and La Porte, L. F. (1970). Congruent fossil communities from Ordovician and Devonian carbonates from New York. *J. Paleontol.* **44,** 928-944.

Watling, L. (1981). An alternative phylogeny of peracarid crustaceans. *J. Crustacean Biol.* **1,** 201-210.

Whittington, H. B. (1975). Trilobites with appendages from the Middle Cambrian, Burgess Shale, British Columbia. *Fossils Strata* **4,** 97-136.

Whittington, H. B. (1979). Early arthropods, their appendages and relationships. *In* "The Origin of Major Invertebrate Groups" (M. R. House, ed.), pp. 253-268. Academic Press, New York.

Whittington, H. B., and Rolfe, W. D. I., eds.(1963). "Phylogeny and Evolution of Crustacea." Mus. Comp. Zool., Cambridge, Massachusetts.

Wills, L. J. (1963). *Cyprilepas holmi,* a pedunculate cirripede from the Upper Silurian of Oesel, Esthonia. *Palaeontology* **6,** 161-165.

Woodward, H. (1879). On the occurrence of *Branchipus* (or *Chirocephalus*) in a fossil state, associated with *Eosphaeroma* and with numerous insect remains in the Eocene freshwater limestone of Gurnet Bay, Isle of Wight. *Q. J. Geol. Soc. London* **35,** 342-350.

Yaeger, J. (1981). A new class of Crustacea from a marine cave in the Bahamas. *J. Crustacean Biol.* **1,** 328-333.

Yochelson, E. L., Flower, R. H., and Webers, G. F. (1973). The bearing of the new Late Cambrian monoplacophoran genus *Knightoconus* upon the origin of the Cephalopoda. *Lethaia* **6,** 275-310.

5

Evolution within the Crustacea

ROBERT R. HESSLER, BRIAN M. MARCOTTE,
WILLIAM A. NEWMAN, AND
ROSALIE F. MADDOCKS

THE BIOLOGY OF CRUSTACEA, VOL. 1

ISBN 0-12-106401-8

Part 1: General: Remipedia, Branchiopoda, and Malacostraca (Hessler)

I. INTRODUCTION

The remarkable variety of crustaceans offers endless opportunity for pursuit of phylogeny. Regrettably, very few critical questions have been solved with finality. The task of this chapter is to describe theories about the paths crustacean evolution has taken and to highlight significant points of contention. Priority is given to current controversy. By and large, the fossil record will not be considered because it is covered in other chapters (Cisne, Chapter 3, and Schram, Chapter 4 of this volume).

Of necessity, discussion begins with the nature of the ancestral crustacean. This is followed by an analysis of the ways the primary crustacean taxa, the classes, might be related. It ends with a consideration of evolution within each of the major classes, generally to the ordinal level. Several class analyses are authored by specialists of those groups: B. M. Marcotte on copepods, W. A. Newman on cirripeds, R. F. Maddocks on ostracodes.

I extend heartfelt appreciation to Erik Dahl, Frederick R. Schram, William A. Newman, Rolf Elofsson, Howard L. Sanders and the students of Scripps Institution of Oceanography for the many dialogues of which this contribution is a result.

II. THE PHYLOGENETIC METHOD

While the goal of discerning the evolutionary relationship between taxa is clear-cut, the means for achieving it are still a matter of considerable debate.

Most frequently, controversy involves the question of whether a similarity in two or more taxa is a homology or the product of convergence.

In the present discussion, I have tried to stay close to the principle that similarities of any sort reflect phylogenetic affinity (homology) unless there is evidence for believing that the similarity is a product of convergence. Unless we adhere to this principle, we have abandoned our only tool for doing systematics. Any other approach makes the evaluation of similarities a matter of personal preference and is therefore no longer science. This does not mean convergence is automatically ruled out. In fact, the possibility of convergence is always present, even with homologies that have survived decades of testing. There is always the chance that some new fact will demonstrate that a supposed homology is actually a product of convergence.

Homology can never be proven; it can only be falsified. The demonstration of convergence is the falsification of the hypothesis of homology. Convergence can be proven using, for example, the criteria of Remane (1952). Because of this, in order to accept convergence instead of homology, we must have a reason for doing so. For example, the homology of the mandible among the various taxa where it is found has never been proven, yet the influential taxon Mandibulata was based primarily on its validity. There was no reason to doubt the validity of the Mandibulata until Manton (1964) demonstrated that the mandible in crustaceans had a fundamentally different origin from that in uniramian taxa (Onychophora, Myriopoda, Insecta). By contrast, the fusion of the thoracic segments to the carapace unites the Euphausiacea, Amphionidacea, and Decapoda into the Eucarida. Without proof, we have no reason to reject the reality of this combination. Here, the hypothesis remains unfalsified.

At the same time, even when no existing data suggest convergence, not all similarities imply homology with equal strength. Experience teaches us where to be suspicious. For example, single characters must always be viewed with care. Thus, with unfalsified homologies, there is always the opportunity to discuss the degree of confidence with which the homology is held. The example just given of the Eucarida could be considered in this way.

III. THE ANCESTRAL CRUSTACEAN

Any inquiry into evolution within a taxon must include an attempt to determine what kind of ancestral form stood at the beginning of the group's radiation. This activity has been attacked as being unwarranted speculation and unrealistically typological (Manton, 1973; Schram, Chapter 4 of this volume), yet it remains one of the essential duties of phylogeny. Much of the criticism is justified; numerous yet-to-be-discovered ancestral organisms are the result of extrapolation beyond what available facts can support. Often, pure, "reasonable" logic is substituted for facts. Nevertheless, if done with proper caution, recognizing the potential for convergence, it is possible to deduce the nature of ancestral organisms that are not even known as fossils. The methodology is the same as that which results in classification of known organisms, and the pursuit of the form of ancestral organisms is no more typological than the process of systematics itself. We seek features whose distribution among the crustacean classes suggests that they are symplesiomorphies.

Speculation on the nature of the urcrustacean has focused on two general possibilities: an ancestor with an abbreviated trunk, like that of a larva or maxillopodan (Section IV,G), or one with an extended series of trunk segments, as with the thoracopodans (Section IV,B).

The influence of the Recapitulation Theory encouraged the hypothesis that the nauplius, being found in all the classes, must represent the ancestral crustacean (Packard, 1833). The zoea was endowed with similar phylogenetic importance (Packard, 1833). Copepods have been regarded as the most primitive living crustaceans (Packard, 1833; Sars, 1887; Hartog, 1888), in part, because as adults they resemble eucarid malacostracan zoeal larvae and in part, because of the well-developed biramous second antenna and mandible (Gurney, 1942). The latter consideration stimulated even greater belief in the primitiveness of mystacocarids (Pennak and Zinn, 1943; Dahl, 1952, 1956b; Ax, 1960; Hessler, 1969).

The idea of an ancestor with an abbreviated trunk was not durable because only an organism with a considerable number of trunk segments could give continuity with an acceptable annelid ancestry (Calman, 1909; Zimmer, 1926–1927). The nauplius retrenched to a secure position as ancestral crustacean larva (Jägersten, 1972), as did the zoea for the eucarids (Calman, 1909).

Prior to the discovery and full appreciation of the Cephalocarida (Fig. 1), the Branchiopoda were generally regarded the most primitive known crustaceans by virtue of their large, but variable number of trunk segments and their long row of serially similar appendages (Claus, 1876; Grobben, 1893; Calman, 1909; Zimmer, 1926–1927; Tiegs and Manton, 1958). The Notostraca (Fig. 1) have been considered most similar to the stem form because of their more annelidan trunk, parapodium-like phyllopodia, and unspecialized carapace fold. A more complicated phylogeny is that of Grobben (1893), who derived other crustaceans from the branchiopods, but in three separate lineages, beginning with the notostracans, conchostracans, or anostracans.

A malacostracan body form has also been considered the basic crustacean type (Siewing, 1956, 1960, 1963). The Leptostraca (Fig. 1), as the best known representative of the Phyllocarida, has received most attention. That

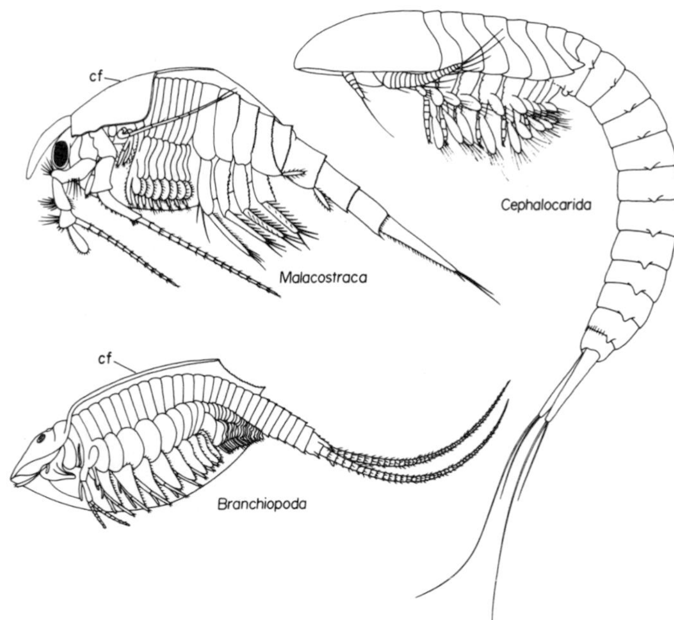

Fig. 1. The "thoracopodan" classes, as exemplified by primitive members. The malacostracan is a leptostracan, and the branchiopod is a notostracan. In both of these, the carapace fold (cf) has been removed from the near side to reveal appendages. (From Hessler and Newman, 1975.)

phyllocarids bridge the gap with other classes is seen in controversy over their placement in the Branchiopoda (Sars, 1887) rather than the Malacostraca (Claus, 1888). The malacostracan conception is not greatly different from the branchiopodan one, involving many trunk segments, a maxillary carapace fold, serially similar phyllopodia, and stalked eyes.

Hypothetical urcrustaceans are not merely based on data derived from the Crustacea, but are influenced by the investigator's perceptions of the taxon from which the Crustacea must have come or to which they are related. As already seen, belief in an annelidan ancestry required a multisegmented urcrustacean (Calman, 1909) or a paddlelike trunk appendage (Cannon and Manton, 1927), and it encouraged the notion of a cephalon with fewer than four post-oral segments (Snodgrass, 1938; Sharov, 1966). Belief in a trilobitomorph ancestry stimulated Hessler and Newman (1975) to endow the urcrustacean with more similarity of the second antenna and mandible to the trunk limb series than could be justified on the basis of data from the crustaceans alone. In parallel logic, Snodgrass (1938) was influenced by the concept of the Mandibulata, comprising the Myriapoda, Insecta, and Crustacea, when he postulated an ancestor with stenopodia and a centipede-like body. With his framework, the Malacostraca had to be the most primitive extant crustacean. Siewing's (1960) analysis of the origin of the Crustacea was influenced by the belief that the Burgess Shale Pseudocrustacea were related to crustaceans.

The discovery of the Cephalocarida (Sanders, 1955) had a profound effect on views of the urcrustacean as well as evolution within the class. The influence of this discovery has been less to introduce new alternatives, than to give firm support for some of the old ones. The cephalocarids (Fig. 1) show that much of the similarity of branchiopods and malacostracans is plesiomorphic, particularly in the limbs (Sanders, 1963a,b), as will be discussed below. The apparent fulfillment of Cannon's (1933) predictions about the nature of the primitive branchiopod limb is remarkable and highlights Cannon's (1927) error in not recognizing the fundamental similarity of the leptostracan plan. The ancestor postulated by Sanders (1955) was a small, benthic, trunk-limb detritivore with strong serial homeomorphy and gradual development, much as conceived by Claus (1876), Calman (1909), Cannon (1933), Tiegs and Manton (1958), Dahl (1963), and others.

Hessler and Newman's (1975) comprehensive attempt to delineate the attributes of the ancestral crustacean argued the affinity of crustaceans with the Trilobitomorpha *sensu lato* and secondarily, the Chelicerata. They endowed the urcrustacean (Cisne, Chapter 3 of this volume; Figs. 1-3) with a morphology not greatly at variance with those conceived by the majority of previous investigators: cephalon with five pairs of appendages—pre-oral first antenna, post-oral second antenna, mandible, and first and second maxillae; adult second antenna not involved in nutrition, having sensory functions; mandible and first maxilla dominating food handling immediately prior to ingestion; second maxilla with thoracic functions; trunk many-segmented, with most segments bearing limbs; all postantennal segments possessing a high degree of serial homogeneity with regard to morphology and functions of body shape, limbs, heart, gut, and nerve cord; post-oral limbs that were multiramous mixopodia, with flattened protopod, foliaceous epipod and exopod, and stenopodial endopod; metachronal activity of post-oral limbs that resulted in concentration of food along the ventral midline and subsequent transport forward to the mouth by protopodal endites; the same limbs participated in locomotion and respiration; the mouth posteriorly directed, floored by posteriorly directed labrum; perhaps having a carapace fold, but if not, all anterior trunk segments with well-developed pleura; compound eyes stalked; tubular heart with paired segmental ostia; nerve cord with well-defined segmental ganglia; digestive tract with midgut diverticula; excretory glands in second antennal and second maxillary segments; first free-living stage a nauplius.

The ancestral nature of these features is based on their presence in all the classes, or at least in those that are regarded as most primitive. The latter point is most subject to contention; how do we decide what taxa are most primitive without becoming involved in circularity?

The belief that primitive metameric animals possess a large number of trunk segments with a high degree of serial homogeneity has always influenced our thinking. It is a reasonable conclusion because the selective pressures for tagmosis are obvious, while those for the elimination of serial specialization are not so compelling. The evolutionary record supports this with examples of fish with relation to mammals, annelids to arthropods, myriapods to insects, trilobitomorphs to chelicerates, and within the chelicerate line. On this basis, the Branchiopoda with a high, variable number of similar segments, had been regarded as most primitive until the discovery of the Cephalocarida, although Calman (1909) issued strong words of caution on this subject.

Equally influential has been the principle that the form of a primitive animal is such that it could easily have given rise to other taxa. The condition of appendages is especially important. Here, the Branchiopoda prove inadequate because it is difficult to derive the endopod-dominated stenopodium of malacostracans (see Fig. 8) from a phyllopodium where the endopod is virtually absent (Figs. 2 and 6) (what portion of the branchiopod limb represents the endopod, if any, has never been established unequivocally).

The importance of the Cephalocarida lies in its satisfying both criteria (Sanders, 1963a,b). It is unique among the living Crustacea that in its long limb series, even the second maxilla is identical to the thoracopods; initially this caused serious confusion about limb identification (Dahl, 1956b; Tiegs and Manton, 1958). Further, the larval stages reveal a similar form in the first maxilla, and to a lesser extent, the mandible and second antenna (Sanders, 1963a). Of particular interest is the continuity in function of thoracic limb endites with those of the maxillae, the gnathic endite of the mandible, and the naupliar process of the larval second antenna.

The basic cephalocarid thoracic limb (Fig. 2) is the most generalized of the Crustacea (Sanders, 1963a). Each complex limb (called a mixopodium by Sanders) serves in locomotion, feeding, and respiration, and, with some modification of posterior appendages, in reproduction as well. Through a process of "specialization, simplification, and reduction" (Sanders, 1963a, p. 42), a large suite of appendages of other Crustacea can be derived from this basic type.

An understanding of cephalocarid functional morphology yields a convincing solution to the questions of the phyllopodium versus the stenopodium and the phylogenetic relationship between branchiopods and

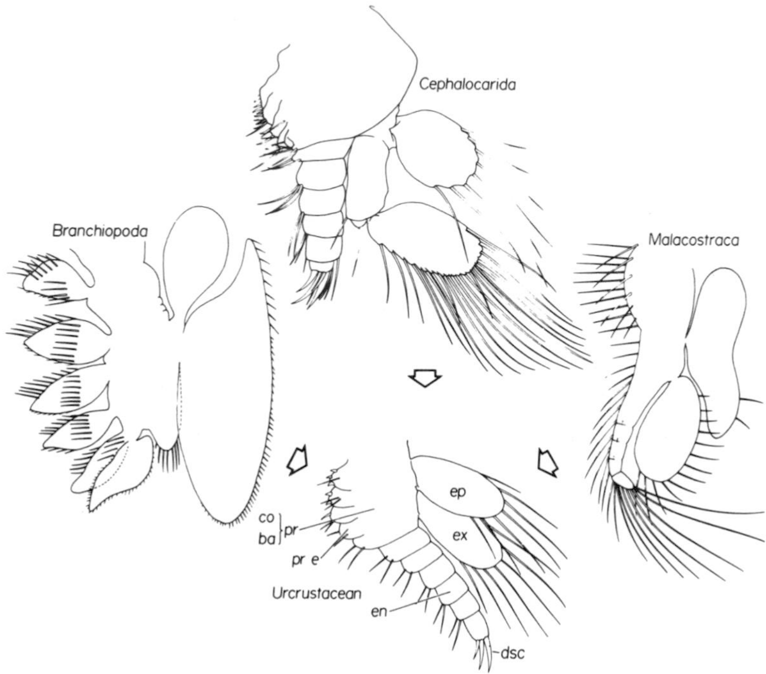

Fig. 2. The trunk limbs of the three "thoracopodan" classes, whose common characteristics are used to deduce the ancestral crustacean limb form. The branchiopod is a notostracan and the malacostracan is a leptostracan. Symbols: ba, basis; co, coxa; en, endopod; ep, epipod; ex, exopod; pr, protopod; pr e, protopodal endite. (From Hessler and Newman, 1975.)

malacostracans. The branchiopods (Cannon, 1933; Eriksson, 1934) and leptostracan malacostracans (Cannon, 1927) display the same basic mode of trunk-limb feeding as do cephalocarids (Fig. 3), and the body and limb structures that make this possible are fundamentally alike. That is, the flattened protopod, foliaceous exites, and enditic lobes with their interdigitating rows of setae function through metachronal limb movement to accumulate food along the ventral midline and pass it forward to the mouth. The latter is ventral and posteriorly directed, being floored by the labrum, to easily receive the food. The primitiveness of this aspect of cephalic morphology was thoroughly documented by Dahl (1956a, 1963).

Branchiopod and leptostracan limbs (Fig. 2) are each more specialized than on the cephalocarid, the first with its reduced or absent endopod, and the second having lost locomotory ability with the food-gathering thoracopods. These are, however, differences in detail in a fundamentally similar plan. In this framework, the branchiopod phyllopodium is comparable to the protopod and exites of the cephalocarid mixopodium. The Leptostraca show that the early malacostracan limb was a mixopodium, although

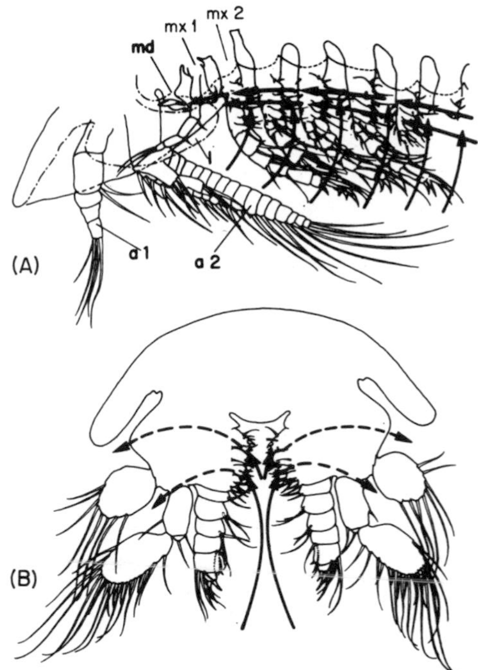

Fig. 3. Primitive system of trunk-limb feeding as seen in the Cephalocarida. (A) Midsagittal view of lower half of head and anterior portion of the thorax. (B) Transverse view of a thoracic segment. Symbols: light solid arrow, water and food particles; light dashed arrow, water from which particles have been removed by enditic setae; heavy solid arrow, movement of food by endites toward the mouth; a 1, first antenna; a 2, second antenna; en, endopod; ep, epipod; ex, exopod; l, labrum; md, mandible; mx 1, first maxilla; mx 2, second maxilla. (Modified from Sanders, 1963a.)

not necessarily of the precise cephalocarid form. The leptostracans are specialized in many ways (Dahl, 1976, 1981), but in this important aspect of their morphology, they show what the urmalacostracan must have been like. Available data are mute on how the stenopodium of eumalacostracans (see Fig. 8) evolved from this origin. However, the leptostracans show that the phyllocarid endopod already possessed the five basic podomeres.

Most of the characters attributed by Hessler and Newman (1975) to the urcrustacean relate to this functional morphological "thoracopodan" (see Section IV,B) facies. To date, no evidence has emerged to suggest that the thoracopodan facies in cephalocarids, branchiopods, and malacostracans is convergent. A surprising, but inescapable corollary of this is that the urcrustacean trunk limb had a broad, flat protopod (Fig. 2) (Hessler and Newman, 1975; Schram, 1978). No data coming from within the Crustacea contradict this conclusion.

Cephalocarids document limb serial homogeneity extending into the cephalic limb series. However, the assumption that all the post-oral cephalic appendages of the ancestral form were essentially like the trunk-limb series is an extrapolation that must be treated with caution. Its validity rests partly on the premise that crustaceans had a trilobitomorph ancestor (Hessler and Newman, 1975); the latter taxon frequently shows this more complete homogeneity. More concrete support is the functional continuity between the larval naupliar process of the second antenna and the masticatory process of the mandible with the basal endites of more posterior limbs. The fact remains that an epipod has not been found on any crustacean second antenna, mandible, or first maxilla. Further, even in thoracopodans with foliaceous exites on the trunk-limb series (and in cephalocarids, the maxillae as well), the exopod of the second antenna and mandible are flagelliform. That these are also the naupliar post-oral appendages may well be more than coincidence. Dahl (1976) emphasizes the existence of two modes of feeding in cephalocarids and branchiopods, the thoracopodan feeding (Fig. 3) of the juvenile and adult, and the cephalic limb feeding of the nauplius or metanauplius larva. He suggests that successive feeding modes probably also existed with the urcrustacean. The special morphology of anterior post-oral cephalic appendages may relate to this (see Cisne, Chapter 3 of this volume). Flagelliform antennal and mandibular exopods are also more suitable than paddles for swimming in an animal that has few limbs (such as a nauplius) because they offer less resistance on the recovery stroke.

Cannon and Manton (1927) proposed that the crustacean limb is primitively biramous, using the evidence of *Lepidocaris,* where posterior biramous paddles grade forward into the more familiar phyllopodium (see Fig. 6). The primarily biramous nature of the second antenna and mandible is consistent with Cannon and Manton's hypothesis. If this view is correct, the Remipedia (Yager, 1981; see Section V) might be regarded as the most primitive subclass because of its long series of enditeless, biramous paddles on the trunk. The consequences of this scenario would be: (1) thoracopodan trunk-limb feeding is secondarily evolved within the Crustacea, convergent on the similar mode seen in trilobitomorphs; (2) the epipod is a completely new structure in crustacean trunk limbs; (3) the biramous form of swimming trunk limbs in copepods, branchiurans, ascothoracicans, and malacostracans is possibly plesiomorphic; and (4) the multisegmented form of the ambulatory endopod evolved within the Crustacea and is convergent on that of other arthropods. This senario does not require polyphyly of enditic trunk-limb feeding within the crustaceans.

Sanders (1963a,b) argued that evolution proceeded the other way, with the more posterior limbs gradually losing epipods and becoming modified to emphasize their locomotory function, just as the most anterior trunk limbs often evolve to become mouthparts. This interpretation seems more probable because it avoids the necessity of the convergences which are corollaries of the other hypothesis. However, it does imply that the biramous limb has evolved more than once.

The presence of a biramous first antenna in Remipedia (Yager, 1981) recalls the condition in malacostracans and phosphatocopine ostracodes (Müller, 1979), raising the question of whether we have been correct in thinking that the urcrustacean first antenna was uniramous. The uniramous antennae of trilobitomorphs and uniramians, plus the unbranched sensory palps of annelids, support the traditional viewpoint.

Telsonic appendages called caudal rami have been regarded as being characteristic of the ancestral crustacean because they are present in all classes, albeit in varying form. Bowman (1971), following Sharov (1966), claimed that in most cases these are not products of the telson, but of the preceding segment, the anal somite. Therefore they should be considered uropods. This view is largely based on the variable position of the anus. However, the orientation of the anus could be expected to vary in response to adaptive necessities (Schminke, 1976); it will not be unduly constrained by the topography of its origin.

The true nature of the telson is revealed by embryology and ontogeny. In the development of more primitive crustaceans, the body increases in length by the successive addition of new somites between the last-formed somite and the terminal region which bears the anus. The teloblastic growth zone is situated at this boundary. When the full number of somites has been reached, the unsegmented terminal region forms the telson of the adult (paraphrased from Calman, 1909, p. 4).

By ignoring the embryological basis of the telson (Kumé and Dan, 1968), Bowman has unduly confused the issue. Sharov's (1966) claims notwithstanding, Manton's descriptions of the embryologies of *Hemimysis* (1928a) and *Nebalia* (1934) clearly show the caudal rami to be postteloblastic, i.e., a product of the telson. We do not have sufficient early embryological information to determine the genesis of caudal rami in many classes, but their similar position and their presence in larvae that do not have the full complement of segments suggest homology with the malacostracan condition. Therefore, because of the lack of supporting evidence, the hypothesis of the pretelsonic uropod should be rejected.

IV. THE CRUSTACEAN PHYLOGENETIC TREE

Much of what has been thought about the crustacean phylogenetic tree has been expressed in terms of the taxonomic names given to clusters of subtaxa regarded to have special affinity (Fig. 4). The more important of such groups are evaluated below.

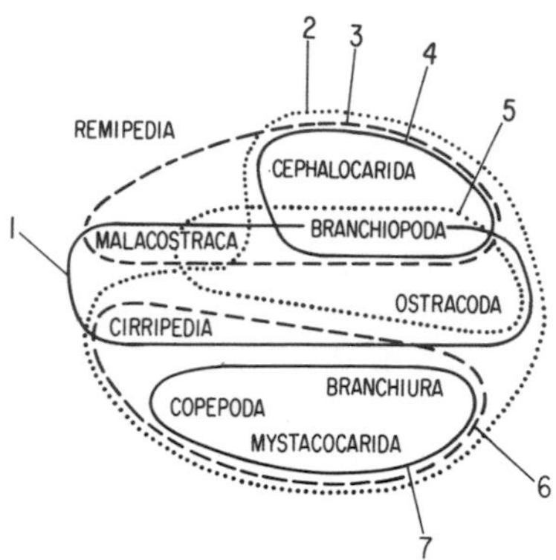

Fig. 4. Clusters of classes that have been used in formal classifications. 1, Palliata; 2, Entomostraca; 3, Thoracopoda; 4, Gnathostraca; 5, Branchiopoda *sensu lato;* 6, Maxillopoda; 7, Copepoda *sensu lato.*

A. Entomostraca

Latreille (1806) opposed to the Malacostraca, all other crustaceans as the Entomostraca. The latter term has convenient uses, but "the Entomostraca, like the Invertebrata, constitute a very heterogeneous group, defined only by negative characters and having no claim to retention in a natural system of classification" (Calman, 1909, p. 27). The Malacostraca have more in common with some entomostracan groups than many entomostracans do with each other.

B. Thoracopoda

Hessler and Newman (1975) proposed the taxon Thoracopoda to include the Cephalocarida, Branchiopoda, and Malacostraca. As already seen, the major unifying features of body and limbs (Figs. 1 and 2) are related to the same fundamental feeding style (Fig. 3). While these and other features give similarity to primitive members of the three classes, they are also basic primitive crustacean traits, assuming that the mixopodium and not the biramous paddle is the primitive crustacean trunk limb. Thus, they tell nothing about any special affinity within the Crustacea. For this reason, the Thoracopoda should not be used except as an informal, convenient way of distinguishing these three classes from those with a more derived "maxillopodan" morphology (see below).

C. Gnathostraca/Anostraca

The Gnathostraca (Dahl, 1956b; Siewing, 1960), combining the Branchiopoda and Cephalocarida, primarily differs from the Thoracopoda in emphasizing the reduction of the first and second maxillae. Siewing (1960) combined cephalocarids, anostracans, and lipostracans in his superorder Anostraca because of similarities in the thoracic limbs. Neither group seems well founded. The former is based on a misunderstanding about the identity of cephalocarid maxillae. The latter fails because the cephalocarid thoracopod is no more like that of branchiopods than phyllocarid malacostracans (Sanders, 1963a,b).

D. Palliata

The conservativeness of the carapace is an issue of continuing importance in crustacean phylogeny. The clustering of classes and orders is strongly influenced by whether this structure evolved singly or multiply.

Lauterbach (1974a,b, 1975) regarded its evolution as a unique event and organized crustacean phylogeny around it. In his view, the urcrustacean lacked a carapace, but possessed a cephalic shield and trunk-segment pleura, as in the Cephalocarida (Fig. 5). This was also the view of Hessler (Hessler and Newman, 1975). From this arose a conchostracan-like form with an almost fully enveloping, bivalved carapace. Subsequent radiation resulted in the phyllopodan Branchiopoda, Malacostraca, Ostracoda, and Cirripedia (including Ascothoracica), which Lauterbach collectively named the Palliata. The remaining taxa are not known to have a true carapace, which Lauterbach regards as a primary condition in Cephalocarida and Copepoda, and probably in anostracan Branchiopoda, Branchiura, and Mys-

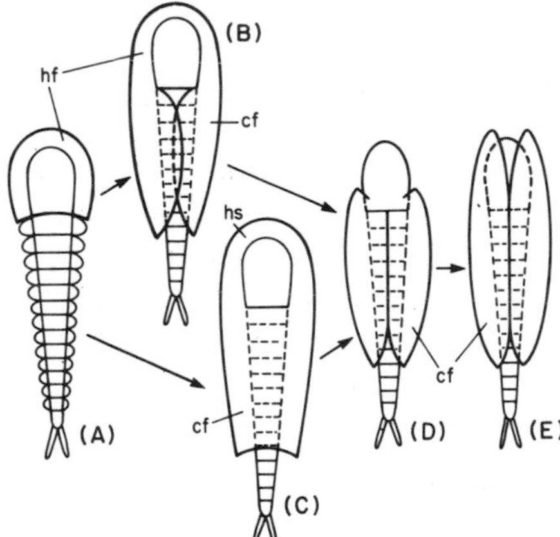

Fig. 5. Evolution of the bivalved carapace: (A–B, D–E) represent Lauterbach's (1974a) hypothesis for the evolution of the Palliata; (C–E) represent the evolution of the conchostracan carapace as espoused herein. (A) Ancestral condition, as seen in cephalocarids. (B) Hypothetical intermediate suggested by conchostracan larvae. (C) Notostracan. (D) Cladoceran or lynceid conchostracan. (E) Limnadiid conchostracan. (In part modified from Lauterbach, 1974a.)

tacocarida, as well. Because its absence is primarily a plesiomorphy, Lauterbach did not attempt to cluster the other taxa or assign collective names. Support of a monophyletic carapace comes from the positioning of the carapace adductor muscle (wherever one is known) in the region of the maxillae (Hessler, 1964).

Dahl (1976, 1981) doubts carapace monophyly and suggests that in view of the multiplicity of important functional demands for this structure, it would be surprising for it not to have evolved more than once. Further, he challenges the assumption that the carapace fold is always a maxillary derivative, citing published (Manton, 1934; Anderson, 1973) and unpublished evidence of other origins, which in some cases involve fusion of branchiostegal folds of several somites.

The Palliata suffers from dependence on a single character. From other points of view, it is a strangely heterogeneous taxon (Dahl, 1976), reducing its plausibility. The concept of a conchostracan-like primitive palliatan is not convincing. The enveloping carapace, as seen in conchostracans and leptostracans, strongly hinders the locomotory ability of thoracic appendages. Thus, it is an unlikely configuration for a malacostracan ancestor.

E. Copepoda *Sensu Lato*

The Branchiura was once thought to belong to the Copepoda (Claus, 1875; Wilson, 1902) on the basis of similarity to the Siphonostoma, including having a maxilliped whose body segment is fused to the cephalon. Martin (1932) demonstrated the true identity of the branchiuran cephalic appendages and concluded that a maxilliped did not exist. Thus, the tagmata of the two groups are different, and the general similarity of branchiurans to siphonostomes is best explained as convergent adaptation to their ectoparasitic existence.

When first discovered, the Mystacocarida was regarded by some as a new group of copepods (Armstrong, 1949), primarily because of similar cephalic appendages. The independence of the maxilliped segment from the cephalon convinced most (Pennak and Zinn, 1943; Dahl, 1952; Buchholz, 1953; Ax, 1960; Hessler, 1969) that the Mystacocarida is a separate but closely related taxon that is even more primitive (see Maxillopoda, Section II,G).

F. Branchiopoda *Sensu Lato*

Ostracodes have been combined with branchiopods (Gerstaecker, 1863, 1866–1879; Packard, 1879, 1883) because they share with conchostracans and cladocerans a bivalved carapace and natatory second antenna, and they include members with a ventrally flexed posterior trunk and similar caudal furca (Calman, 1909). The presence in ostracodes of a well-developed mandibular palp, first and second maxillae (Maddocks, Part 4 of this chapter), and trunk limb endopods testifies to the wide gulf separating the two taxa. Most of the similarities are probably simply accommodations to the bivalved carapace. The presence of a carapace may well be a plesiomorphy, but its bivalved form in these two taxa is likely to be convergent.

Leptostracan malacostracans have also been united with branchiopods (see Malacostracan Evolution).

G. Maxillopoda/Copepodoidea

The Copepoda, Cirripedia, Branchiura, and Mystacocarida have relatively abbreviated trunks (eleven or fewer segments) and feeding mechanisms which, at least primitively, rely on cephalic appendages. Because of these and other similarities, they have been combined as the Copepodoidea (Beklemischev, 1952, 1969) or the more commonly used Maxillopoda (Dahl, 1956b). The Ostracoda has also been regarded as a potential member (Siewing, 1960). Some form of these clusterings has received subsequent, recurrent support (Ax, 1960; Birstein and Novozhilov, 1960; Newman *et al.*, 1969; Brown, 1970; Bowman and Abele, Chapter 1 of this volume).

While these four or five classes do show resemblances, few similarities are common to them all. Avoiding specialized subtaxa, the following synapomorphies have been suggested:

(1) The trunk is abbreviated, but only copepods, cirripeds, and mystacocarids have the same number of segments including the telson (11). In copepods and cirripeds, the trunk is supposed to be subdivided into six thoracic and five abdominal segments; however, the presence of a seventh trunk limb in several harpacticoid copepod families (Marcotte, Part 2 of this chapter) casts doubt on the validity of this similarity.

(2) In copepods, branchiurans, mystacocarids, and ostracodes the maxillae are large and, except in branchiurans, play an important role in food gathering. However, the details of the trunk-limb feeding mode show no uniformity.

(3) Concomitant with (2), the primitive thoracopodan mode of trunk-limb feeding that includes passing food forward in the food groove is absent. This is true in cirripeds as well, and in all five classes, thoracic endites are absent.

(4) At the same time, the first thoracopod has frequently been enlisted to some extent in cephalic feeding, constituting a maxilliped in copepods and mystacocarids. Some cirripeds have a maxilliped, but not the more primitive ones; Newman *et al.* (1969) regard the cirripedian maxilliped as being independently derived.

(5) A large, biramous mandibular palp is present in adult copepods, mystacocarids, and ostracodes. A small uniramous palp is seen in cirripeds, but a palp occurs in malacostracans as well.

(6) The female genital pore shows no intertaxic constancy of position. However, with males, it is on the fourth trunk segment in branchiurans and mystacocarids and on the seventh (first abdominal) in cirripeds and copepods.

(7) Cirripeds, branchiurans, and mystacocarids are alone among crustaceans in having flagellate sperm, wherein the elongate nucleus lies alongside the proximal portion of the axoneme (Brown, 1970; Grygier, 1981).

(8) Copepods and cirripeds have six naupliar stages.

(9) Elofsson (1963, 1965, 1966) has documented a cluster of uniquely similar features of naupliar eyes and frontal organs in copepods, cirripeds, branchiurans, and ostracodes. The three naupliar eyes of all possess tapetal cells, and except for the cirripeds, all have lens cells. Further, none of the four taxa display any sort of frontal organ *sensu stricto*.

Gurney (1942) proposed that progenesis played a major role in the evolution of copepods. This line of reasoning could be extended to explain much of the morphology of maxillopodans in general; i.e., the abbreviated trunk, mandibular palp, and well-developed food-gathering maxillae accompanied by lack of the primitive type of trunk-limb trophic function are all aspects of larval crustaceans. From this point of view, any apparently primitive aspect of maxillopodans must be viewed with caution. For example, the seemingly primitive mandible in mystacocarids is a naupliar appendage in every respect. Further, if progenesis is the driving force in the evolution of these classes, then the above naupliar features are in reality plesiomorphies and do not argue against the possibility that the maxillopodan condition could have evolved through progenesis more than once. The similarity of copepods and mystacocarids (Buchholz, 1953) may be explained simply as being their mutual resemblance to naupliar larvae.

Other features sometimes used to document the unity of the Maxillopoda are also plesiomorphic. Among these are the presence of a compound eye and flagellar sperm. The former is reasonably regarded as an attribute of the urcrustacean (Hessler and Newman, 1975; Paulis, 1979), although Dahl (1963) and Elofsson and Dahl (1970) contest its monophyly. Flagellar sperm is a primitive feature of metazoans in general.

A few similarities remain that are not apparently primitive and that are sufficiently specific to imply synapomorphy. These include the condition of naupliar eyes, the lack of frontal organs, and the trunk segment count. Naupliar eyes and frontal organs would unite nearly all the maxillopodan taxa and ostracodes as well; they would exclude of the mystacocarids whose blindness makes them mute on this point.

The common position of the copepod and cirriped male genital pores constitutes a reasonable synapomorphy, as does the total trunk segment count shared by mystacocarids. There is no factual justification for claiming that there are six thoracic segments in mystacocarids, of which the sixth is limbless (Armstrong, 1949; Ax, 1960; Newman *et al.*, 1969). The sixth trunk somite is identical to those following it; singling it out as having undergone limb reduction does little to solidify the case for maxillopodan unity.

These data suggest that copepods, cirripeds, and perhaps mystacocarids may have some special phylogenetic affinity which could justify taxonomic formalization as Maxillopoda or Copepodoidea. Evidence for joining branchiurans or ostracodes exists, but is not strong. There appears to be no more reason for maintaining the independence of ostracodes, as has been preferred in the past (Dahl, 1956b, 1963; Ax, 1960), than for branchiurans, where past inclusion may be based largely on its having once been regarded as a subgroup of copepods (Calman, 1909). In summary, while there is some evidence for a natural unit containing a subset of these five classes, there is not sufficient concurrence of similarities to decide which classes should be included. Because of the uncertainty about its boundaries, use of Maxillopoda or Copepodoidea as a formal taxon should be avoided. Until more is known, their adoption will yield confusion and potential misconceptions.

The problematical Y nauplius and cypris (Hansen, 1899; T. Schram, 1970a,b, 1972) are maxillopodan by virtue of the six free, limbed thoracic segments of the cyprid instar, but until adults are found, it cannot be resolved whether these larvae belong to the cirripeds or to a separate class. The minute ectoparasites *Basipodella* (Becker, 1975) and *Deoterthron* (Bradford and Hewitt, 1980) also have six limbed thoracic segments, and *Basipodella* has the requisite five abdominal segments.

H. Conclusion

The preceding analysis leads to the unsatisfying conclusion that, except possibly for the two or three classes that form the maxillopodan core, one cannot identify any phylogenetic branchings between the nine crustacean classes. Thus, the situation differs little from Dahl's perception in 1963.

V. REMIPEDIA

The monotypic Remipedia, found in a Bahamian marine cave, is the most recently described class (Yager, 1981). *Speleonectes lucayensis* shows no close alignment with any other class. The homomorphic, multisegmented trunk with appendages on each segment is unlike that of any maxillopodan. The laterally directed, biramous, paddlelike trunk limbs lack endites and are devoted entirely to swimming. The first and second maxillae and the maxilliped are uniramous and raptorial. This condition of trunk and posterior head appendages differs greatly from that of any thoracopodan.

The Pennsylvanian *Tesnusocaris* bears some resemblance to *Speleonectes.* It may indeed be a remipedian, but poor preservation of the fossil precludes absolute determination (Yager, 1981).

Aspects of *Speleonectes* also recall *Lepidocaris,* particularly the biramous swimming paddles and the raptorial first maxilla. These similarities are probably superficial. The small protopod, multisegmented rami, and lateral direction of the speleonectid trunk limb are quite different from the lepidocarid condition. Further, the raptorial first maxilla of *Lepidocaris* occurs only on some individuals and is probably a male clasper, whereas it is clearly a feeding structure on *Speleonectes.*

VI. BRANCHIOPODA

As recognized today, the Branchiopoda contains six well-defined subgroups: Anostraca, Notostraca, Conchostraca, Cladocera, Lipostraca (Schram, Chapter 4 of this volume; Fig. 3A), and Kazacharthra (Schram, Chapter 4; Fig. 3D), of which the last two are known only from fossils. Lehmann (1955) included the Devonian *Vachonisia* in a seventh, the Acercostraca, but these are now regarded as trilobitomorphs (Stürmer and Bergström, 1976).

Most investigators regard the Branchiopoda as monophyletic, at the same time recognizing that the class subsumes an unusual range of morphologies. Unifying features are the endite- and exite-bearing foliaceous trunk limb whose endopod is reduced or absent, the large number of trunk segments (except in cladocerans), and reduced first antenna and first and second maxillae.

The alignment of taxa within the class has been the subject of controversy. Early workers and some recent ones combined Notostraca, Conchostraca, and Anostraca within a taxon usually called Phyllopoda, a group with equal rank to the Cladocera (Milne Edwards, 1840; Packard, 1883; Sars, 1896; Pennak, 1978). Claus (1860) used the same conception, but reversed the terms Branchiopoda and Phyllopoda. This division emphasized the difference in the number of trunk segments and limbs and the degree of metamorphosis in larval development.

The conchostracans and cladocerans are nevertheless quite similar: they both possess a bivalved carapace; large, biramous, natatory second antenna; and posterior abdomen and furca developed as an abreptor. The major differences result from exposure of the cephalon (already apparent in lynceid conchostracans), reduction of segments, and serial specialization in the latter group. The abbreviated trunk and exposed cephalon are features seen in larval conchostracans, suggesting that progenesis played a major role in cladoceran evolution (Brooks, 1959). Modern classifications combine conchostracans and cladocerans as Diplostraca (Gerstaecker, 1866) or Onychura (Eriksson, 1934).

Recent investigations tend to emphasize the isolation of anostracans with respect to other living branchiopods (Linder, 1945; Preuss, 1951, 1957; Tasch, 1969). Anostracans have pedunculate eyes which are not enclosed in a secondary chamber, lack a carapace, bear dorsal sensory setae on the trunk but not on the telson, have genital openings on somites other than the eleventh, and have additional epipods on the trunk limbs. Preuss (1951, 1957) lists other features, and emphasizing the difference in trunk limb musculature, he concludes that anostracans are completely separate from the other branchiopods. Such extreme separation is not warranted. Branchiopodan unity is demonstrated by the reduced maxillae, trunk limb morphology, and details of naupliar morphology (Sanders, 1963a).

Some investigators (Eriksson, 1934; Borradale, 1958) divide living branchiopods into three units: Anostraca, Notostraca, and Diplostraca/Onchyura. This ignores several apomorphies shared by the latter two groups: i.e., carapace, encapsulated eyes, trunk limb musculature, and other features mentioned by Linder (1945) and Preuss (1951).

The Devonian *Lepidocaris rhyniensis* (Scourfield, 1926, 1940) has had extraordinary impact on crustacean phylogeny, especially within the branchiopods. It is generally agreed that this species, for which Scourfield erected the order Lipostraca, is most closely allied to the Anostraca (Scourfield, 1926; Calman, 1926; Eriksson, 1934; Linder, 1945; Preuss, 1951; Borradaile, 1958; Tasch, 1969). However, the only feature that substantiates this special affinity is the absence of a carapace fold. The generally similar appearance stems from this and the basically pelagic life style of the two groups.

The apomorphies of *Lepidocaris* include no eyes, pleural scales, and male claspers on the first maxilla. The "lateral caudal process" (Scourfield, 1926) is probably also a specialization and may represent the appendages of a teloblastic segment fused to the telson. The large, biramous natatory second antenna is primitive or progenetic.

Most significant phylogenetically is the change in morphology along the trunk-limb series (Fig. 6E–G). Posterior limbs are simple biramous paddles clearly adapted for swimming. Anterior limbs have five setose endites, including a large "gnathobase." The medial ramus is developed as a large, articulated endite, while the lateral ramus is reduced. Scourfield (1926) and

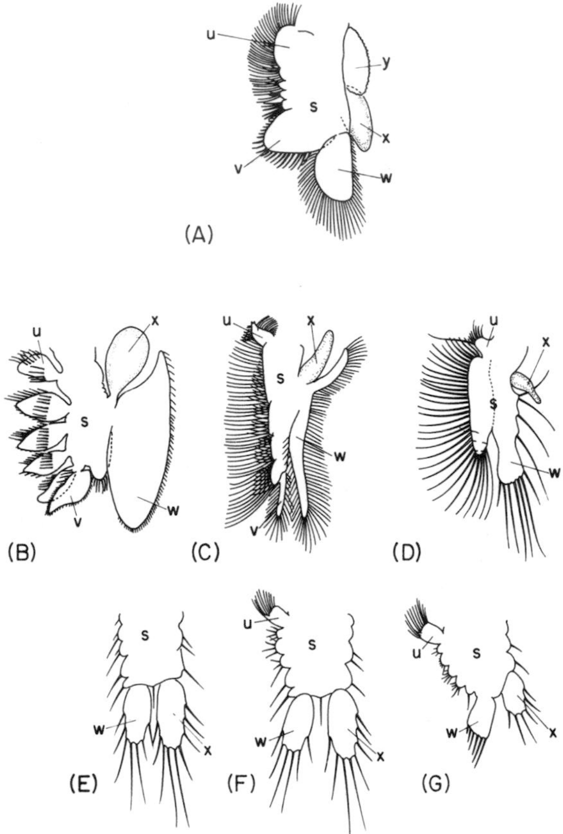

Fig. 6. Branchiopod appendages. (A) Anostraca (*Branchinecta*). (B) Notostraca (*Apus*). (C) Conchostraca (*Limnadia*). (D) Cladocera (*Sida*). (E–G) Lipostraca (*Lepidocaris*), from posterior, middle, and anterior parts of the limb series, respectively. Symbols: s, stem; u, proximal endite; v-y, structures whose identities are subject to controversy (see text). The letters indicate the present author's view of homologies. (A,C, modified from Sars, 1896; B, modified from Eriksson, 1934; E–G, modified from Scourfield, 1926.)

Sanders (1963a) believe the anterior limbs display a more primitive condition, while Calman (1926) and Cannon and Manton (1927) hold the opposite view (see Section III). The anterior trunk limbs are most branchiopodan, but are not more similar to those of anostracans than to notostracans or diplostracans, as is shown for example, in Cannon's (1933) diagram of branchiopod limb evolution.

These comparisons do not disprove the affinity of the Lipostraca and Anostraca, but highlight the tenuousness of this taxonomic union.

The Kazacharthra (Tasch, 1969) are clearly related to notostracans.

Packard (1883) considered the Cladocera to be the most primitive branchiopod because it is the only one with marine representatives and because its low segment count makes it the best link with what he believed to be the nauplius-like crustacean ancestor. Current belief in a multisegmented ancestor makes one of the other taxa a better candidate for resembling the primitive branchiopod.

Cannon (1933) regards the notostracan as most primitive in terms of limb structure and feeding mechanisms, but emphasizes that certain specializations remove it from the direct line to other taxa. Within the Calmanostraca [coined by Tasch (1969) to include notostracans and diplostracans because Preuss' (1951) re-use of Phyllopoda for this same group results in a confusing shift from the original composition], the Notostraca seem to stand nearest to the ancestral condition, with its head shield giving rise posteriorly to a large dorsal carapace fold (Fig. 5C). The conchostracan condition would have resulted from reduction in the head shield (Fig. 5D), while the carapace fold became bivalved and grew forward laterally to even enclose the head (Fig. 5E). Evidence for this shift is seen in lynceids, whose cephalon is not enclosed by the bivalved carapace fold and which have naupliar stages with an undivided dorsal shield like that of notostracans (Linder, 1945). The metanauplii of other conchostracans only have the carapace fold, and the head is uncovered. This view of conchostracan evolution differs from that of Lauterbach (1974a), who suggests a pair of lateral folds grew from the head shield (Fig. 5B) and subsequently fused dorsally over the trunk to form the bivalved carapace. The progenetic origin of cladocerans has already been discussed.

Evidence from within the branchiopods is inadequate to document the relationship of the Sarsostraca [Tasch's (1969) term for anostracans and lipostracans] to calmanostracans. Cannon (1933) regards the anostracan trunk limb and its functions as highly derived. The large stalked eyes on anostracans suggest that the lack of a carapace is primary, yet those who would derive the branchiopods from a carapaced ancestor (Lauterbach, 1974a; Hessler and Newman, 1975) would consider the anostracan condition to be secondarily attained.

A most perplexing aspect of branchiopod morphology is the composition of its trunk limbs (Fig. 6). Structurally, it is a foliaceous stem with setose enditic lobes and two or more lateral or terminal flaps, often called exites. The central question concerns the identity of the endopod, if one exists at all.

If an endopod exists, its development varies between taxa. In notostracans and conchostracans, the terminal endite (Fig. 6B,C:v), which is articulated with the stem, has been called the telopod (roughly equivalent to endopod) (Eriksson, 1934; Preuss, 1957; Pennak, 1978). It has also been regarded simply as the sixth endite (Packard, 1883; Sars, 1896; Borradaile, 1958). If it is an endopod, it follows that the large distolateral flabellum (Fig. 6B,C:w) is the exopod and that the more proximal branchial exite (Fig. 6B,C:x), called the bract, is an epipod (Huxley, 1877; Sars, 1896).

In anostracans, the large, distal endite (Fig. 6A:v) is an integral part of the stem, and therefore has been regarded as simply the sixth endite by Packard (1883) and Calman (1909). Others consider it to be the fused endopod (Sars, 1896; Borradaile, 1926; Calman, 1926; Cannon, 1928; Kaestner, 1970). For Calman (1926) and Borradaile (1926), *Lepidocaris* yielded the solution because the endopod (Fig. 6E–G:w) of its posterior swimming legs becomes the distal, articulated endite of anterior trophic limbs, equivalent to the distal, fused endite in anostracans. Thus, the distolateral exite in anostracans (Fig. 6A:w) becomes the exopod (flabellum), and the one next proximal to it the epipod (bract) (Fig. 6A:x). In anostracans, there are extra exites proximal to the epipod. *Lepidocaris* has a flabellum (exopod) (Fig. 6E–G:x), but no bract (epipod) in Calman and Borradaile's scheme.

Other hypotheses regard yet more of the stem as being part of the endopod (Lankester, 1881; Behning, 1912; Hansen, 1925; Siewing, 1960).

All these interpretations must be viewed with caution because there are few firm reference points for limb homology within the Branchiopoda. The portion of the stem that constitutes coxa and basis cannot be determined with confidence, Hansen (1925) notwithstanding. The identity of the exites cannot be determined from position because they may shift, as suggested by the epipod in cephalocarids. One is inclined to accept the bract and flabellum in notostracans and conchostracans as epipod (Fig. 6B,C:x) and exopod (w), respectively, because their position corresponds so well to that of the exites in cephalocarids and malacostracans (Hansen, 1925; Sanders, 1963a; Hessler and Newman, 1975). The extra exites in anostracans complicates determination, although the branchial function of the penultimate exite (bract) correlates it with the bract in other branchiopods (Fig. 6A–D:x) (Wagler, 1927), and even the epipods of eumalacostracans. Thus, the anostracan flabellum would indeed be the exopod, as many have claimed.

Lepidocaris is not as helpful as has been thought. There is no objective reason to regard the rami of the swimming legs as exopod and endopod. This determination gained its force from the old idea that arthropod limbs had their ancestry in the annelid parapodium (Cannon and Manton, 1927). The rami might both be exites, which is especially reasonable if the anterior trophic limbs are the primitive ones (Sanders, 1963a). Further, the articulated distal endite (Fig. 6G:w) of anterior trophic limbs is not necessarily a homologue of the sixth endite in anostracans (Fig. 6A:v). Contrary to the Calman (1926) and Borradaile (1926) claims, it could just as well be equivalent to the anostracan flabellum (Fig. 6A:w); the exite proximal to it would then be equivalent to the anostracan bract (Fig. 6A:x).

In the final analysis, the identity of the branchiopod endopod remains unresolved, especially in anostracans, where there is no way of knowing whether it exists as a fused or unfused rudiment, or is completely absent. In short, there seems to be as much uncertainty about the homologies of the branchiopod limb as existed one century ago.

VII. MALACOSTRACA

Malacostracans, along with remipedians, possess a nearly complete set of abdominal limbs, a characteristic which is surely plesiomorphic to the class. The limbless abdomen of *Canadaspis* (Briggs, 1978) (Cisne, Chapter 3 of this volume; Figs. 4 and 5) must be regarded as a specialization, assuming that this animal is indeed a malacostracan. (If what Briggs identified as the mandible actually involves the reduced first and second maxillae as well, placement in the Branchiopoda might be in order.) Perhaps the most significant specialization attending the origin of the Malacostraca was the tagmatization of the trunk appendages; the abdominal limbs lost their epipods and became specialized for swimming, while the thoracopods retained trophic, ambulatory, and respiratory functions (Lauterbach, 1975). The loss of natatory function by thoracopods was gradual; thoracic exopods are used for swimming by anaspidaceans, euphausiaceans, mysidaceans, cumaceans, and some carideans (Hessler, 1981a).

Among the living Malacostraca, the Leptostraca (Fig. 1) represent in many ways the most primitive grade of evolution, as demonstrated by a suite of plesiomorphies shared with entomostracan classes. The general function and morphology of the thoracic limbs (Fig. 2) and the orientation of the mouth are quite close to those of cephalocarids and branchiopods, reflecting a primitive crustacean condition (Hessler and Newman, 1975). The abdomen has a seventh pretelsonic segment, and the telson bears large caudal rami. These features are absent among eumalacostracans or linger there only sporadically: e.g., caudal rami in some adult bathynellacean syncarids (Siewing, 1959) and developing mysids (Manton, 1928a), and hints of the seventh abdominal segment in lophogastrid mysidaceans (Manton, 1928b). Aspects of embryology (Manton, 1934) and the long, tubular heart are also primitive features. These entomostracan similarities, particularly to the branchiopods, led to the alignment of leptostracans with the latter group (Milne Edwards, 1840; Sars, 1896). The studies of Metschnikov (1968) and Claus (1888) established *Nebalia*'s malacostracan affinities.

Some aspects of leptostracans must be regarded as apomorphic. Most important is the loss of ambulatory function for thoracic limbs, which now serve only for feeding, respiration, and brood protection (Dahl, 1976). The complete enclosure of the thorax by the laterally compressed carapace fold, the flattening of the thoracopodal endopods with concomitant loss of articulation and musculature and the unique feeding-current direction are surely derived. The reduction of pleopods V–VI, the uniramous second antenna, the scale-like ramus of the first antenna and the animal's epimorphic development are other specializations. Because the cephalic kinesis is not seen in non-malacostracans, its presence in leptostracans may also be apomorphic. In *Nebalia,* it serves to close facultatively the thoracic chamber, probably for protection during mud burrowing (Cannon, 1927).

The most serious challenge to the primitiveness of leptostracans among malacostracans has been the claim (Cannon, 1927; Tiegs and Manton, 1958) that its feeding mechanism must be derived from a system like that of mysids. This claim preceded Cannon's (1933) work on evolution of feeding mechanisms in branchiopods and predates the discovery of the Cephalocarida (Sanders, 1955). It ignores the leptostracan's more plesiomorphic morphology without adequate defense for why the mysid condition must be basic to all Malacostraca (Sanders, 1963a).

The Eumalacostraca have lost the seventh pretelsonic abdominal segment, probably through fusion with the sixth, as seen in the development of *Hemimysis* (Manton, 1928a) and adult *Lophagaster* (Manton, 1928b). Since the seventh abdominal segment is the only one not known to bear appendages in any pre-eumalacostracan (see *Nebalia,* for example), the presence of limbs on all abdominal segments in lower eumalacostracans suggests that it is the seventh which is "missing" (Rolfe, 1981). This adds to the evidence (Shiino, 1942; Burnett, 1973) against the idea that in stomatopods it is indeed the first two segments that fused (Komai and Tung, 1931; Siewing, 1963; Schram, 1969b).

Eumalacostracans lack the thoracopodan mechanisms of thoracic limb fine particle feeding (Dahl, 1976). Instead, the recurring mode of fine particle feeding among lower eumalacostracans relies on protopodal, enditic, setal filters and brushes on the maxillae and first maxilliped (Cannon and Manton, 1927, 1929; Manton, 1928b; Dennell, 1934, 1937). Dahl (1976) proposes this is a neotenic elaboration of the primitive larval mode of cephalic feeding, as seen in cephalocarids. In this view, the stenopodium (see Fig. 8) is an apomorphy of eumalacostracan genesis.

Within the Eumalacostraca, there recurs in several taxa a morphology that has been labeled the "caridoid facies" (Calman, 1904). Its features, going beyond Calman's original list, are:

(1) Carapace extends posteriorly to cover and enclose the thorax.
(2) Movably stalked eyes.
(3) Biramous first antenna.
(4) Exopod of second antenna scale-like.

(5) All thoracic limbs with well-developed flagelliform exopods.

(6) Abdomen well developed, with complex and massive musculature, all designed for strong ventral flexion of the tail fan.

(7) Uropod with paddle-like rami, in conjunction with flattened telson forming a strong tail fan.

(8) Pleopods I–V alike, with two flagelliform rami.

(9) Internal organs mainly excluded from the abdomen.

This morphology is seen most fully in lophogastrid mysidaceans, euphausiaceans, and except for maxillipedal exopods, many natantian decapods (Fig. 7). It is well developed in anaspidacean syncarids, except for the total absence of a carapace.

The close adherence to caridoid morphology of the benthic *Anaspides* (Fig. 7) demonstrates that a "shrimp-like" habitus *sensu stricto* (as in euphausiids and penaeids) is not a necessary quality of the caridoid facies in

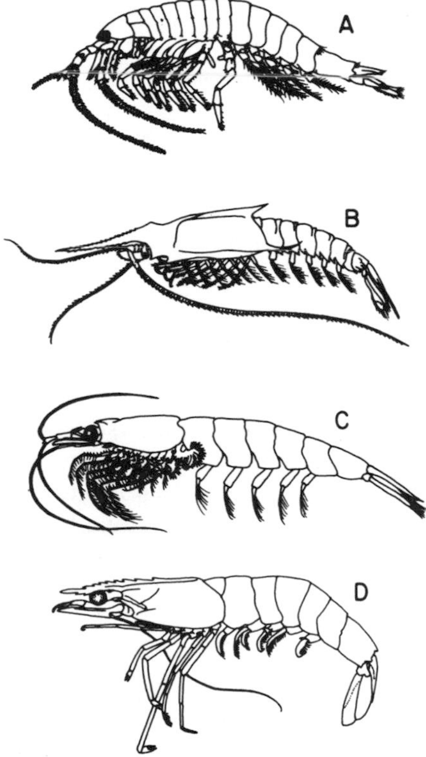

Fig. 7. Living taxa which largely conform to the caridoid facies. (A) Syncarida, Anaspidacea. (B) Mysidacea, Lophogastrida. (C) Euphausiacea. (D) Decapoda, Penaeidea. (A, from Brooks, 1962; B, from Hessler, 1969; C, from Hessler, 1969; D, from McLaughlin, 1980.)

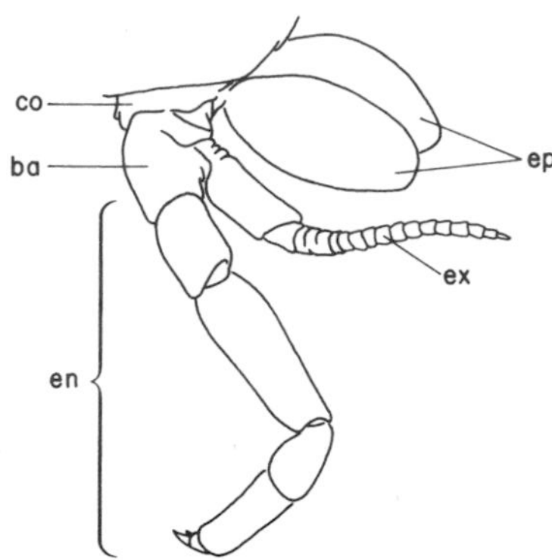

Fig. 8. An unspecialized malacostracan stenopodium from the syncarid *Anaspides*. Setae not included. Symbols: ba, basis; co, coxa; en, endopod; ep, epipod; ex, exopod. (Modified from Hessler, 1981b.)

its most meaningful sense. Some of the reluctance to accept the unity of the caridoids (Dahl, 1976; Schram, 1981b) may stem from using the too restrictive concept. A shrimp-like habitus is readily evolved, as seen from the more pelagic anaspidacean *Paranaspides,* whose close resemblance to shrimps is surely secondary.

Because the caridoid habitus is characteristic of members of three (Syncarida, Eucarida, Peracarida) eumalacostracan divisions (Calman, 1904), it was concluded that the caridoid features are synapomorphies clustering those divisions into a natural unit (although points 1–3 are really plesiomorphies). The common, perfect possession of the caridoid facies by mysidaceans and euphausiaceans was the basis for Latreille's (1817) Schizopoda, which Claus (1885) regarded as a central, primitive group. Today, however, this taxon has given way to a vertical classification, in which caridoid taxa are gathered with more derived forms on the basis of non-caridoid features (Calman, 1904). Thus, mysidaceans join other oostegite-bearing orders in the Peracarida, and euphausiaceans are grouped with decapods in the Eucarida by virtue of their cephalothorax. Similarly, the primarily caridoid anaspidaceans are joined with the more derived bathynellaceans. Hoplocarids were an early eumalacostracan offshoot of the developing caridoid line (Calman, 1909; Siewing, 1956, 1960, 1963).

This conception of relationships within the Eumalacostraca has prevailed for most of this century (see Schram, 1969b, for earlier interpretations), but important concerns have emerged in recent years. Are hoplocarids truly related to other eumalacostracans? Has the caridoid facies evolved more than once, particularly with respect to the carapace?

Hoplocarids possess a number of features of the caridoid facies: scale-like expod on second antenna, five pairs of natatory pleopods, well-developed uropod forming tail fan in conjunction with flattened telson, and abdomen capable of strong ventral flexure. However, the abdominal trunk musculature is not fully caridoid, being only spiraled (Hessler, 1964) and without the special caridoid elaborations (Daniel, 1933, and earlier). Nor does it fill the abdomen, leaving room for viscera (Dahl, 1963). Hessler (1964) regards this as a precaridoid condition, which is compatible with the traditional view.

Alternatively, it has been argued that hoplocaridan evolution is completely separate from that of the caridoids (Schram, 1969a,b, 1973; Kunze, 1981). This view emphasizes the possibility that the hoplocaridan abdomen resulted from fusion of the first two somites (Komai and Tung, 1931; Siewing, 1956, 1963), and it points to the many major unique features of hoplocarids: triflagellate first antenna; pereiopods with a three-segmented protopod and four-segmented endopod; pleopodal gills; and numerous aspects of the digestive system in adaptation to carnivory. Here, caridoid similarities are considered superficial: the scale of the second antenna is the distal of two exopodal segments rather than the whole ramus; the abdomen has evolved to permit reversal in tight spaces; and the caridoid escape reaction is not well developed. In this scenario, the hoplocarids evolved independently from a fine-particle feeding (Schram, 1969b) or carnivorous (Kunze, 1981) phyllocarid with respiratory epipods on the pleopods (Schram, 1969a).

The long, tubular heart with a pair of ostia in each segment, even in the abdomen, is quite primitive (Siewing, 1963), although its extension into the abdomen may be secondary if abdominal gills are a secondary acquisition, as in isopods. If abdominal gills are primary, as Schram suggests, they may not be an apomorphy of the hoplocaridan line, but rather an extreme plesiomorphy reflecting the state that preceded urmalacostracan subdivision of trunk appendages into pereiopods and pleopods, that is, the stage when all limbs had epipods. Special affinity with phyllocarids is documented by the common presence of the cephalic kinesis, although this could be convergent (Calman, 1909). Burnett and Hessler (1973) argue that while hoplocarids do show many unique features, these are not known to occur in any precursor and thus must have evolved within hoplocaridan evolution. If that is the case, there is no *a priori* reason to reject the homology of characters hoplocarids share with caridoids, although as with all similarities the possibility remains that some future discovery will provide evidence of convergence.

Schram (1973) regards the phyllocarid *Sairocaris* as just this sort of evidence. He segregates this genus as the Hoplostraca, a distinct preeumalacostracan group which could have given rise to the Hoplocarida independent of the other eumalacostracans. Rolfe (1981) mentions additional hoplocaridan similarities in the Hoplostraca, but places them away from the common stem of the Hoplocarida and other eumalacostracans. The poor preservation of these fossils obscures most of the information required for a meaningful evaluation. The presence of a relatively long abdomen is not sufficient grounds for such an important conclusion.

The absence of a carapace in an animal apparently so primitive as the Anaspidacea has stimulated much discussion on whether the carapace is truly plesiomorphic. Dahl (1976, 1981) expresses doubt that the ancestral eumalacostracan possessed one and suggests its multiple origin within the group. In this view, not only syncarids, but amphipods and perhaps even isopods have no carapace in their ancestry (Schram, 1981; Watling, 1981). This idea has less strength regarding isopods than amphipods, as will be discussed shortly.

A generally primitive body plan does not preclude the possession of derived traits as well; the lack of eyes in cephalocarids (Elofsson, 1966) is an example. [Burnett's (1981) documentation of eyes in cephalocarids is too superficial to be convincing.] Therefore, the primitiveness of anaspidaceans does not demand the conclusion that the caridoid ancestor lacked a carapace. The universal presence of a carapace on all other lower eumalacostracans (including eocarids and hoplocarids) and all phyllocarids makes it more parsimonious to conclude that a full carapace is plesiomorphic to the Eumalacostraca and that it was secondarily lost in the ancestry of the Syncarida. Anaspidaceans are no more primitive than the Devonian carapace-bearing eocarid *Waterstonella* and the euphausiaceans, whose first thoracopod is also unmodified as a maxilliped.

Disregarding the question of the carapace, the synapomorphy of the rest of the caridoid facies seems secure with respect to syncarids, eucarids, peracarids, and pancarids. It follows that within each division, the general course of evolution is a pattern of disintegration of the caridoid form.

Little need be said with respect to the Syncarida. From the free-living caridoid paleocaridaceans and anaspidaceans, evolution involved the suite of adaptations related to interstitial life, as seen in bathynellaceans and stygocaridaceans. Fragments of the same trend are apparent for the Pancarida as well.

The relationship of the eucaridan orders Euphausiacea and Decapoda appears simple. Euphausiaceans are nearly perfect caridoids, whose primary apomorphies are luminescent organs and loss of ambulatory ability of the thoracic endopods, both of which are related to a purely pelagic existence. In lower decapods, the primary specialization is the initial conversion of the first three thoracopods into maxillipeds.

Most of the resemblances of euphausiaceans to decapods (Fig. 7) are plesiomorphies and are thus inadequate indicators of affinity. Here we include the elements of the caridoid facies, naupliar and zoeal larvae, and a two-segmented protopod on the second antenna. Brinton (1966) regards many special similarities between euphausiaceans and sergestids as independent adaptations to their common pelagic existence. Fusion of all thoracic segments to the carapace, being but a single synapomorphy, must be regarded as a shaky indicator of affinity.

The Amphionidacea has been included in the Caridea, but is currently recognized as an independent eucaridan order (Williamson, 1973; McLaughlin, 1980; Bowman and Abele, Chapter 1 of this volume). Still, the single species is too poorly known to define the order's position with confidence.

The Peracarida contains eumalacostracans with a thoracopodal oostegal marsupium, a lacinia mobilis in adults, a single maxilliped (two in pygocephalomorphs), and a second antenna with a three-segmented protopod. In conjunction with the thoracic marsupium, the pereiopodal coxae are more or less immobilized, and the flexibility so necessary for the base of the limb is by and large taken over by a monocondylic articulation between coxa and basis (Hessler, 1981b). The peracaridan system for ambulation is in an incipient condition in mysidaceans, and it possesses some exceptional features in amphipods. A further synapomorphy is seen in the sperm, which is whip-like, but immobile and devoid of fibrils and mitochondria (Brown, 1970; PochonMasson, 1978; Baccetti, 1979). As a diagnostic feature, the lacinia mobilis should receive less emphasis than it has; it is found in some adult bathynellaceans (Siewing, 1963) and thermosbaenaceans, and in some young euphausiaceans and natatian decapods (Weigmann-Haass, 1977; Gurney, 1942; Dahl and Hessler, 1981). As a functionally important derivative of the spine row (Dahl and Hessler, 1981, in contrast to Manton, 1928b, and Gordon, 1964), it appears to have evolved more than once.

The classical view of evolution of the peracaridan orders (Mysidacea, Cumacea, Tanaidacea, Spelaeogriphacea, Isopoda, and Amphipoda) begins with a caridoid habitus as seen in lophogastrid mysidaceans (Fig. 7), and involves (1) reduction of the carapace, (2) immobilization and loss of the eyestalk, (3) loss of the antennal scale, (4) changes in respiratory patterns related to (5) reduction, modification, or loss of thoracopodal exites, (6) incorporation of thoracic coxae into the body, and (7) reduction of the abdomen, including reduction or even loss of uropods and pleopods (Calman, 1909; Siewing, 1963). Underfolding of the abdomen, a process so important in the evolution of decapods, does not occur in peracarids, probably because it would interfere with the subthoracic marsupium. Except in the amphipods (Dahl, 1977), peracarids beyond the lophogastrids have lost epipods on all thoracopods except the maxilliped. Even in lophogastrids the maxillipedal epipod is developed as a bailer of the branchial chamber. This function is preserved in cumaceans, spelaeogriphaceans, and tanaids. These three orders show intermediate degrees of deviation from the caridoid plan, as displayed in having carapaces of varying length, loss of more posterior thoracic exopods, reduction and loss of eye stalks and antennal scale, and in some cases, reduction of the abdomen and uropod and loss of pleopods. Each order shows its own unique specializations as well, although such features are at a minimum in spelaeogriphaceans.

Cumaceans and tanaids share with isopods the manca larva and oostegites lacking marginal setae, as well as similarities of the digestive and circulatory systems (Siewing, 1963). These document their belonging to a single evolutionary complex. The spelaeogriphacean condition of these features is unknown. Isopods (even disregarding their parasitic constituents) are extreme in lacking a carapace and thoracic exopods, and they have totally sessile eyes. The maxillipedal epipod has lost its respiratory function and now serves as a cheek to the mouth field; respiration, in a shift whose intermediate stages are unknown, is now entirely pleopodal. Tanaids have been thought to display the likely precursor condition of isopod evolution (Calman, 1909), but the Spelaeogriphacea is a more suitable possibility because of its well-developed abdomen, uropods, and strong, natatory pleopods, and its unmodified second thoracopod.

In many ways, amphipod morphology deviates from the caridoid facies in a fashion similar to the isopods. Most conspicuous is the loss of the antennal scale, eye stalks (except in the ingolfiellids), the carapace, and thoracic exopods. The resulting similarity of the two orders is the basis of Burmeister's (1834) Arthrostraca and Schram's (1981) Acaridea. However, the many differences between the two justify rejection of the Arthrostraca (Calman, 1909; Siewing, 1956, 1960, 1963). Differences in the excretory organs (antennal gland in adult amphipods), circulatory (no subneural artery) system, brooding structures (setose oostegites with blade retained after brooding), development (dorsally flexed embryo, no manca), structure of abdominal appendages (three pairs of "uropods"), and morphology of the digestive track (many details; Siewing, 1963) indicate that the Amphipoda is the end product of its own distinct line, unfortunately one which is illuminated neither by living nor fossil intermediates. Possibly the myriad differences document an origin so isolated that its peracaridan similarities are convergent (Dahl and Hessler, 1981). Even its maxilliped is somewhat different, and the unique features in the skeletomusculature of posterior pereiopods may reflect a similar, but independent solution to the need for coxal immobilization (Hessler, 1981b).

The implications of potential isolation of the amphipods are great. Within such a framework, none of the available data is adequate to demonstrate that amphipods had a caridoid ancestor (although at the same time, no existing data disallow one). Moveable stalked eyes (ingolfiellids), biramous antennules, and possession of some natatory pleopods are all basic to the known Malacostraca as a whole and could have been present in a noncaridoid precursor. Further, there is no evidence that the immediate ancestor had a carapace (Dahl, 1977); it might well have been more like an anaspidacean than a lophogastrid.

Proponents of the polyphyly of the carapace (Dahl, 1976, 1981; Watling, 1981; Schram, Chapter 4, of this volume) have suggested that even isopods may represent a primitive rather than derived condition within the Peracarida. However, except for the lack of a carapace, there is little to relate them to the basic condition represented by anaspidaceans. The unity of isopods with the other "mancoid" (Schram, Chapter 4 of this volume) orders seems sound, so if isopods were primitive, it would follow that much of the caridoid facies would have to be polyphyletic. This is unlikely; the caridoid facies has so many facets that it is difficult to envision the coincident multiple evolution of them all. Particularly difficult would be the re-evolution of thoracic exopods as the carapace evolved.

Thermosbaenacea is a caridoid derivative at an intermediate level, much as with the lower mancoids. Its lacinia mobilis and single maxilliped with bailing epipod have been used as evidence for placement within the Peracarida in spite of the lack of oostegites and use of the carapace fold as a brood chamber (Gordon, 1958; Barker, 1962; Fryer, 1964). Another peracaridan similarity is the infolded monocondylic articulation of pereiopodal coxa and basis, although its orientation is unusual (Hessler, 1981b). Since the lacinia mobilis is not an exclusively peracaridan trait, its use in discerning affinities loses its force (Dahl and Hessler, 1981). Other than having the epipodal bailer, the maxilliped is not particularly similar to those of unquestioned peracarids. Thus, the argument (Siewing, 1958) that the very different brooding system warrants placement in an independent order, Pancarida, must be taken seriously.

REFERENCES

Anderson, D. T. (1973). "Embryology and Phylogeny in Annelids and Arthropods." Pergamon, Oxford.

Armstrong, J. C. (1949). The systematic position of the crustacean genus *Derocheilocaris* and the status of the subclass Mystacocarida. *Am. Mus. Novit.* **1413,** 1-6.

Ax, P. (1960). "Die Entdeckung neuer Organisationstypen im Tierreich." A. Ziemsen Verlag, Wittenberg, Lutherstadt.

Baccetti, B. (1979). Ultrastructure of sperm and its bearing on arthropod phylogeny. *In* "Arthropod Phylogeny" (A. P. Gupta, ed.), pp. 609-644. Van Nostrand-Reinhold, New York.

Barker, D. (1962). A study of *Thermosbaena mirabilis* Monod (Malacostraca, Peracarida) and its reproduction. *Q. J. Microsc. Sci.* [N.S.] **103,** 261-286.

Becker, K.-H. (1975). *Basipodella harpacticola* n. gen., n. sp. (Crustacea, Copepoda). *Helgol. Wiss. Meeresunters.* **27,** 96-100.

Behning, A. (1912). Studien über die vergleichende Morphologie sowie über temporale und Lokalvariation der Phyllipodextremitaten. *Int. Rev. Gesamten Hydrobiol. Hydrogr.* Suppl. 4, Part 1, 1-70.

Beklemishev, V. N. (1952). "Osnovy sraviteljnoi anatomii bespozvonocnych," 2nd ed. Nauka, Moscow.

Beklemishev, V. N. (1969). "Principles of Comparative Anatomy of Invertebrates" (transl. of 3rd (1964) ed. by J. M. MacLennan), Vol. 1. Oliver & Boyd, Edinburgh.

Birstein, J. A., and Novozhilov, N. I. (1960). Class Crustacea. Crustaceans [J.A.B.]; Subclass Maxillopoda [N.I.N]. *In* "Osnovy Paleontologii" (Yu. A. Orlov, ed.), Chlenistonogie Trilobitoobraznie i Rakoobraznie (N. E. Cheryshova, ed.), 201-216, 253-260. State Scientific Technical Press, Moscow (in Russian).

Borradaile, L. A. (1926). Notes upon crustacean limbs. *Ann. Mag. Nat. Hist.* [4] **17,** 193-213.

Borradaile, L. A. (1958). The Class Crustacea. *In* "The Invertebrata" (L. A. Borradaile, F. A. Potts, L. E. S. Eartham, and J. T. Saunders, eds.) (revised by G. A. Kerkut), 3rd ed., pp. 340-419. Cambridge Univ. Press, London and New York.

Bowman, T. E. (1971). The case of the nonubiquitous telson and the fradulent furca. *Crustaceana* **21,** 165-175.

Bradford, J. M., and Hewitt, G. C. (1980). A new maxillopodan crustacean, parasitic on a myodocopid ostracod. *Crustaceana* **38,** 67-72.

Briggs, D. E. G. (1978). The morphology, mode of life, and affinities of *Canadaspis perfecta* (Crustacea: Phyllocarida), Middle Cambrian, Burgess Shale, British Columbia. *Philos. Trans. R. Soc. London, Ser. B* **281,** 439-487.

Brinton, E. (1966). Remarks on euphausiacean phylogeny. *Proc. Symp. Crustacea, Ernakulum, 1965* Vol. 1, pp. 255-259.

Brooks, J. L. (1959). Cladocera. *In* "Fresh Water Biology" (W. T. Edmondson, ed.), 2nd ed., pp. 587-655. Wiley, New York.

Brown, G. G. (1970). Some comparative aspects of selected crustacean spermatozoa and crustacean phylogeny. *In* "Comparative Spermatology" (B. Baccetti, ed.), pp. 183-204. Academic Press, New York.

Buchholz, H. A. (1953). Die Mystacocarida. Eine neue Crustaceenordnung aus dem Lückensystem der Meeressande. *Mikrokosmos* **43,** 13-16.

Burmeister, H. (1834). Beiträge zur Naturgeschichte der Rankenfüsser (Cirripedia). Reimer, Berlin.

Burnett, B. R. (1973). Notes on the lateral arteries of two stomatopods. *Crustaceana* **23,** 303-305.

Burnett, B. R. (1981). Compound eyes in the cephalocarid crustacean *Hutchinsoniella macracantha*. *J. Crustacean Biol.* **1,** 11-15.

Burnett, B. R., and Hessler, R. R. (1973). Thoracic epipodites in the Stomatopoda (Crustacea): A phylogenetic consideration. *J. Zool.* **169,** 381-392.

Calman, W. T. (1904). On the classification of the Crustacea Malacostraca. *Ann. Mag. Nat. Hist.* [7] **13,** 144-158.

Calman, W. T. (1909). Crustacea. *In* "Treatise on Zoology" (E. R. Lankester, ed.), Part VII, Fasc. 3. Adam & Black, London.

Calman, W. T. (1926). The Rhynie crustacean. *Nature* (*London*) **118,** 89-90.

Cannon, H. G. (1927). On the feeding mechanism of *Nebalia bipes*. *Trans. R. Soc. Edinburgh* **55,** 355-369.

Cannon, H. G. (1928). On the feeding mechanism of the fairy shrimp, *Chirocephalus diaphanus* Prevost. *Trans. R. Soc. Edinburgh* **55,** 807-822.
Cannon, H. G. (1933). On the feeding mechanism of the Branchiopoda. *Philos. Trans. R. Soc. London, Ser. B* **222,** 267-352.
Cannon, H. G., and Manton, S. M. (1927). On the feeding mechanism of a mysid crustacean, *Hemimysis lamornae. Trans. R. Soc. Edinburgh* **55,** 219-253.
Cannon, H. G., and Manton, S. M. (1929). On the feeding mechanism of the syncarid Crustacea. *Trans. R. Soc. Edinburgh* **56,** 175-189.
Claus, C. (1860). "Beiträge zur Kenntniss der Entomostraken," Part 1. Schriften d. Gesellsch. zur Beförderung der gerammt. Naturwiss. zu Marburg, Marburg.
Claus, C. (1875). Über die Entwicklung, Organisation und systematische Stellung der Arguliden. *Z. Wiss. Zool.* **25,** 217-284.
Claus, C. (1876). "Untersuchungen zur Erforschung der genealogischen Grundlage des Crustaceensystems." Carl Gerold, Wien.
Claus, C. (1885). Neue Beiträge zur Morphologie der Crustaceen. *Arb. Zool. Inst. Univ. Wien* **6,** 1-108.
Claus, C. (1888). Über den Organismus der Nebaliden und die systematische Stellung der Leptostraken. *Arb. Zool. Inst. Univ. Wien* **8,** 1-148.
Dahl, E. (1952). Mystacocarida. *Lunds Univ. Arsskr.* [N. S.] **48,** 1-41.
Dahl, E. (1956a). On the differentiation of the topography of the crustacean head. *Acta Zool. (Stockholm)* **37,** 123-192.
Dahl, E. (1956b). Some crustacean relationships. *In* "Bertil Hanström: Zoological Papers in Honour of His Sixty-Fifth Birthday, November 20th, 1956" (K. Wingstrand, ed.), pp. 138-147. Lund Zool. Inst., Lund, Sweden.
Dahl, E. (1963). Main evolutionary lines among recent Crustacea. *In* "Phylogeny and Evolution of Crustacea" (H. B. Whittington and W. D. I. Rolfe, eds.), Spec. Publ., pp. 1-15. Mus. Comp. Zool., Cambridge, Massachusetts.
Dahl, E. (1976). Structural plans as functional models exemplified by the Crustacea Malacostraca. *Zool. Scr.* **5,** 163-166.
Dahl, E. (1977). The amphipod functional model and its bearing upon systematics and phylogeny. *Zool. Scr.* **6,** 221-228.
Dahl, E. (1982). "Alternatives in Malacostracan Evolution." Rec. Australian Museum (in press).
Dahl, E., and Hessler, R. R. (1982). The lacinia mobilis: A reconsideration of its origin, function and phylogenetic implications. *J. Linn. Soc. London, Zool.* (in press).
Daniel, R. J. (1933). Comparative study of the abdominal musculature in Malacostraca. Pt. III. The abdominal muscular systems of *Lophogaster typicus* M. Sars, and *Gnanthophausia zoea* Suhm, and their relationships with the musculatures of other Malacostraca. *Proc. Trans. Liverpool Biol. Soc.* **47,** 71-133.
Dennell, R. (1934). The feeding mechanism of the cumacean crustacean *Diastylis bradyi. Trans. R. Soc. Edinburgh* **58,** 125-142.
Dennell, R. (1937). On the feeding mechanism of *Apseudes talpa,* and the evolution of the peracaridan feeding mechanisms. *Trans. R. Soc. Edinburgh* **59,** 57-78.
Elofsson, R. (1963). The nauplius eye and frontal organs in Decapoda (Crustacea). *Sarsia* **12,** 1-68.
Elofsson, R. (1965). The nauplius eye and frontal organs in Malacostraca (Crustacea). *Sarsia* **19,** 1-54.
Elofsson, R. (1966). The nauplius eye and frontal organs of the non-Malacostraca (Crustacea). *Sarsia* **25,** 1-128.

Elofsson, R., and Dahl, E. (1970). The optic neuropiles and chiasmata of Crustacea. *Z. Zellforsch. Mikrosk. Anat.* **107,** 343-360.

Eriksson, S. (1934). Studien über die Fangapparate der Branchiopoden nebst einigen phylogenetischen Bemerkungen. *Zool. Bidr. Uppsala* **15,** 23-287.

Fryer, G. (1964). Studies on the functional morphology and feeding mechanism of *Monodella argentarii* Stella (Crustacea: Thermosbaenacea). *Trans. R. Soc. Edinburgh* **66,** 49-90.

Gerstäcker, A. (1863). "Handbuch der Zoologie," Vol. 2, Arthropoda. Engelmann, Leipzig.

Gerstaecker, K. E. A. (1866-1879). Gliedenfüssler (Arthropoda). *Bronn's Klassen* **5**(1), 1-1320.

Gordon, I. (1958). A thermophilous shrimp from Tunisia. *Nature* (*London*) **182,** 1186.

Gordon, I. (1964). On the mandible of the Stygacaridae (Anaspidacea) and some other Eumalacostraca, with special reference to the lacinia mobilis. *Crustaceana* **7,** 150-157.

Grobben, C. (1893). A contribution to the knowledge of the geneology and classification of the Crustacea. *Ann. Mag. Nat. Hist.* [7] **11,** 440-473.

Grygier, M. J. (1981). Sperm of the ascothoracican parasite *Dendrogaster,* the most primitive found in Crustacea. *Int. J. Invertebr. Reprod.* **3,** 65-73.

Gurney, R. (1942). "Larvae of Decapod Crustacea." Ray Society, No. 129, London.

Hansen, H. J. (1899). Die Cladoceren und Cirripedien der Plankton-Expedition. *Ergeb. Plankton-Exped. Humboldt-Stift.* **2,** 1-58.

Hansen, H. J. (1925). On the comparative morphology of the appendages in the Arthropoda. A. Crustacea. *In* "Studies on Arthropoda II." Gyldendalske, Copenhagen.

Hartog, M. M. (1888). The morphology of *Cyclops* and the relations of the Copepoda. *Trans. Linn. Soc. London, Zool.* **5,** 1-46.

Hessler, R. R. (1964). The Cephalocarida—comparative skeleto-musculature. *Mem. Conn. Acad. Arts Sci.* **16,** 1-97.

Hessler, R. R. (1969). Mystacocarida. *In* "Treatise on Invertebrate Paleontology" (R. C. Moore, ed.), Part R, Arthropoda 4, Vol. 1, pp. R192-R195. Geol. Soc. Am., Boulder, Colorado, and the Univ. of Kansas Press, Lawrence.

Hessler, R. R. (1981a). Evolution of arthropod locomotion: A crustacean model. *In* "Locomotion and Exercise in Arthropods" (C. F. Herreid and C. R. Fourtner, eds.), pp. 9-30. Plenum, New York.

Hessler, R. R. (1981b). The structural morphology of walking mechanisms in eumalacostracan crustaceans. *Philos. Trans. R. Soc. London, Ser. B* **296,** 245-298.

Hessler, R. R., and Newman, W. A. (1975). A trilobitomorph origin for the Crustacea. *Fossils Strata* **4,** 437-459.

Huxley, T. H. (1877). "A Manual of the Anatomy of Invertebrated Animals." London.

Jägersten, G. (1972). "Evolution of the Metazoan Life Cycle." Academic Press, New York.

Kaestner, A. (1970). Crustacea. *In* "Invertebrate Zoology" (transl. by H. W. Levit and L. R. Levi), Vol. 3. Wiley, New York.

Komai, T., and Tung, Y. M. (1931). On some points of the internal structure of *Squilla oratoria. Mem. Coll. Sci., Kyoto Imp. Univ.* **6,** 1-15.

Kumé, M., and Dan, K. (1968). "Invertebrate Embryology." NOLIT Publ. House, Belgrade.

Kunze, J. C. (1981). The functional morphology of stomatopod Crustacea. Ph.D. Thesis, University of Sydney, Sydney, Australia.

Lankester, E. R. (1881). Observations and reflections on the appendages and on the nervous system of *Apus cancriformis. Q. J. Microsc. Sci.* [N. S.] **21,** 343-376.

Latreille, P. A. (1806-1809). "Genera Crustaceorum et Insectorum secundum ordinem naturalem in familias disposita, iconibus exemplisque plurimis explicata." Parisiis and Argentorati, König.

Latreille, P. A. (1817). Les crustacés, les arachnides et les insectes. *In* "Le règne animal distribué d'après son organization, pour servier de base à l'histoire naturelle des animaux et d'introduction à l'anatomie comparée," (G. Cuvier, ed.). Déterville, Paris.

Lauterbach, K.-E. (1974a). Über die Herkunft des Carapax der Crustaceen. *Zool. Beitr.* [N. S.] **20,** 273-327.

Lauterbach, K.-E. (1974b). Die Muskulatur der Pleurotergite im Grundplan der Euarthropoda. *Zool. Anz.* **193,** 70-84.

Lauterbach, K.-E. (1975). Über die Herkunft der Malacostraca (Crustacea). *Zool. Anz.* **194,** 165-179.

Lehmann, W. M. (1955). *Vachonia rogeri,* n. gen. n. sp., ein Branchiopod aus dem unterdevonischen Hünsruckschiefen. *Paläontol.* **29,** 126-130.

Linder, F. (1945). Affinities within the Branchiopoda, with notes on some dubious fossils. *Ark. Zool.* **37A,** 1-28.

McLaughlin, P. A. (1980). "Comparative Morphology of Recent Crustacea." Freeman, San Francisco, California.

Manton, S. M. (1928a). On the embryology of the mysid crustacean *Hemimysis lamornae. Philos. Trans. R. Soc. London, Ser. B* **216,** 363-463.

Manton, S. M. (1928b). On some points in the anatomy and habits of the lophogastrid Crustacea. *Trans. R. Soc. Edinburgh* **56,** 103-119.

Manton, S. M. (1934). On the embryology of the crustacean *Nebalia bipes. Philos. Trans. R. Soc. London, Ser. B* **223,** 163-238.

Manton, S. M. (1964). Mandibular mechanisms and the evolution of arthropods. *Philos. Trans. R. Soc. London, Ser. B* **247,** 1-183.

Manton, S. M. (1973). Arthropod phylogeny—a modern synthesis. *J. Zool.* **171,** 111-130.

Martin, M. F. (1932). On the morphology and classification of *Argulus* (Crustacea). *Proc. Zool. Soc. London,* pp. 771-806.

Metschnikov, E. (1868). The history of the development of *Nebalia. Zap. Imp. Akad. Nauk, St.-Petersb.* **13,** 1-48 pp (in Russian).

Milne Edwards, H. (1840). "Histoire naturelle des Crustacés comprenant l'anatomie, la physiologie et la classification de ces animaux," Vol. III. Paris.

Müller, K. J. (1979). Phosphatocopine ostracodes with preserved appendages from the Upper Cambrian of Sweden. *Lethaia* **12,** 1-27.

Newman, W. A., Zullo, V. A., and Withers, T. H. (1969). Cirripedia. *In* "Treatise on Invertebrate Paleontology" (R. C. Moore, ed.), Part R, Arthropoda 4, Vol. 1, pp. R206-R295. Geol. Soc. Am., Boulder, Colorado, and the Univ. of Kansas Press, Lawrence.

Packard, A. S. (1879). "Zoology for High Schools and Colleges." New York.

Packard, A. S. (1883). A monograph of the North American phyllopod Crustacea. *Annu. Rep. (1878), U.S. Geol. Geogr. Sur. Territories* **12,** 295-592.

Paulis, H. F. (1979). Eye structure and the monophyly of the Arthropoda. *In* "Arthropod Phylogeny" (A. P. Gupta, ed.), pp. 299-383. Van Nostrand-Reinhold, New York.

Pennak, R. W. (1978). "Fresh-water Invertebrates of the United States," 2nd ed. Wiley, New York.

Pennak, R. W., and Zinn, D. J. (1943). Mystacocarida, a new order of Crustacea from intertidal beaches in Massachusetts and Connecticut. *Smithson. Misc. Collect.* **103,** 1-11.

Pochon-Masson, J. (1978). Les différenciations infrastructurales liées a la perte de le motilité chez les gamètes mâles des Crustacés. *Arch. Zool. Exp. Gen.* **119,** 465-470.

Preuss, G. (1951). Die Verwandtschaft der Anostraca und Phyllopode. *Zool. Anz.* **147,** 49-64.

Preuss, G. (1957). Die Muskulatar der Gliedmassen von Phyllopoden und Anostraken. *Mitt. Zool. Mus. Berl.* **33,** 221-257.

Remane, A. (1952). "Die Grundlagen des naturlichen Systems, der vergleichenden Anatomie und der Phylogenetik," 2nd ed. Akad. Verlags ges., Leipzig.

Rolfe, W. D. I. (1981). Phyllocarida and the origin of the Malacostraca. *Géobios* **14,** 17-27.

Sanders, H. L. (1955). The Cephalocarida, a new subclass of Crustacea from Long Island Sound. *Proc. Natl. Acad. Sci. U.S.A.* **41,** 61-66.

Sanders, H. L. (1963a). The Cephalocarida: Functional morphology, larval development, comparative external anatomy. *Mem. Conn. Acad. Arts Sci.* **15,** 1-80.

Sanders, H. L. (1963b). Significance of the Cephalocarida. *In* "Phylogeny and Evolution of the Crustacea" (H. B. Whittington and W. D. I. Rolfe, eds.), Spec. Publ., pp. 163-175. Mus. Comp. Zool., Cambridge, Massachusetts.

Sars, G. O. (1887). Report of the Phyllocarida collected by H.M.S. Challenger during the years 1873-76. *Challenger Rep., Zool.* **19,** 1-38.

Sars, G. O. (1896). "Phyllocarida and Phyllopoda," Fauna Norvegiae, Vol. 1. Aktie-Bogtrykkeriet, Christiania.

Schminke, H. K. (1976). The ubiquitous telson and the deceptive furca. *Crustaceana* **30,** 292-300.

Schram, F. R. (1969a). Some Middle Pennsylvanian Hoplocarida (Crustacea) and their phylogenetic significance. *Fieldiana, Geol.* **12,** 235-289.

Schram, F. R. (1969b). Polyphyly in the Eumalacostraca? *Crustaceana* **16,** 243-250.

Schram, F. R. (1973). On some phyllocarids and the origin of the Hoplocarida. *Fieldiana, Geol.* **26,** 77-94.

Schram, F. R. (1978). Arthropods: A convergent phenomenon. *Fieldiana, Geol.* **39,** 61-108.

Schram, F. R. (1981). On the classification of the Eumalacostraca. *J. Crustacean Biol.* **1,** 1-10.

Schram, T. A. (1970a). Marine biological investigations in the Bahamas. 14. Cypris Y, a later developmental stage of nauplius Y Hansen. *Sarsia* **44,** 9-24.

Schram, T. A. (1970b). On the enigmatical lava nauplius Y type I Hansen. *Sarsia* **45,** 53-68.

Schram, T. A. (1972). Further records of nauplius Y type IV Hansen from Scandinavian waters. *Sarsia* **50,** 1-24.

Scourfield, D. J. (1926). On a new type of crustacean from the Old Red Sandstone (Rhynie Chert Bed, Aberdeenshire)—*Lepidocaris rhyniensis* gen. and sp. nov. *Philos. Trans. R. Soc. London, Ser. B* **214,** 153-187.

Scourfield, D. J. (1940). Two new and nearly complete specimens of the Devonian fossil crustacean *Lepidocaris rhyniensis. Proc. Linn. Soc. London* **152,** 290-298.

Sharov, A. G. (1966). "Basic Arthropodan Stock." Pergamon, Oxford.

Shiino, S. M. (1942). Studies on the embryology of *Squilla oratoria* de Haan. *Mem. Coll. Sci., Kyoto Imp. Univ., Ser. B* **17,** 77-173.

Siewing, R. (1956). Untersuchungen zur Morphologie der Malacostraca (Crustacea). *Zool. Jahrb., Abt. Anat. Ontog. Tiere* **75,** 39-176.

Siewing, R. (1958). Anatomie und Histologie von *Thermosbaena mirabilis. Abh. Math-Naturwiss. Kl., Akad. Wiss. Lit. Mainz, pp. 195*-270.

Siewing, R. (1959). Syncarida. *Bronn's Klassen* **5,** 1-121.

Siewing, R. (1960). Neuere Ergebnisse der Verwandtschaftsforschung bei den Crustaceen. *Wiss. Z. Univ. Rostock, Math.-Naturwiss. Reihe* **9,** 343-358.

Siewing, R. (1963). Studies in malacostracan mophology: Results and problems. *In* "Phylogeny and Evolution of the Crustacea" (H. B. Whittington and W. D. I. Rolfe, eds.), Spec. Publ., pp. 85-103. Mus. Comp. Zool., Cambridge, Massachusetts.

Snodgrass, R. E. (1938). Evolution of the Annelida, Onychophora, and Arthropoda. *Smithson. Misc. Collect.* **97,** 1-159.

Stürmer, W., and Bergström, J. (1976). The arthropods *Mimetaster* and *Vachonisia* from the Devonian Hunsrück Shale. *Paläontol. Z.* **50,** 78-111.

Tasch, P. (1969). Branchiopoda. *In* "Treatise on Invertebrate Paleontology" (R. C. Moore, ed.), Part R, Arthropoda 4, Vol. 1, pp. R128-191. Geol. Soc. Am., Boulder, Colorado, and the Univ. of Kansas Press, Lawrence.

Tiegs, O. W., and Manton, S. M. (1958). The evolution of the Arthropoda. *Biol. Rev. Cambridge Philos. Soc.* **33,** 255-337.

Wagler, E. (1927). 1. Ordnung: Branchiopoda, Phyllopoda-Kiemenfüsser. *In* "Handbuch der Zoologie" (W. Kükenthal and T. Krumbach, eds.), Vol. 3, No. 1, pp. 305-398. de Gruyter, Berlin.
Watling, L. (1981). An alternative phylogeny of peracarid crustaceans. *J. Crust. Biol.* **1,** 201-210.
Weigmann-Haass, R. (1977). Die Calyptopis und Furcilia-Stadien von *Euphausia hanseni* (Crustacea: Euphausiacea). *Helgol. Wiss. Meeresunters.* **29,** 315-327.
Williamson, D. I. (1973). *Amphionides reynaudii* (H. Milne Edwards), representative of a proposed new order of eucaridan Malacostraca. *Crustaceana* **25,** 35-50.
Wilson, C. B. (1902). North American parasitic copepods of the family Argulidae, with a bibliography of the group and a systematic review of all known species. *Proc. U.S. Natl. Mus.* **25,** 635-742.
Yager, J. (1981). A new class of Crustacea from a marine cave in the Bahamas. *J. Crustacean Biol.* **1,** 328-333.
Zimmer, C. (1926-1927). Crustacea = Krebse Allgemeine Einleitung in die Naturgeschichte der Crustacea. *in* "Handbuch der Zoologie" (W. Kükenthal and T. Krumbach, eds.), Vol. 3, No. 1, pp. 275-304. de Gruyter, Berlin.

Part 2: Copepoda (Marcotte)

The Subclass Copepoda is composed of over 7500 free-living and parasitic species. The adult body of free-living copepods is divided into a wide anterior end, the prosome (cephalon + metasome), and a narrow posterior end, the urosome. The adult body of parasitic copepods is variously modified; segmentation is often reduced. Adult free-living and ectoparasitic copepods usually have six feeding appendages (A_1, A_2, Md, Mx_1, Mx_2, Mxp) and five (six) walking or swimming legs. The abdomen is limbless except for a pair of rami on the telson. Bowman (1971) has argued that copepods have no telson. His argument failed to recognize that the telson is defined by embryological characters, not by adult ones (Calman, 1909). It is homologous to the pygidium (telotrochal plus anal regions) of protostomal embryos. The telson bears the anus and has teloblastic cells on its anterior edge from which body segments proliferate. Copepods (and other crustaceans) have a telson (cf. Schminke, 1976).

Thorell (1859) recognized three groups of copepods based on the structure of their oral appendages. (1) The Gnathostoma were copepods with open buccal cavities formed by a large labrum and a small labium (paragnaths). Their mandibles were biramous and had large, broad, denticulate gnathobases (Figs. 1B and C). (2) The Poecilostoma (poecil = small cup) had an open buccal cavity and mouthparts similar to those of the Gnathostoma, but Thorell thought that they lacked mandibles. (3) The Siphonostoma had a tubular buccal cavity formed by the fusion of the labrum and labium (Figs. 1E and 3D). The mandibular gnathobases were formed into long, thin stylets,

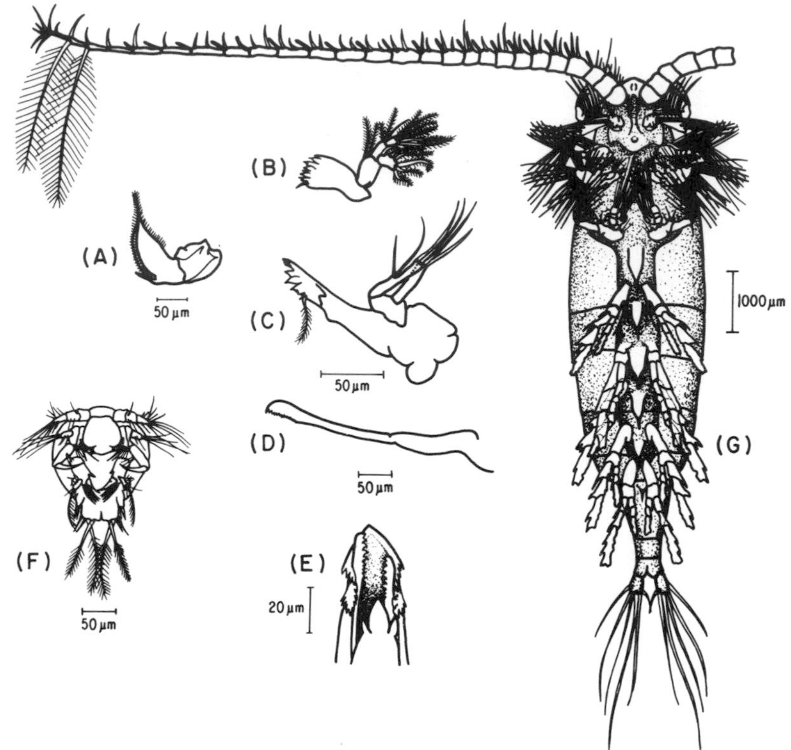

Fig. 1. (A) Poecilostomatoid mandible (*Paeudanthissus ferox*) (after Humes and Ho, 1967). (B) Gnathostomous mandible (*Longipedia coronata*) (after Sars, 1901–1904) (No scale available). (C) Gnathostomous mandible (*Tisbe furcata*). (D) Siphonostomatoid mandible (*Trebius candatus*) (after Kabata, 1979). (E) Siphonostomatoid buccal complex (*Pandarus bicolor*) (after Kabata, 1979). (F) Fifth naupliar stage (*Harpacticus littoralis*) (after Castel, 1976). (G) Planktonic calanoid (*Calanus hyperboreus*) (after Giesbrecht, 1892).

either with or without distal teeth (Fig. 1D). The gnathostoma were free-living, but the others were mostly commensal and parasitic.

Claus (1862) showed that the Poecilostoma indeed had a mandible (e.g., Fig. 1A). He redistributed the species in this group among the Gnathostoma and Siphonostoma. Claus' work notwithstanding, Brady (1880) continued to use Thorell's system because he thought its division of the Copepoda was "a very natural one, the three groups presenting characters which, though differing in degree in various species, do point, on the whole, to habits of life very remarkably different and deserving of expression in any natural classification."

Giesbrecht (1892) abandoned Thorell's system and divided the Copepoda into two groups: (1) The Gymnoplea were copepods in which the prosome-urosome articulation occurred between the sixth and seventh thoracic somites, i.e., behind the fifth pair of thoracic legs. The urosome had no legs (gymno = naked). (2) The Podoplea had their prosome-urosome articulation between the fifth and sixth throacic somites, i.e., in front of the fifth pair of thoracic legs. Thus, their urosome had legs (podo = foot).

Giesbrecht's system divided most planktonic copepods (Gymnoplea) from all other copepods (Podoplea). Later workers found this division unsatisfactory. However, this system was seen to emphasize the role of tagmosis in copepod evolution and the importance of body fusion in promoting hydrodynamically efficient swimming among planktonic copepods.

Sars (1901-1904) divided the Copepoda into seven groups defined by their type genera: *Calanus, Harpacticus, Cyclops, Notodelphys, Monstrilla, Caligus,* and *Lernaea*. Each taxon was of equal value (orders or suborders). Recognition of these groups was based on apparent morphological discontinuities among taxa. A truly phylogenetic classification awaited description of more species. Later, Wilson (1910) added an eighth order (suborder), the Lernaeopodoida, to Sars' system by amalgamating two morphologically similar parasitic families.

As more species were found and described, it became clear that most of the orders of Sars and Wilson were paraphyletic or polyphyletic groups. Lang (1948a,b) tried to resolve some of the emerging phylogenetic and taxonomic problems in his consideration of harpacticoid systematics and notodelphyoid oral morphology. Kabata (1979) provided a comprehensive re-examination of the issues involved and a new scheme of copepod phylogeny and nomenclature, again based mostly on mandibular modifications.

The following diagnoses are derived from Kabata (1979) by making explicit the taxonomic descriptions implicit in his discussion of copepod orders and their phylogenetic relationships. Here, Kabata's gymnoplean and podoplean taxa are not given superordinal status. These supertaxa numerically divide the copepods very unequally. The Calanoida are separated as the Gymnoplea from all other orders on the basis of a single morphological character. Their prosome-urosome articulation is between the sixth and seventh (genital complex) thoracic somites. This articulation in podopleans is between the fifth and sixth thoracic somites. Embryologically, the prosome-urosome articulation becomes clear in the first copepodid stage behind the third free thoracic somite (Calman, 1909). It moves back one somite with each subsequent molt. Its position is fixed in the second copepodid stage for podopleans and in the third for gymnopleans. This character in gymnopleans is probably apomorphic and units a monophyletic assemblage. However, two views of podoplean evolution are possible. First, if podoplean tagmosis is plesiomorphic, the assemblage may be polyphyletic. In this view, no single supertaxonomic name is warranted. Second, if podoplean tagmosis is apomorphic, the assemblage can be regarded as monophyletically derived from an ancestor common to podopleans and gymnopleans, which lacked prosome–urosome differentiation (Kabata, 1979). This second hypothesis has not been supported by the discovery of either living or fossil copepods with the requisite characteristics. Until such data are available, the "Gymnoplea" must be considered taxonomically equal to the "Calanoida," and the "Podoplea" must be considered, at best, of dubious empirical reality and, at worse, inadequately defined by a plesiomorphic character and polyphyletic.

Order Calanoida, (Figs. 1G and 2D): Free-living, mostly planktonic; A_1 very

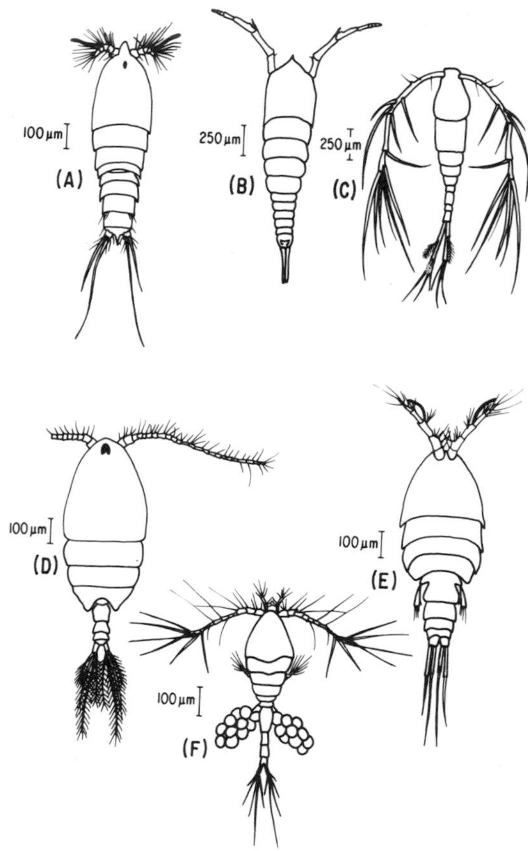

Fig. 2. (A) Epibenthic (demersal planktonic) harpacticoid (*Longipedia weberi*) (after Sewell, 1940). (B) Planktonic harpacticoid (*Aegisthus aculeatus*) (after Boxshall, 1979). (C) Planktonic mormonilloid (*Mormonilla phasma*) (after Boxshall, 1979). (D) Near bottom, planktonic calanoid (*Pseudophaenna typica*) (after Sars, 1901–1904). (E) Epibenthic (semiplanktonic) harpacticoid (*Tisbe lancii*) (after Marcotte, 1974). (F) Planktonic cyclopoid (*Oithona nana*) (after Giesbrecht, 1892).

long with 16-26 articles; buccal cavity open; A_2, Md, and Mx_1 biramous; Md gnathostomous; Mx_2 and Mxp uniramous; first thoracic legs biramous, multi-articulate, with plumose setae for swimming; last thoracic leg uniramous, may be modified or missing; heart present in many but not all [= Calanoida (Sars, 1901-1904)].

Order Misophrioida (Fig. 3A): Free-living on and above benthos; A_1 shorter with 11-16 articles; buccal cavity open; A_2, Md, and Mx_1 biramous; Md gnathostomous; thoracic legs biramous, multi-articulate, with plumose setae for swimming; heart present [= Misophriidae (Brady, 1880), = Misophrioida (Lang, 1948a)].

Order Harpacticoida (Figs. 2A,B, and E): Mostly free-living, benthic, epibenthic, planktonic; A_1 short with fewer than 10 articles; buccal cavity open; A_2 and Md biramous; Md gnathostomous; Mx_1 usually biramous; various degrees of fusion, reduction, and loss of rami in cephalic and thoracic appendages; heart absent [= Harpacticoida (Sars, 1901-1904; Lang, 1948b)].

Order Monstrilloida (Fig. 3B): Nauplii at first, swimming, then sac-like endoparasites of marine polychaetes, prosobranch gastropods, and, infrequently, echinoderms; adults planktonic, without A_2 mouthparts, or functional gut; adult thoracic legs biramous for swimming [= Monstrilloida (Sars, 1901-1904)].

Order Siphonostomatoida (Figs. 3C-F) (Note: this diagnosis is necessarily broad and emphasizes the order's possible diphyletic origin, as will be discussed below): Adults ecto- or endoparasitic on freshwater and marine fish and on various invertebrates; adult segmentation often reduced or lost; A_1 reduced or elongate and multi-articulate; A_2 may end in single massive claw for attachment to host; labrum and labium prolonged into siphon or tube sometimes with some fusion; Md enclosed in buccal siphon, uniramous, gnathobasis styliform, with or without distal teeth, palp present in some forms; Mx_1 ancestrally biramous, modified or reduced in derived forms; Mx_2 subchelate or brachiform (like human arm) for attachment to host; Mxp subchelate or absent, sometimes absent in female only; adult thoracic limbs may be normal swimming appendages in some, in majority variously modified and reduced [= Siphonostoma (Thorell, 1859) = Caligoida (Sars, 1901-1904) = Lernaeopodoida partim (Wilson, 1910) = Lernaedoida partim (Sars, 1901-1904) = Lernaeidae partim (Wilson, 1917)].

Order Cyclopoida (Fig. 2F): Free-living in plankton and benthos, commensal and ectoparasitic; A_1 short with 10-16 articles; buccal cavity open; A_2 uniramous; Md and Mx_1 usually biramous; Md gnathostomous [= Cyclopoida and Notodelphyoida (Sars, 1901-1904) = Lernaeoida partim (Sars, 1901-1904) = Lernaeidae partim (Wilson, 1917)].

Order Poecilostomatoida, Figs. 3G-I): Adults parasitic on mostly marine invertebrates and fishes; adult segmentation often lost with copepodid

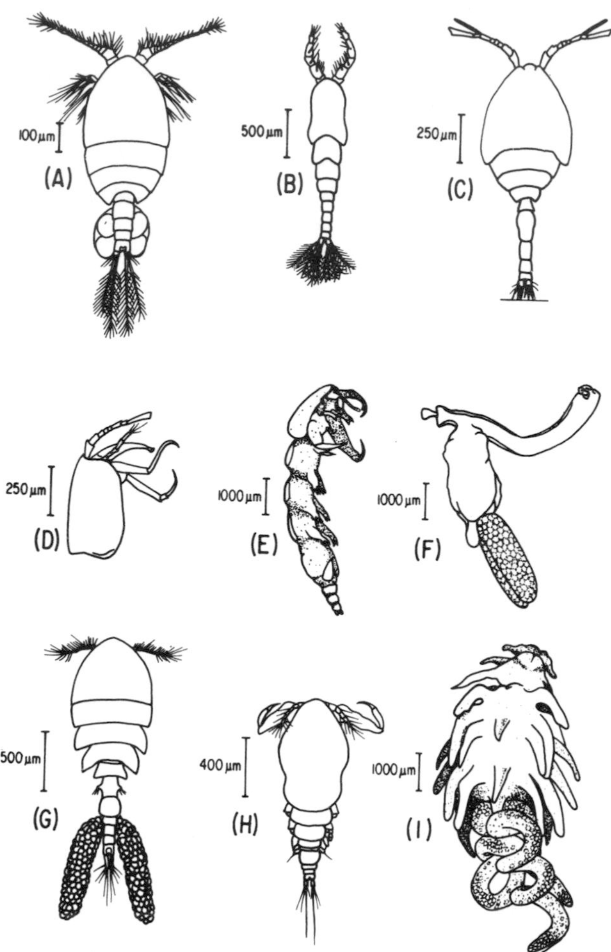

Fig. 3. (A) Epibenthic or semiplanktonic misophrioid (*Misophria pallida*) (after Sars, 1901–1904). (B) Planktonic (dispersing) stage of monstrilloid life cycle (*Monstrilla anglica*) (after Wilson, 1932). (C) Siphonostomatoid parasite probably of invertebrates (Pontoeciella abyssicola) (after Boxshall, 1979). (D) Same as Fig. 3C, lateral view of head. (E) Siphonostomatoid parasite of fish (*Nemesis lamna vermi*) (after Kabata, 1979). (F) Siphonostomatoid parasite of fish (*Clavella adunca* f. *deliciosa*), lateral view (after Kabata, 1979). (G) Poecilostomatoid parasite of polychaetes (*Pseudanthessius ferox*) (after Humes and Ho, 1967). (H) Poecilostomatoid parasite of fish (*Ergasilus sieboldi*) (after Kabata, 1979). (I) Poecilostomatoid parasite of fish (*Chondracanthus neali*) (after Kabata, 1979).

metamorphosis; A_1 often insignificant in size; buccal cavity slit-like; A_2 often ends in many small claws for attaching to host; Md with falcate (falcatus = sickle-shaped) gnathobasis, rami missing; Mx_1 much reduced; Mx_2 reduced with denticulate inward pointing claw or slender, armed grasping claws; Mxp subchelate in males, often missing in females; adult thoracic limbs variously modified and reduced [= Poecilostoma (Thorell, 1859) = Lernaeopodoida partim (Wilson, 1910) = Chondracanthidae (Oakley, 1930)].

Boxshall (1979) has added the Order Mormonilloida to Kabata's list (Fig. 2C). Its diagnosis is based on two species in a single genus. They are gnathostomous podopleans resembling the cyclopoid genus *Oithona*. Their A_1 has three or four, long articles; A_2 is biramous. A heart and their fifth thoracic legs are completely absent. Their systematic position is uncertain. Their mixture of cyclopoid and calonoid characteristics may be due to convergence (Boxshall, 1979).

The original copepod probably dwelled on or near the bottom of bodies of water using simple, biramous, postcephalic appendages synchronized to a metachronal beat for locomotion. It was probably not fully planktonic since swimming characteristically requires large locomotory appendages with elaborate setation. Further, swimming requires specialization of the skeletomuscular system. For example, *Calanus,* a swimming copepod *par excellence,* has its swimming legs internally coupled in order that they move in a single, rapid, wide-angular stroke when huge, dorsal retractor muscles are contracted (Perryman, 1961). Adaptations to a fully swimming planktonic life cannot be regarded as ancestral.

Commensurate with a benthic or semibenthic existence, the original copepod probably had a flexible body with clearly separate segments. Tagmosis in the service of swimming, i.e., enlargement and coalescence of anterior body segments to form a rigid, fusiform prosome, had not occurred. This ancestor was probably podoplean if the early embryological occurrence of this feature is instructive. The tagmotic pattern of this ancestor may have superficially resembled that of the contemporary harpacticoid families Chappuisiidae, Darcythompsoniidae, or Phyllognathopodiidae, which are freshwater benthic and moss-dwelling copepods. This ancestor could have been freshwater-dwelling since limnic environments are centers for diversification of many Harpacticoida, Cyclopoida, and Siphonostomatoida. However, divergence of the Calanoida, Misophrioida, most Harpacticoida, the Poecilostomatoida, and many Siphonostomatoida clearly took place in the sea.

The first copepod probably had six adult appendages adapted to feeding: pre-oral A_1 amd A_2, post-oral Md, Mx_1, and Mx_2, and Mxp. This feature is common to all copepods. The A_1 of the hypothetical ancestor may have had 13 articles, a number which can be morphologically divided according to the patterns of homologies given by Claus (1862) and Giesbrecht (1892) to generate the 25 articles (26 in *Ridgewayia*) of a calanoid A_1 or fused to form the A_1 of harpacticoids which have at most 9 articles. The mouth had an open buccal cavity with a well-defined labrum at its anterior edge. The mandibles were biramous and had large, denticulate gnathobases which extended into the space below the labrum. The gnathostomous mandible can be seen in many contemporary copepod families, for example the calanoid Metridiidae, the harpacticoid Canuellidae and Cerviniidae, and the cyclopoid genera *Notodelphys* and *Cyclopina*. The Mx_1 was biramous and the Mx_2 and Mxp were uniramous. This original copepod was a raptorial feeder.

The ancestral copepod may have had a dorsal heart like other crustaceans. A heart is seen in some contemporary Calanoida and in the (podoplean) Misophrioida. It is unlikely that the anatomical complexities of a heart and associated structures could have developed independently twice in copepod evolution, and it may therefore be considered plesiomorphic. Conversely, loss of the heart could easily have occurred more than once. Whatever its origin, the presence of a heart implied large body size and/or a high rate of physical (metabolic) activity.

Finally, this original copepod probably had five pairs of locomotory legs on its metasome. These limbs were probably biramous and each ramus probably had three articles. These characters are widespread among the Copepoda and must be considered ancestral. The last metasomal segment probably also had a pair of limbs, but these were only inarticulate setose plates resembling limb buds seen in naupliar development. This agrees with the generally held conception that the original copepod evolved through neoteny of some pre-existing crustacean (e.g., Lang, 1946).

From this ancestor the Order Harpacticoida arose. Harpacticoids remained for the most part in the ancestral habitat and have repeatedly diverged from the plesiomorphic condition through specific radiations into most of the adaptive zones occupied more fully by the other copepod orders. Harpacticoids are variable in body shape but are podoplean. They have a gnathostomous oral morphology and they are raptorial feeders (e.g., Marcotte, 1977). However, they lack a heart. One can look to the harpacticoids to understand both the evolutionary robustness and ecological plasticity of the plesiomorphic condition.

Probably convergence has led to resemblances between harpacticoids and other copepod orders. For example, the sucker-like body of the epiphytic family Peltididae is like that of epizoic caligid siphonostomatoids. The buccal cavity and oral appendages of the harpacticoid *Tisbe* (labrum, Md, Mx_2) is reminiscent of the caligid siphonostomatoid buccal apparatus. The fusiform body and the structure of oral and thoracic appendages in the harpacticoid families Canuellidae and Cerviniidae resemble those of the calanoids. Finally, the podoplean tagmosis and structure of thoracic appendages in the semiplanktonic harpacticoid genus *Tisbe* resemble those of planktonic and invertebrate epizoic (parasitic) Cyclopoida. Although these examples are probably due to convergence, they can be instructive when reconstructing copepod phylogenies.

From the ancestral copepod, fully planktonic copepods arose. The first were the (gymnoplean) Calanoida. They differed from the ancestral copepod and from the other copepods in the position of their prosome-urosome articulation. The first calanoids may have been like members of the contemporary calanoid family Centropagidae (Gurney, 1931). Lowndes (1935) has shown that these copepods are raptorial feeders. From this beginning, all other calanoids arose through specializations for planktonic life. For example, the A_1 became long and the caudal rami became fitted with plumose setae, which enhanced flotation (Giesbrecht, 1892; Brehm, 1927). The thoracic legs became large and setose and were internally coupled by specializations of the skeletomusculature, allowing rapid swimming, and the rami of the Md became large and plumose, permitting a slow, gliding motion (Lowndes, 1935; Marshall and Orr, 1955; Perryman, 1961). Plumose maxillary setae permitted filter feeding within the context of raptorial movements (Cannon, 1928; Marshall and Orr, 1955; Conover, 1966; Alcaraz *et al.*, 1978). Predator avoidance was facilitated by sensory adaptations of the first antenna (Strickler and Bal, 1973; Friedman, 1978). Calanoids have retained the plesiomorphic heart.

The Misophrioida, like the Calanoida, are planktonic and have a gnathostomous oral morphology. They differ from the calanoids in tagmotic pattern; the Misophrioida are podoplean. They resemble the other epibenthic and planktonic podopleans, especially the Harpacticoida, except in having a heart. The Misophrioida may have played a pivotal role in a possibly monophyletic, gymnoplean-podoplean divergence. However, too little is known about their biology to even speculate about their phylogenetic significance, and monophyly for these groups cannot be assumed.

Also problematic is the phylogenetic position of the Monstrilloida. The nauplii of this small group ontogenetically become endoparasitic exclusively in marine invertebrates, indicating an early origin for the order. Its exact relationship to the hypothetical ancestor and other copepod orders is unknown because the adult lacks mouth parts. As an adult, it is semiplanktonic with well-developed natatory thoracic legs. Perhaps neoteny of a semiplanktonic ancestor related to the harpacticoids brought its less specialized, benthic-dwelling nauplius into contact with the mantle cavity and hemal system of adult gastropods and polychaetes. Naupliar processes on the oral appendages of developing harpacticoids are often used for attachment to substrates, and for the monstrilloid ancestor these processes would have been pre-adapted for attachment to a potential host. Such heterochronic adaptations are common in other animal groups (Gould, 1977), and their occurrence among the Harpacticoida is well known (Noodt, 1974). A parasitic life habit might have evolved with subsequent reduction in body parts. The "adult" phase of the present monstrilloid life cycle may have evolved secondarily for dispersion. This may have been necessary because the adult hosts of the Monstrilloida were not widely dispersing. The hosts' planktonic larvae were dispersive, but these were not infested with the monstrilloids. Copepod parasites of fish (e.g., see Siphonostromatoida below) do not require this mechanism, since the freely motile host provides adequate dispersal. Building upon its semiplanktonic ancestry, strong natatory legs are present, but oral appendages and a functional gut did not secondarily evolve.

The final great radiation of the Copepoda was into planktonic and parasitic habitats. The three orders involved may have arisen from a common ancestor resembling, in sundry features, the harpacticoid genus *Tisbe*. *Tisbe,* itself, is too apormorphic to be ancestral to these derived forms (Marcotte, 1977). Nevertheless, it does embody many of the morphologies and behaviors which could have been ancestral to the three remaining copepod orders: Cyclopoida, Poecilostomatoida, and Siphonostomatoida.

The first Cyclopoida probably lived swimming among bottom-dwelling weeds. Its swimming legs were well developed and its first antenna was short, though still many articulated. Its second antenna was uniramous, and it had well-developed gnathostomous mouthparts. It was probably a predator eating microcrustaceans (Fryer, 1957; Lane, 1979). These characters are best seen in the contemporary cyclopoid genera *Cyclopina* and *Macrocyclops.* From this ancestor, the other free-living and parasitic cyclopoids arose.

Some contemporary cyclopoids feed on large food items, even fish (Fryer, 1957). After feeding, they undergo a period of rest. From such feeding behavior, Fryer (1957) concluded that one can visualize "a copepod attacking a fairly large organism which would not be killed as a result of the attack, and after completing its meal tending to pass its period of quiescence still attached to its host, which would then be able to provide another meal later." Such might have been the origin of cyclopod parasites (e.g., Notodelphyidae).

The Poecilostomtoida probably branched off from the Cyclopoida relatively recently (Kabata, 1979). *Hemicyclops* best represents the connecting link. At this point in cyclopod evolution, the labrum was wide and short and covered only the anterior edge of the buccal cavity. The gnathobasis of the mandible was falcate and the mandibular palps were reduced. In subsequent poecilostomatoid evolution, hypertrophy and paedomorphosis led to reduction of the fifth thoracic legs to projections resembling limb buds, and adult body segmentation was lost. The second antenna and other head appendages in some cases became adapted to clinging to hosts.

No single phylogenetic scenario can be reasonably framed to account for the origin of all the families in the last copepod order, Siphonostomatoida. Although the members of this order possess somewhat similar oral morphologies, when other morphological characters are considered, the order seems to have had at least a diphyletic origin.

The siphonostomatoid parasites of invertebrates (e.g., Rataniidae and Pontoeciellidae) are cyclopoid-like in adult body segmentation, antennula articulation, and in having typical swimming legs. Their buccal siphon is only loosely connected near its distal tip. The remaining siphonostomatoids are parasites of vertebrates. Their bodies are not cyclopoid-like in the features listed above, and their mouthparts consist of an elaborate buccal cone with a tube formed by the distal part of the labium.

The origin of the siphonostomatoid parasites of invertebrates seems intimately tied to a *Tisbe*-like ancestor of the Cyclopoida. The original siphonostomes' mouth could only be used when stationary and in contact with a firm substrate, at least temporarily, and not when swimming (Kabata, 1979). Such an ancestor could have been epibenthic, epiphytic, or epizoic, behaviors occurring in *Tisbe*. Also in *Tisbe,* the gnathobasis of the mandible is often thin and elongate and only distally dentate. The labrum is long and distally setose or serrate. The gnathobasis of the mandible and the labrum together reach far below and behind the mouth. There is a strong tendency for the labrum-mandible complex to be semicircular in frontal section. Finally, the Mx_2 of *Tisbe* are medially recurved hooks. All these features are at least reminiscent of morphologies and behaviors seen in the siphonostomatoid parasites of invertebrates. Subsequent diversification of these siphonostomatoids involved specializations of cephalic appendages as holdfasts and a general tendency toward a convexo-concave body in longitudinal section promoting hydrodynamic adhesion to flat surfaces on the host's body.

Too little is known about the biology of the siphonostomatoid parasites of vertebrates to permit a meaningful phylogeny for them. Their origin seems obscurely embedded in cyclopoid evolution, and the direction(s) of their subsequent diversification is not well understood. Whatever the outcome of future research into the biology of this group, their placement in the Order Siphonostomatoida should be regarded as tentative at best.

In summary, of the seven orders of Copepoda, the Calanoida, Misophrioida, Harpacticoida, Monstrilloida, Cyclopoida, and Poecilostomatoida, are probably monophyletic assemblages, although their exact relationships to one another are not precisely known. The Siphonostomatoida is probably polyphyletic for the reasons outlined above. Of the seven orders, the Misophrioida, Monstrilloida, and Siphonostomatoida are less well understood. Only with more information about their biology can a convincing phylogentic scheme of copepod evolution be developed.

REFERENCES

Alcaraz, M., Paffenhöfer, G.-A., and Strickler, J. R. (1978). Catching the Algae: A first account of visual observations on filter-feeding Calanoids. *Spec. Symp.—Am. Soc. Limnol. Oceanogr.* **3,** 241-248.

Bowman, T. E. (1971). The case of the nonubiquitous telson and the fraudulent furca. *Crustaceana* **21** (2), 165-175.

Boxshall, G. A. (1979). The planktonic copepods of the northeastern Atlantic Ocean: Harpacticoida, Siphonostomatoida and Mormonilloida. *Bull. Br. Mus. (Nat. Hist.), Zool.* **35** (3), 201-264.

Brady, G. S. (1880). "A Monograph of British Copepoda." Ray Society, London.

Brehm, V. (1927). Copepoda. *In* "Handbuch der Zoologie" (W. Kükenthal, ed.), pp. 435-496. W. de Gruyter, Berlin.

Calman, W. T. (1909). Crustacea. *In* "A Treatise on Zoology" (R. Lankester, ed.), Vol. VII, pp. 1-346. Adam & Black, London.

Cannon, H. G. (1928). On the feeding mechanism of the copepods *Calanus finmarchicus* and *Diaptomus gracilis. J. Exp. Biol.* **6,** 131-144.

Castel, J. (1976). Developpement larvaire et biologie de *Harpacticus* littoralis Sars 1910 (Copepod, Harpacticoide) dans les étangs saumâtres de la région D'Arachon. *Cah. Biol. Mar.* **17,** 195-212.

Claus, C. (1862). Untersuchungen uber die Organisation und Verwandtschaft der Copepoden. *Wurzburg. Naturw. Z.* **3.**

Conover, R. J. (1966). Feeding on large particles by *Calanus hyperboreus* (Kroyer). *In* "Some Contemporary Studies in Marine Science" (H. Barnes, ed.), pp. 187-194. Allen & Unwin, London.

Friedman, M. M. (1978). Comparative morphology and functional significance of copepod receptors and oral structures. *Spec. Symp.—Am. Soc. Limnol. Oceanogr.* **3,** 195-197.

Fryer, G. (1957). The feeding mechanism of some freshwater cyclopoid copepods. *Proc. Zool. Soc. London* **129,** 1-25.

Giesbrecht, W. (1892). Systematik und Faunistik der pelagischen Copepoden des Golfes von Neapel. *Fauna Flora Golfes Neapels* **19.**

Gould, S. J. (1977). "Ontogeny and Phylogeny." Harvard Univ. Press, Cambridge, Massachusetts.

Gurney, R. (1931). "British Freshwater Copepods," Vol. 1. Ray Society, London.

Humes, A. G., and Ho, J. S. (1967). New cyclopoid copepods associated with polychaete annelids in Madagascar. *Bull. Mus. Comp. Zool.* **135** (7).

Kabata, Z. (1979). "Parasitic Copepoda of British Fishes." Ray Society, London.

Lane, P. (1979). Relative roles of vertebrates and invertebrate predation in structuring zooplankton communities. *Nature* (*London*) **280,** 391-393.

Lang, K. (1946). A contribution to the question of the mouth parts of Copepoda. *Ark. Zool.* **38A**(5), 1-24.

Lang, K. (1948a). Copepoda Notodelphyoida from the Swedish West-coast with an outline on the systematics of the copepods. *Ark. Zool.* **40**(14), 1-36.

Lang, K. (1948b). "Monographie der Harpacticiden." Ohlssons Bokryckeri, Lund.

Lowndes, A. G. (1935). The swimming and feeding of certain calanoid copepods. *Proc. Zool. Soc. London* pp. 687-715.

Marcotte, B. M. (1974). Two new harpacticoid copepods from the North Adriatic and a revision of the genus *Paramphiascella*. *J. Linn. Soc. London, Zool.* **55**(1), 65-82.

Marcotte, B. M. (1977). An introduction to the architecture and kinematics of harpacticoid (Copepoda) feeding: *Tisbe furcata* (Baird, 1837). *Mikrofauna Meeresboden*. **61,** 183-196.

Marshall, S. M., and Orr, A. P. (1955). "The Biology of a Marine Copepod." Oliver & Boyd, Edinburgh and London.

Noodt, W. (1974). Anspassung an interstitielle Bedingungen: Ein Faktor in der Evolution Hoherer Taxa der Crustacea? *Fauna. -Oekol. Mitt.* **4,** 445-452.

Oakley, C. L. (1930). The Chondrocanthidae (Crustacea: Copepoda); with a description of five new genera and one new species. *Parasitology* **22,** 182-201.

Perryman, J. C. (1961). The functional morphology of the skeleto-muscular system of the larval and adult stages of the copepod *Calanius* together with an account of the changes undergone by this system during larval development. Ph.D. Thesis, University of London.

Sars, G. O. (1901-1904). "An Account of the Crustacea of Norway," Vols. 1-4. Bergen Museum, Bergen, Norway.

Schminke, H. K. (1976). The ubiquitous telson and the deceptive furca. *Crustaceana* **30**(3), 292-300.

Sewell, R. B. S. (1940). "Copepoda, Harpacticoida," John Murray Exped. Sci. Rep. 7(2). British Museum (Natural History), London.

Strickler, J. R., and Bal, A. K. (1973). Setae of the A_1 of the copepod *Cyclops scutifer* (Sars). Their structure and importance. *Proc. Natl. Acad. Sci. U.S.A.* **70,** 2656-2659.

Thorell, T. (1859). Bidrag till Kannedomen om Krustaceer som lefva i Arter av Slagter Ascidia. *K. Sven. Vetenskaps akad. Handl.* **3**(8).

Wilson, C. B. (1910). The classification of the copepods. *Zool. Anz.* **35,** 609-620.

Wilson, C. B. (1917). North American parasitic copepods belonging to Lernaeidae, with a revision of the entire family. *Proc. U.S. Natl. Mus.* **53,** 1-150.

Wilson, C. B. (1932). The copepods of the Woods Hole region Massachusetts. *Smithson. Bull.* **158.**

Part 3: Cirripedia (Newman)

I. INTRODUCTION

Numerous advances have been made in our knowledge of the Cirripedia since the subclass was reviewed last (Newman *et al.,* 1969). The present treatment extends or reinterprets the broader evolutionary issues covered in that work. One should attempt to treat the subject at several levels; general systematics, the fossil record, the adaptive advantages of morphological change, and ultimately biogeography and ecology. However, because of the nature of the present volume, the former rather than latter aspects will be emphasized.

II. ORIGIN OF THE CIRRIPEDIA

By the time of Darwin's (1851, 1854) incomparable monographs on living members of the subclass, it was widely recognized that barnacles were crustaceans. In comparing the grosser morphology of a simple stalked barnacle with that of the somewhat aberrant shrimp, *Lucifer,* Darwin (1851) provided insights into how the stalked barnacles (Lepadomorpha), acknowledged ancestors of the sessile barnacles (Brachylepadomorpha, Verrucomorpha, and Balanomorpha) evolved from a free-living stock. The ancestors must have had a normal component of five sensory and feeding cephalic appendages plus compound eyes, six natatory thoracic limbs, and an appendageless abdomen of at least three segments, the last of which bore furcal rami. Despite the notable facies similarities between the cyprid larva of cirripeds and the Ostracoda, an ancestor with a bivalved carapace was not envisaged at that time.

The nature of the urcirriped remained obscure until Lacaze-Duthiers (1880) described the parasite *Laura*, recognized its affinities with the Cirripedia, and established the new order Ascothoracica, the most generalized members of which have five abdominal segments. Other works (cf. Beklemischev, 1952; Dahl, 1956, 1963) noted similarities between the cirripeds, copepods, mystacocarids, and branchiurans, and proposed that they be united under a taxon: the Copepodoidea or Maxillopoda. The latter name has come to prevail among workers recognizing the affinities of its members, probably because it was proposed in the English language. Newman *et al.* (1969) emphasized the 5-6-5 (head-thorax-abdomen) maxillopodan affinities of the Cirripedia.

Beurlen (1930) and Gurney (1942) recognized that the copepods had a reduced body plan. The latter noted similarities with the protozoea larva and suggested that the group evolved by paedomorphosis from a malacostracan ancestor (cf. Newman, 1982). The ascothoracicans, despite their specializations for parasitism, are in many ways more generalized than copepods, and it was noted that clues to maxillopodan origins might better be sought in them (Newman, 1974). However, similarities between appendages presumed to unite the maxillopodan taxa may be plesiomorphies reacquired through neoteny or progenesis, and therefore it is possible that the taxon is not wholly monophyletic, as noted under the discussion of the Maxillopoda (Hessler, Part I of this chapter). Nonetheless, the Maxillopoda can be defined by a number of characteristics, many of which appear to be apomorphies.

(1) First antennae uniramous, well developed; second maxillae normal or reduced but not vestigial (except where modified or lost in parasitic forms). (The first antenna separates the Maxillopoda from the Malacostraca and the

Remipedia, but it is apparently plesiomorphic. The second maxillae on the other hand, while plesiomorphic as compared to the branchiopods, are apparently apomorphic as compared to the cephalocarids.)

(2) Trunk (thorax and abdomen) never more than 11 segments, including telson. (An apomorphy that distinguishes the maxillopodans from all free-living crustaceans except the Cladocera among the Branchiopoda.)

(3) Thoracic limbs never more than six pairs (see below); biramous and without gnathobases or epipods. (The low number is highly apomorphic as is the lack of epipods and gnathobases.)

(4) First thoracic segment free or fused to the head; limbs normal (plesiomorphic) or modified as maxillipeds (apomorphic). [The plesiomorphic condition of the first thoracic limb in some maxillopodans (cirripeds, ascothoracicans) renders the concept underlying the name Maxillopoda unfortunate.]

(5) Abdomen no more than five segments (see below); genital apertures seven segments or less from the cephalon. [The short abdomen is an apomorphy. Genital apertures of both sexes opening on the same segment is apparently the plesiomorphic condition (copepods, branchiurans, mystacocarids, and ostracodes), whereas their separation is derived (cirripeds and ascothoracicans)]. The abdomen distinguishes the Maxillopoda from all other crustaceans, as does the genital being the seventh trunk segment.

When the first abdominal segment supports genital apertures, as in copepods and cirripeds, limbs or limb rudiments may be present and variously modified for reproductive purposes. The so-called "seventh trunk" limbs of certain harpacticoid copepods are apparently very plesiomorphic in this regard (Marcotte, Part 2 of this chapter). But such limbs are not inconsistent with the maxillopodan plan (cf. Hessler, Part I of this chapter, and Grygier, 1982b) and by convention are those of the first abdominal segment.

The sixth trunk segment of mystacocarids has been variously interpreted (cf. Hessler, Part 1 of this chapter). Since it is limbless, its homology with the sixth thoracic segment can be doubted, and those who take this stand note that it is identical in appearance to the segment that follows. It should be noted, however, that it is also identical to the preceding one (including having toothed grooves) except for the uniramous, vestigial limbs found there. Considering the other remarkable similarities with cirripeds and copepods and the fact that vestigiality commonly precedes from the rear of a tagma forward, the most parsimonious explanation for the missing limbs in mystacocarids is that they have been lost. The cumaceans, bathynellaceans, and thermosbenaceans, which also have reduced or lost eyes as well as posterior trunk limbs, illustrate comparable changes in other groups inhabiting fine sediments.

The sum of these adult characters, many of which appear synapomorphic, distinguishes the Maxillopoda from the other crustacean classes. It is instructive that larval characters serve to do so as well. The naupliar appendages of Branchiopoda and Cephalocarida (and Malacostraca) are distinct from each other (Sanders, 1963), and certainly would not be confused with those of cirripeds, ascothoracicans, copepods, or mystacocarids, which in turn are similar among themselves. Therefore, the Maxillopoda appears to be, for the most part, a natural grouping and is recognized here (cf. Grygier, 1982a,b). However, it is not the purpose of this review to champion the Maxillopoda but rather to draw attention to the inescapable conclusion that the free-living ancestors of the cirripeds would have had a body plan very similar to that of the Ascothoracica; in fact, the urascothoracican plan is one from which all the maxillopodan taxa could have been derived. Furthermore, if, in addition to the Ascothoracica, the Rhizocephala eventually are removed from the Cirripedia, the need to recognize sister groups within the Maxillopoda will certainly be required.

A. Apoda

The status of the Apoda, represented by the parasite *Proteolepas bivincta* Darwin, has been in question (Krüger, 1940), and while it was retained *incertae sedis* in the Cirripedia, it seemed more likely to be a copepod or an epicaridean (Newman *et al.*, 1969). It has since been demonstrated to be an epicaridean (Bocquet-Védrine, 1972a, 1979) and therefore need not concern us further here.

B. Ascothoracica

The most primitive living crustaceans attributed to the Cirripedia are parasites of echinoderms and corals included in the Ascothoracica. While it is generally agreed that the Ascothoracica is closer to the Cirripedia than to any other maxillopodan taxon (Newman, 1974), some authors (Wagin, 1976 and earlier; Krüger, 1940; Grygier, 1982b) have argued for separate, coordinate status (Fig. 1). The Ascothoracica differ from the Cirripedia *sensu lato* in having nauplii lacking frontolateral horns; a relatively anamorphic rather than highly metamorphic development; an ascothoracid larva capable of feeding at all stages rather than a single cyprid stage lacking a complete gut and functional mouthparts; wholly prehensile first antennae rather than ones also provided with cement glands; and natatory thoracic limbs that show no indication of ever having been used as cirri. Furthermore, it has recently been shown that the ascothoracican sperm, the most generalized known for the Crustacea, is distinct from that of the remainder of the cirripeds (Grygier,

1982a). Therefore, it would seem reasonable to separate the Ascothoracica from the Cirripedia, provided that they and related coordinate taxa are grouped under the Maxillopoda.

C. Cirripedia *Sensu Lato*

If the Ascothoracica is not included with the cirripeds, we are left with the Thoracica, Acrothoracica, and Rhizocephala (Fig. 1). The last will be taken up first because its affinities can be questioned.

D. Rhizocephala

The Rhizocephala is represented by highly specialized parasites of primarily decapod crustaceans. Members are recognizable as cirripeds by their larval stages (nauplii with frontolateral horns and/or a typical cyprid) and comparable spermatozoa (cf. Baccetti, 1979). The cyprid attaches to the host and, in presumed primitive forms (Akentrogonida), undergoes metamorphosis into an external reproductive sac nourished by roots extending into the host at the site of infection (Fig. 2C). In more advanced forms (Kentrogonida), the attached cyprid injects its cellular contents into the host through a hypodermic device, the kentrogon, and the internal phase (interna) subsequently develops roots and eventually produces a reproductive sac (externa) that erupts on the surface (Fig. 2E). It should be noted that the kentrogonid system enables the reproductive body of the parasite to be located at a site remote from the site of infection, typically under the abdomen of the host where it mimics the host's egg mass.

It was first observed that the kentrogon passed inside one of the cyprid's first antennae and entered the host where the antennae attach at the base of a seta (Delage, 1884). Since two pedunculate barnacles, *Anelasma* and *Rhizolepas,* are known to have atrophied feeding appendages and to gain nourishment by sending roots from the peduncles into their hosts (a shark and polychaete worm, respectively), it seemed reasonable to infer that the Rhizocephala had evolved from a lepadomorph stock (Newman *et al.*, 1969). The attachment area of the peduncle centers on the place where the cyprid antennae become permanently attached (Fig. 2A). Therefore, a pedunculate barnacle with roots could have given rise to the akentrogonid (Fig. 2C), which in turn would have given rise to the kentrogonid level of organization.

It has recently been shown (Ritchie and Høeg, 1981), in the kentrogonid *Lernaeodiscus,* that the cyprid settles on the smooth surface of the gills rather than at the base of a seta of the host crab *Petrolisthes.* The dart forms at the position of the cyprid's mouth field, and rather than penetrating the host by

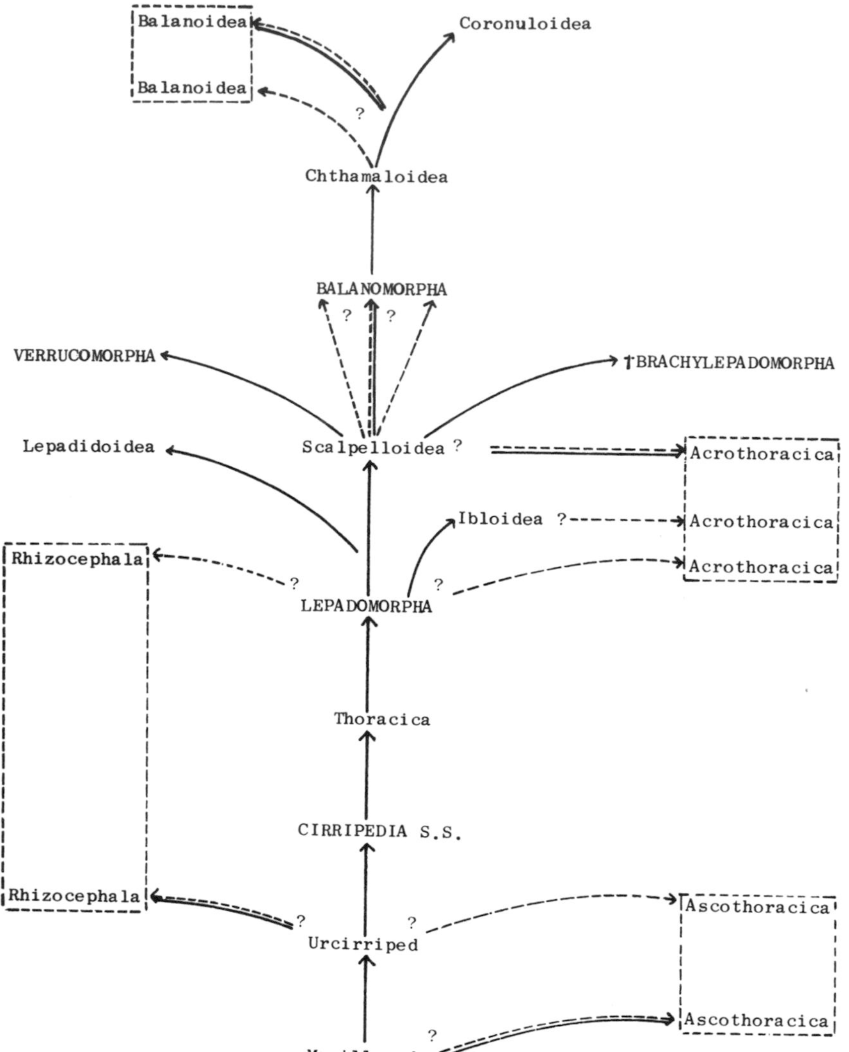

Fig. 1. Phylogeny of the Cirripedia: The first two orders, the Ascothoracica and Rhizocephala, are represented by wholly parasitic forms. Gross morphology of the ascothoracicans is typically maxillopodan, but prehensile first and reduced second antennae, a bivalved carapace, and the position of the genital apertures places them closer to the Cirripedia than to any other maxillopodan group. Whether or not they should be included in the cirripeds has long been a point of contention, but in the final analysis they will likely become firmly established as a coordinate taxon within the Maxillopoda (cf. Grygier, 1982a). The Rhizocephala, on the other hand, have generally been considered derivable from the lepadomorph cirripeds. However, some evidence for an urcirriped origin has recently come to light (cf. text and Fig. 2) that may eventually require removing the Rhizocephala from the Cirripedia. The remaining orders, the Thoracica and Acrothoracica, constitute the Cirripedia *s.s.* A lepadomorph origin for the acrothoracicans has been variously interpreted, and while a scalpelloid origin is favored in the present report (cf. text and Fig. 5), the alternatives cannot be

first passing through one of the first antennae of the cyprid (Fig. 2E2), it is forced in directly from the position where it was formed (Fig. 2E1). An intermediate between these two orientations is seen in *Gemmosaccus* (cf. Viellet, 1964). It should be noted that while rhizocephalan nauplii lack gnathobases and are unable to feed, they retain a "v"-shaped labrum, reminiscent of the oral cone of ascothoracicans, and it is likely that the kentrogen forms in relationship to it in *Lernaeodiscus* (L. E. Ritchie, personal communication). This suggests an alternative hypothesis concerning the origin of the Rhizocephala. The ancestral stock may have been an ectoparasite that bit its host (Fig. 2B), as do the generalized ascothoracicans such as *Synagoga* and the siphonostome copepods, but that developed absorptive roots from the mouth field comparable to those of copepods such as "*Rhizorhina*" (Calman, 1909).

If further investigations substantiate this inference, the model offered by *Rhizolepas* and *Anelasma* for a lepadomorph ancestry of the Rhizocephala will have been falsified; that is, the ur-rhizocephalan was a free-swimming epizoite that formed a parasitic mode of feeding by its mouth field. In this model, the kentrogonid ancestor would have developed the hypodermic dart or kentrogon from the oral cone. Injection through the antennae (Fig. 2E2) would be a specialization in higher kentorogonid rhizocephalans whose larvae settle at the base of setae surrounded by a hard exoskeleton rather than on soft structures such as the gills of the host.

Such an origin would preclude a lepadomorphan ancestor, since it is highly improbable that a cirral-feeding form could become a biting parasite. A prethoracican capable of temporary attachment by prehensile first antennae and facultative feeding, including ectoparasitism, seems most likely. The Rhizocephala then would have to have been derived from some prethoracican stock that has become extinct (Fig. 2B). It would have led to the akentrogonid Rhizocephala on one hand and the thoracican cirripeds on the other. The Rhizocephala, considering the similarities of their spermatozoa and their larvae to those of ordinary cirripeds, are closer to the stem of the cirripeds than are the ascothoracicans (Fig. 1).

---

discounted. The Thoracica forms the core of the Cirripedia, with the more highly evolved scalpelloids among the Lepadomorpha having given rise to the three sessile orders, the Brachylepadomorpha, Verrucomorpha, and Balanomorpha (cf. text and Fig. 3). There is a possibility that the superfamilies constituting the Balanomorpha are polyphyletic, but the evidence is not compelling (Newman and Ross, 1976). If monophyletic, the problem of whether the balanoids stem from the chthamaloids or coronuloids remains to be resolved. In any event, the balanids among the Balanoidea represent the most advanced of the benthic forms with regard to the structure of their wall, method of attachment to the substratum, and specializations in the opercular plates (cf. text and Fig. 4). (Solid arrows indicate reasonable certainty and dashed arrows where possible alternative interpretations exist. A dashed arrow accompanied by a parallel line indicates, in the view of the author, the most likely alternative.)

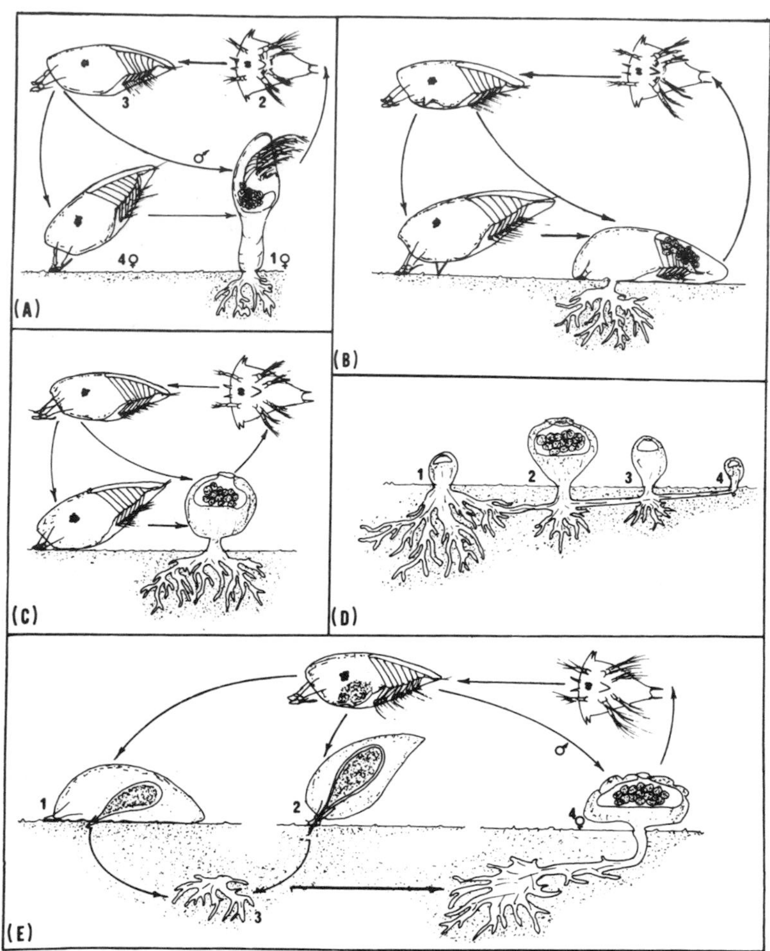

Fig. 2. Origin of the Rhizocephala and evolution of the Kentrogonida. (A) Thoracican origin: The lepadomorphs *Analasma* and *Rhizolepas* have given up setose feeding for parasitism by extending roots into their hosts from their peduncles and thereby provide a plausible model for the origin of the akentrogonid Rhizocephala (Newman *et al.,* 1969). (B) Urcirriped origin: A free-living clade that gave rise to the cirripeds could also have included an ectocommensal that became a biting parasite and that subsequently developed roots from its mouth field, as have such copepod parasites as "*Rhizorhina*" (cf. Calman, 1909). When all vestiges of crustacean structure had been lost in the reproductive body, the akentrogonid level of organization would have been achieved (C). (D) Some akentrogonids, such as *Thompsonia,* produce additional reproductive bodies by budding, apparently in connection with the ability of regenerating lost or damaged ones (1). The result, in addition to increasing reproductive output, allows reproductive bodies to develop at a distance from the site of infection (1-4) and, concomitantly, at sites where they may be less likely damaged. This advance would have been an essential step in the evolution of the kentrogonids where the externa appears remote from the site of infection, beneath the abdomen of the host decapod crustacean, where it is cared for as the host's brood rather than, as a foreign object, being removed by the host's cleaning appen-

E. Thoracica

1. LEPADOMORPHA

The pedunculate or stalked barnacles are the most generalized thoracicans, and it is from them that the sessile forms (Brachylepadomorpha, Verrucomorpha, and Balanomorpha) can be derived (Figs. 1 and 3). The Lepadomorpha also has an extensive fossil record, dating back at the time of the last review to the Silurian (Wills, 1963). *Cyprilepas* Wills (1962), occurring on eurypterids, apparently had a bivalved, uncalcified carapace forming the capitulum, rather than a penta-valved carapace, as did the next oldest genus, *Praelepas* Chernyshev (1930), from the Carboniferous. A bivalved carapace occurs in ascothoracicans and in the cyprid larval stage of all contemporary cirripeds, and consequently in the inferred free-living ancestors of the cirripeds. Therefore, a bivalved carapace in the earliest known thoracican cirriped added significantly to understanding their origins.

Since the discovery of *Cyprilepas*, a larger, more impressive form, *Priscansermarinus* Collins & Rudkin (1981) from the Cambrian (Burgess Shale), has been identified as a lepadomorph. It too appears to have a bivalved or globular carapace, further indicating an early radiation of such forms, but in the earliest rather than middle Paleozoic. Such evidence lends support to the suggestion that members of the extant genus *Heteralepas s.l.* may never have had a carapace protected by calcareous plates (Foster, 1978), rather than the generally held view that their calcareous plates have been lost (Broch, 1922; Newman *et al.*, 1969).

Evidence that subdivision of the carapace valves had been achieved by the Carboniferous, provided by *Praelepas* from the U.S.S.R., was substantiated by discovery of a *Praelepas*-like form of approximately the same age from the United States (Schram, 1975). But these are all uncalcified and lepadoid-like forms. Therefore, it was particularly interesting when Whyte (1976) described what could be a scalpelloid from the Carboniferous of the

---

dages (Ritchie and Høeg, 1981; L. E. Ritchie, personal communication). (E) The kentrogonid rhizocephalans differ from the akentrogonids in establishing an internal phase (3, interna) which subsequently produces a reproductive body (4, externa) remote from the site of infection. The kentrogon evolved in connection with establishment of the interna, and in *Sacculina* (2) it penetrates the host through one of the cyprid larva's first antennae, where it attaches at the base of a seta of the host (Delage, 1884). However, in *Lernaeodiscus* (1) the kentrogon forms in the position of the mouth field, on the ventral side of the closely attached cyprid, and directly penetrates the gill membrane of the host (Ritchie and Høeg, 1981). The latter situation appears to be the plesiomorphic condition, since the "v"-shaped oral cone (labrum) is seen in the nauplii of such forms, and it is likely that the kentrogon evolved in connection with the oral cone. Therefore, the system seen in *Sacculina* would have been derived. The development of the kentrogon from the oral cone favors the hypothetical uncirriped origin of the Rhizocephala (B).

United Kingdom, which, as will be discussed below, may have a significant bearing on our understanding of the affinities of the acrothoracicans.

The Lepadomorpha has undergone some rather extensive revisions, particularly at the generic level, primarily among the Scalpellidae (Newman and Ross, 1971; Zevina, 1978a,b, 1980, 1981). These changes, while enhancing our appreciation of a tremendous diversity, have had little impact on our understanding of lepadomorphan evolution in general. On the other hand, the discovery of the first barnacle associated with an abyssal hydrothermal spring, the scalpelloid *Neolepas* (Newman, 1979), has aided substantially in bridging the gap between the Triassic-Cretaceous genera *Eolepas* and *Archaeolepas* (scalpelloids without latera in the armature of the capitulum) and the scalpelloid *Scillaelepas s.l.* (with three pairs of latera), which in turn is inferred to have given rise to the Balanomorpha (Darwin, 1851; Aurivillius, 1894; Newman *et al.,* 1969; Anderson, 1980b) (Fig. 3).

Another advance bearing on the evolution of the scalpelloids was the discovery of subrostral complemental males in *Scillaelepas* by Bocquet-Védrine (1971). As noted by Darwin (1851, 1854), lepadomorphs are normally hermaphrodites; however, in some species the hermaphrodite is accompanied by a so-called "complemental" male, with reduced armament but still capable of feeding, while in others, the male is further reduced and dwarfed, and the normally hermaphroditic individual has become wholly female (cf. Ghiselin, 1974). In all cases except in *Scillaelepas,* the males were situated suprarostrally, in the aperture between the scuta or in specialized pockets formed in the interior of the scuta. A subrostral complemental male, situated between the rostrum and the subrostrum in *Scillaelepas*, was unique among scalpelloids. Its significance became apparent when, in a species of *Scillaelepas* that is plesiomorphic in lacking subrostra, the male was found situated even further away from the aperture, between the subrostral peduncular scales. This is the most primitive form of association between complemental males and hermaphrodites known in scalpelloids (Newman, 1980).

Scillaelepas is an old genus, dating back to Upper Jurassic, and numerous species are known from the Cretaceous of Europe and Australia. It subsequently declined and disappeared from shallow water, but it is presently represented by a few species in deep water (340-3000 m) in the North Atlantic, southwestern Indian Ocean, and south of New Zealand. The acquisition of complemental males appears to be correlated with this decline, which apparently included a reduction in population densities to the point that established hermaphrodites were commonly too far apart to ensure cross-fertilization. The transfer of males from the subrostral situation to between the scuta, as seen in *Calantica,* apparently paved the way for the more

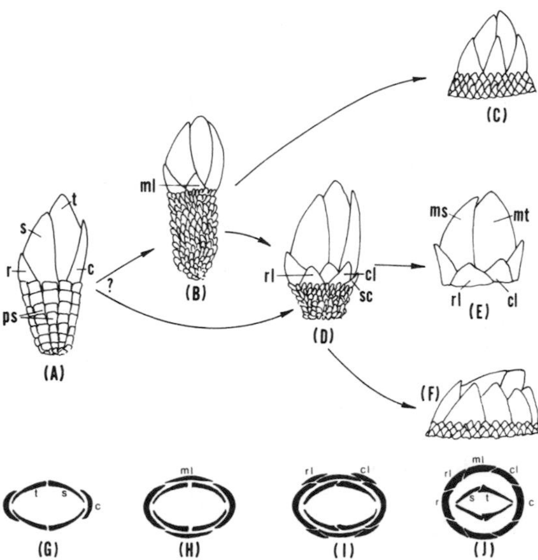

Fig. 3. Radiation of the scalpelloids leading to the evolution of the three suborders of sessile barnacles: Brachylepadomorpha, Verrucomorpha, and Balanomorpha. Early radiation of the scalpelloids included an increase in the number of capitular plates beyond that seen in the extinct genera, *Archaeolepas* (A) and *Eolepas*. *Neolepas* (B), in which a pair of median latera has been added, illustrates the grade of construction seen in the earliest of the sessile barnacles contained in the Brachylepadomorpha (C, *Brachylepas*). The next higher level, seen in the organization of *Scillaelepas* (D) includes three pairs of latera and forms a plausible model from which the Verrucomorpha (E, *Proverruca* or *Eoverruca*) and the Balanomorpha (F, *Catophragmus*) could have been derived. There is no evidence that forms like *Eoverruca* ever had median latera, but since the rostro- and carinolatera of the fixed side have been lost, and since those of the movable side also have been lost in *Verruca*, it is plausible that the median latera were lost before *Proverruca* and *Eoverruca* appeared.

Since a "*Neolepas*" stage occurs in the ontogeny of *Scillaelepas,* it appears most likely that its level of capitular organization was achieved by the addition of the rostro- and carinolatera to a *Neolepas*-like form (B to D), rather than from a *Archaeolepas*-like form by the incorporation of rosto-, medio-, and carinopeduncular scales into the capitulum (A to D).

The four-plated capitular organization (excluding terga and scuta) of *Neolepas* (H) is passed through not only in the ontogeny of higher related scalpelloids (Newman, 1979), but in balanomorphs as well. During the ontogeny of the chthamaloid *Chionelasmus* (a genus which has six plates making up the adult wall), an eight-plated stage is passed through in which the plates are arranged as in *Scillaelepas* (I). But this arrangement differs from the definitive arrangement in *Catophragmus* (J) in which the carinolatera are overlapped by the median latera. This suggests that *Chionelasmus* has not descended from *Catophragmus s.l.,* as has been supposed. Abbreviations: c, carina; cl, carinolatus; ml, median latus; ms, movable scutum; mt, movable tergum; ps, peduncular scales; r, rostrum; rl, rostrolatus; s, scutum; sc, subcarina; t, tergum.

advanced and presently diverse deep-sea scalpelloids (*Arcoscalpellum s.l.*), which are commonly females accompanied by dwarf, nonfeeding males.

It is important to realize that the late Mesozoic forms like *Neolepas, Scillaelepas*, and *Stramentum* (cf. Hattin, 1977), represent a dramatic increase in skeletal armament, and that this was a time marked by an equally dramatic increase in predation pressure, particularly in shell-crushing forms such as the true crabs (Stanley, 1974; Vermeij, 1977). It is also the time when the sessile barnacles first appeared; the very intermediate Brachylepadomorpha (Upper Jurassic to Miocene), the Verrucomorpha [Middle and Upper Cretaceous (three genera) to Holocene (one genus)], and Balanomorpha (Upper Cretaceous (1 genus) to Holocene (three superfamilies)] (Fig. 3).

When one considers how relatively easy it is to crush a scalpelloid capitulum or tear the whole animal off a rock with a pair of pliers, as compared to the difficulty encountered using the same instrument on a sessile barnacle, the impact of true crabs, which first appeared in the Upper Cretaceous, can readily be appreciated. It is interesting, therefore, that according to ease of removal among the sessile barnacles, the tenure of the Brachylepadomorpha was short and extinction occurred in the Miocene. Generic diversity of Verrucomorpha was highest in the Cretaceous and, as in the scalpelloids, most species today are in deep water (where true crabs are rare). The Balanomorpha, appearing in the Upper Cretaceous, are presently the dominant forms in shallow water, especially at low and mid latitudes (Spivey, 1981; Dayton *et al.*, 1982).

2. BRACHYLEPADOMORPHA AND VERRUCOMORPHA

The Brachylepadomorpha, little more than lepadomorphs with very short peduncles, can readily be derived from a *Neolepas*-like ancestor (Fig. 3B–C). The Verrucomorpha, or wart barnacles, are truly sessile. They are not only distinguishable but remarkable, compared to all other barnacles, in having the opercular valves of one side forming part of the box-like wall and the other pair forming the hinged lid. The evolutionary sequence required in achieving this arrangement was inferred by Newman *et al.* (1969). However, a plausible explanation for the adaptive advantage conferred by the asymmetry has come from studies on feeding by Anderson (1980a), who suggested that it allowed the cirri of *Verruca* (*Verruca*) to sweep over the substratum rather than through the water above the animal. One can only agree with the plausibility of this suggestion, and if it bears out, the nearly vertical rather than horizontal arrangement in *Verruca* (*Altiverruca*) could be the result of a reversion to feeding above the animal rather than an intermediate stage between the vertical and horizontal orientation suggested in the Newman *et al.* (1969) model. Actually the fossil record may lend some support to this view because, while all subgenera of *Verruca* appear in the Eocene (Newman and Schram, 1982), the oldest fossil inferred to be a *Verruca* (M. Cretaceous) appears to be of the horizontal type (Schram and Newman, 1980).

3. BALANOMORPHA

The acorn barnacles encompass the greatest diversity of the sessile barnacles today, and the order was recently revised (Newman and Ross, 1976, 1977). Three superfamilies were proposed: the Chthamaloidea, Coronuloidea, and Balanoidea. However, there have been suspicions that these superfamilies may be polyphyletic, stemming from different scalpelloid stocks, as did the Brachylepadomorpha and Verrucomorpha (Figs. 1 and 3). So far, such suspicions have not been confirmed, and the superfamilies are treated as monophyletic here.

The coronuloids include such commonly known benthic genera as *Tetraclita s.l.* on one hand and the sessile whale and turtle barnacles on the other. The Coronuloidea appears to have arisen from the chthamaloids and to have given rise to the balanoids during ascendency. The fossil record supports the order of increasing complexity with time: Chthamaloidea—Cretaceous; Coronuloidea—Paleocene; Balanoidea—Eocene (Zullo and Baum, 1979; Buckeridge, 1980; Newman and Stanley, 1981).

The chthamaloids for the most part have not faired well during the Tertiary, except in the higher reaches of the intertidal zone where species of *Chthamalus s.l.* (approximately 35 of the 51 chthamaloids species known) have found some refuge from competition and predation (Stanley and Newman, 1980; Paine, 1981). It has become apparent, primarily by work with enzyme electrophoresis, that species richness in high intertidal forms is likely very much greater than previously known (Southward, 1976; Hedgecock, 1979; Dando and Southward, 1980). Nonetheless, both the chthamaloids and the benthic coronuloids are presently represented by relatively few intertidal and deep-sea species, while shallow waters are inhabited by a relatively high diversity of benthic balanoids. Stanley and Newman (1980) and Newman and Stanley (1981) suggest that the balanoids are competitively superior and have replaced the chthamaloids in shallow water and also in the low intertidal in most regions.

The evolutionary advances made by coronuloids and balanoids over chthamaloids have been in the feeding structures (conversion of a bullate to a flat labrum and the third cirri into third maxillipeds) and, in the more advanced benthic forms, improvements in the shell (straightening out of the articulation between the tergum and scutum and development of a tubiferous wall). The modifications of the trophic structures are adaptations to more efficient and/or specialized feeding (Anderson, 1978), whereas the changes in the structure of the opercular parts and wall are, for the most part, adaptations for improved armament.

Darwin (1854) noted that the opercular valves of chthamaloids generally were deeply articulated (dovetailed) together, whereas those of balanoids were not. In the latter, the terga can advance apically in relationship to the scuta, thus explaining how the beaks of the terga, in those species having them (e.g., *Tetraclita, Semibalanus cariosus, Balanus* spp., and *Megabalanus s.l.*), come to extend well beyond the apexes of the scuta. Beaks apparently have developed convergently in these distantly related forms. The opercular valves in such forms can be forcefully ground around inside the orifice, with the beaks sweeping the orificial region in the process (Fig. 4C). These modifications evidently have evolved in response to predators.

It has been inferred that the advances in the wall, including chitinous layers as well as longitudinal tubes separating inner and outer layers, were also antipredation devices (Newman and Ross, 1971). But since tubiferous walls can grow much faster than a solid wall of comparable strength, a significant aspect also must be the competitive advantage of rapid growth to a relatively large size (Stanley and Newman, 1980). It would also permit species requiring large size as a refuge from predation to achieve it rapidly, but tubiferous walls are found throughout the size spectrum in both coronuloids and balanoids, so this is not the only advantage. It is also noteworthy that a tubiferous wall has the distinct disadvantage of being more susceptable to desiccation and vulnerable to relatively rapid deterioration through erosion and boring sponges, etc. Some intertidal coronuloids and balanoids secondarily fill the tubes with more resistant material (e.g., *Tetraclita, Tesseropora, Balanus glandula* and *Semibalanus balanoides*), beginning shortly after their initial period of rapid growth, and many balanoids seal the tubes at intervals by transverse septa.

The tubiferous walls of coronuloids and balanoids have apparently evolved independently from different solid walled ancestors. In addition, balanids have developed a tubiferous calcareous basis, and in *Megabalanus s.l.*, even the radii are tubiferous. An important aspect of such walls is not only their capacity for rapid growth but their rigid construction (Newman *et al.*, 1967). Upon death most barnacle shells, given an opportunity to decompose before burial, disarticulate. Many balanids do not, and this is because the plates of the wall and calcareous basis are intricately locked together by denticulate sutures. The interlocking is precise and the geometry such that there is virtually no play between sutures, even after an extended treatment with sodium hypochlorite to remove the organic matter. The parts can only be disassembled after breaking the shell. This is quite a different system than that discussed by Bourget (1980) for a lower balanoid. Most

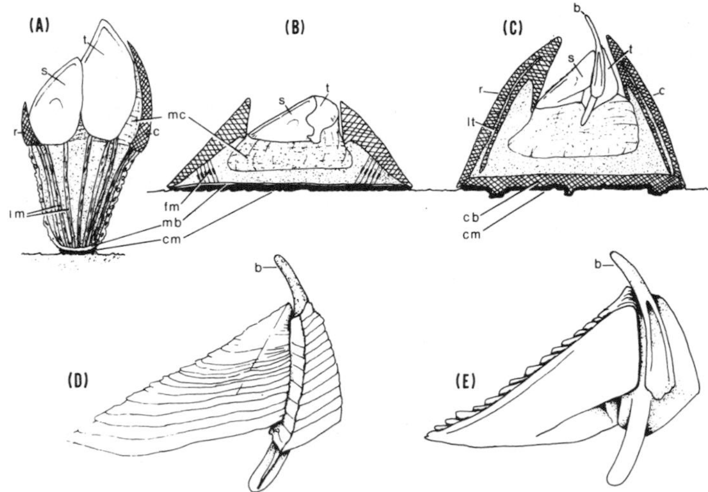

Fig. 4. Evolution of attachment and opercular valves in benthic balanomorphs: (A) the lepadomorph *Pollicipes;* (B) an archeobalanid like *Semibalanus balanoides;* (C) a higher balanid like *Megabalanus;* and (D–E) the opercular valves of the megabalanine *Austromegabalanus psittacus.* The lepadomorph exoskeleton is divided into the capitulum supporting the principal plates and the peduncle, which is in turn cemented to the substratum (A). The peduncle, provided with longitudinal, circular, and oblique muscles, is capable of flecture and changes in length (Darwin, 1851). Consequently a lepadomorph must expend energy to maintain a posture or to remain attached when under disturbance or attack. Balanomorphs differ from lepadomorphs in having dispensed with the peduncle, bringing the capitulum into contact with the substratum. However, primitive forms such as *S. balanoides* (B) apparently still depend upon a longitudinal fibromuscular system for fixation throughout life (Bourget and Crisp, 1975; Bourget, 1980). Some chthamaloids overcome this requirement when mature by forming a permanent calcareous attachment, but at the expense of further growth (Newman, 1961). Higher balanids such as *Megabalanus* (C) form an intricate interlocking mechanism between the calcareous basis and the wall plates (Newman *et al.,* 1967), which allows further growth while the animal is rigidly attached to the substratum. Both of these systems apparently alleviate the need for fibromuscular elements by remaining attached.

The transition from the scalpelloid to the balanomorph (chthamaloid) form included formation of an articulation between the scutum and tergum of each side (Darwin, 1854), while the transition from chthamaloid to balanoid consisted of a 90° rotation of the articulation, changing it from a dovetail to a tongue and groove system (B to C). The impetus for this change is unknown, but the result allows the terga to move apically relative to the scuta and thus to ultimately elevate a beak of relatively resistant material which becomes exposed by erosion (Darwin, 1854; C–E). Beaked terga have evolved independently in coronuloids and different balanoids, apparently in response to predation. Abbreviations: b, beak; c, carina; cb, calcareous basis; cm, cement; fm, fibromuscular tissue; lt, longitudinal tube; mb, membranous basis; mc, mantle cavity; r, rostrum; s, scutum; t, tergum.

lower balanoids, like most coronuloids and chthamaloids, do not form a calcareous basis, and those that do are not intricately locked into it.

Lower balanomorphs, especially those without a calcareous basis of any sort, may grow in the manner suggested by Bourget and Crisp (1975). The shell wall of such forms is apparently held to the substratum by internal fibers provided with contractile elements. Relaxation of the attachment fibers, concomitant with an increase in blood hydrostatic pressure, is inferred to precede an extension of the secretory hypodermis out under the edge of the shell where the calcareous growth increments are to be added. Such a system is compatible with the method of attachment and growth in the lepadomorph barnacles, and it is of course from the lepadomorphs that the balanomorphs evolved (cf. Fig. 4A–B). Lepadomorphs are cemented to the substratum by a peduncle supporting the capitulum that can be shortened by muscles and lengthened by circulatory hydrostatic pressure (Burnett, 1975). This is essentially the same system Bourget and Crisp (1975) described for the balanoid *Semibalanus*. However, connective contractile tissue antagonistic to hydrostatic pressure cannot be the system employed by the higher balanids, at least in mature individuals (Fig. 4C). They have developed an interlocking mechanism that does not depend on contractile elements to hold the shell to the calcareous basis, which in turn is cemented to the substratum; but how such an intricately structured shell grows has not been analyzed.

Darwin (1854) established the homologies between the principal plates of the lepadomorphs, verrucomorphs, and balanomorphs, and he clearly envisaged in a general way how the lepadomorphs gave rise to balanomorphs. He looked to *Pollicipes s.l.* as the type of pedunculate that would have given rise to *Catophragmus*. However, in *Pollicipes,* the median latus extends well up between the terga and scuta, and the remaining capitular plates, other than the rostrum and carina, are too generalized to be pin-pointed as incipient carino- and rostrolatera. A living species of *Scillaelepas* was subsequently discovered, and Aurivillius (1894) noted that it had the three pairs of latera required. It was his model, somewhat modified, that was used by Newman *et al.* (1969) to depict the evolution of the lepadomorph wall (cf. Anderson, 1980b). But how did *Scillaelepas* acquire its latera?

Some species of *Archaeolepas* apparently had three pairs of lateral pedunculate scales, in addition to the single subrostral and subcarinal ones, as illustrated in Newman *et al.* (1969, Fig. 115) and as Hattin (1977) demonstrated for *Stramentum.* This suggests the source of the three pairs of latera needed to reach the *Scillaelepas* level of organization; that is, they could have been acquired by the transfer of the uppermost three pairs from the peduncle to the capitulum (Fig. 3A to D). Alternatively, the rostro- and carinolateral pairs could have been added to *Neolepas,* which already has a small median latera (Fig. 3B to D). This seems more likely since higher extant scalpelloids as well as *Scillaelepas* pass through a "*Neolepas*" stage during their ontogeny (Newman, 1979).

Darwin (1851) observed that the principal five capitular plates of lepadomorphs (carina and paired terga and scuta) were preceded ontogenetically by primordial valves that remain at the apexes of these plates unless eroded away. Wills (1963) noted the structural similarities of the primordial valves to the entire capitular plate in the Silurian genus, *Cyprilepas*. The primordial valves therefore were inferred to be a recapitulation of an uncalcified ancestral shell (Newman *et al.*, 1969). Darwin (1854) also noted that primordial valves occurred in the Verrucomorpha, and their presence on the so-called fixed tergum and scutum was one of the criteria he used in establishing the homology of these plates with the normal terga and scuta of Lepadomorpha. Interestingly, the balanomorphs lacked primoridal valves, but Darwin (1854) thought he saw traces of them in *Chthamalus,* and this was additional evidence that the chthamaloids were primitive. However, to my knowledge his observation has never been confirmed. It was therefore very exciting for us to observe, during a study of the ontogeny of the very primitive chthamaloid *Chionelasmus* (in preparation), obvious primordial valves on the terga and scuta (none could be detected on the carina). This is very informative not only because *Chionelasmus* stands much closer to the stem of the chthamaloids than does *Chthamalus*, but because, through more detailed study of the primordial valves themselves, it offers the possibility of more accurately determining relationships with the scalpelloids. In this connection, it is noteworthy that *Chionelasmus* passes through an eight-plated ontogenetic stage, with the plates arranged as in *Scillaelepas* rather than as in *Catophragmus* (personal observation; Fig. 3I versus 3J). This suggests that the phylogenetic history of *Chionelasmus* may not be preceded by *Catophragmus s.l.,* as had been inferred in the past (Newman and Ross, 1976). If not, it further heightens the possibility that the Balanomorpha is polyphyletic.

Darwin (1854) had also noted what appeared to be a conversion of lepadomorph ovigerous frenae (structures holding the eggs in place while brooding) to balanomorph branchiae (structures inferred to have a respiratory and/or incubatory function). The evidence for homology was their more or less comparable positions in the mantle cavity, and the reduction of supporting function and enlargement of frenae in "some species of *Pollicipes,* the genus which is nearest to the Balanidae." According to G. Walker (1980, personal communication), however, the most primitive living chthamaloid, *Catophragmus,* has both frenae and branchiae. It is still possible that branchiae developed from and replaced frenae by a supernumerary or replication process, but it does seem that the two structures are not strictly homologous, as Darwin supposed.

F. Acrothoracica

Darwin (1854) established the order Abdominalia (= Acrothoracica) to accommodate *Cryptophialus,* but he placed the acrothoracican *Alcippe* (= *Trypetesa*) with the Lepadomorpha, and there has been no doubt that the two orders are closely related (cf. Ghiselin and Jaffe, 1973). Newman *et al.* (1969) suggested that the acrothoracican organization plan was comparable to the bivalved lepadomorphan *Cyprilepas,* which had an uncalcified bivalved shell and is of adequate geologic age. Tomlinson (1969) looked to the lepadomorphan, *Ibla* as a plausible ancestor for the acrothoracicans, because it shows a slight separation between the first and remaining pairs of cirri, has the carapace abductor muscle posterior rather than anterior to the esophagus, has reduced, uncalcified armament, and some species are dioecious (female plus dwarf males), as are all acrothoracicans. However, it was discovered (Newman, 1974) that deep-water species of the most generalized acrothoracican, *Weltneria,* had a calcareous plate secreted in the rostral position and was cemented by the animal to the interior of the burrow during growth (Fig. 5C). A rostral plate is characteristic of the scalpelloid Lepadomorpha, and it is therefore possible that a scalpelloid gave rise to the acrothoracicans.

Turquier (1978) states, however, that the attachment area in *Trypetesa* forms "to the rear of the peduncular vestige, which excludes possible homology of the disc with the lepadomorph rostrum." But, in a previous study of the metamorphosis of *Trypetesa* (Turquier, 1970), he shows clearly that the last rudiment of the "peduncular vestige" (1970, Fig. 9F) occupies exactly the same position as the first deposited adult cement (1970, Fig. 10). Thus, the attachment disc at its inception is in the same position as the ancestral lepadomorph first antennae and ultimately the peduncle. The permanent calcareous plate in *Weltneria* forms between the lower margin of the aperture and the cement and is therefore wholly rostral in position. While the so-called rostral plate of *Weltneria* could be an analagous structure, it is identical to the scalpelloid rostrum in shape, construction, and position, and therefore appears to be homologous with it.

A reasonable model for a scalpelloid ancestry for the acrothoracicans is found in *Lithotrya,* the only burrowing thoracican (Fig. 5B). However, there are two problems: the first, already mentioned, is the pre- rather than post-esophageal position of the carapace adductor in all lepadomorphs except *Ibla;* the second is the greater geologic age of the Acrothoracica. I am concerned but not disuaded by the former; the pre-esophageal position of the carapace adductor is achieved in all thoracicans except *Ibla* during metamorphosis from the cyprid larva to the adult, and therefore a reversion to the ancestral or plesiomorphic condition is possible. On the other hand,

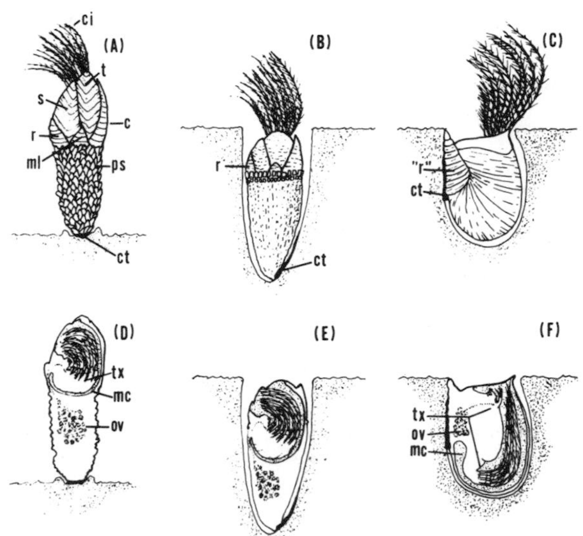

Fig. 5. Scalpelloid origin for the Acrothoracica: A form like the extant genus *Neolepas* (A), with its trophic structures occupying the capitulum (D), would have become associated with depressions in calcium carbonate substrates. The habit led to enlarging such depressions, first simply using mechanical abrasion by the peduncular scales, but eventually assisted by chemical means; such a form must have preceded the development of the only burrowing thoracican, *Lithotrya* (B). In *Lithotrya,* the capitular plates are somewhat reduced, the peduncular scales are not only relatively small and of two types, but can be replaced by periodic molting, and the body bearing the cirri occupies more the peduncle than the capitulum (E) (Darwin, 1851). It is therefore envisaged that a comparable sequence gave rise to the Acrothoracica, since some species of the most generalized genus, *Weltneria* (C), have what appears to be a scalpelloid rostral plate cemented to the wall of the burrow as it grows. The remainder of the capitular armament in acrothoracicans is represented by a pair of chitinous opercular bars (C) closing the aperture and guarding the opening to the burrow. The mantle cavity, surrounding the body and elongate thorax, occupies a sac equivalent to the peduncle of the scalpelloid ancestor (F). Abbreviations: c, carina; ci, cirri; ct, cement; mc, mantle cavity; ml, median latus; ov, ovary; ps, peduncular scales; r, rostrum; s, scutum; t, tergum; tx, thorax.

the plesiomorphic condition may have persisted into the scalpelloid stock that gave rise to the acrothoracicans. Therefore, while this remains a difficulty, it does not preclude a scalpelloid origin. As for geologic age, the acrothoracicans, previously known to range back into the Carboniferous, are now known to occur in the Late Devonian (Rodriguez and Gutschick, 1977), while the scalpelloids, previously known from the late Triassic, are now known from possible remains of Carboniferous age (Whyte, 1976). While the discrepancy in age persists, it should be noted that the acrothoracicans are recognized not by their remains, but by their characteristic burrows in limestone and shells, and therefore may be more easily preserved and/or recognized than the shells of small scalpellids which are most commonly disarticulated in fossil deposits.

III. CONCLUSIONS

The Cirripedia have been considered to consist of five orders: the parasitic Apoda, Ascothoracica, and Rhizocephala, and the cirral setose-feeders Thoracica and Acrothoracica. However, it is now known that the Apoda was based on a cryptoniscid isopod (Bocquet-Védrine, 1972a), and previous reservations about including the Ascothoracica in the cirripeds have received support through studies of their spermatozoa (Grygier, 1982a). In addition, the question of including the Rhizocephala in the Cirripedia is raised herein, since it has recently been discovered that the so-called kentrogon is likely a modified oral cone (Ritchie and Høeg, 1981; L. E. Ritchie, personal communication). Thus, the Thoracica and Acrothoracica can be considered to constitute the Cirripedia *sensu stricto* (Fig. 1).

The Rhizocephala are very closely related to the Thoracica by larval characters, and a free-living, urcirriped, ascothoracican-like form may have given rise to both of these taxa (Figs. 1 and 2B). This and the notable affinities of the cirripeds with the rhizocephalans, ascothoracicans, copepods, branchiurans, mystacocarids, and likely ostracodes further emphasizes the need for a taxon encompassing them, and more so now than before (Newman *et al.*, 1969), the Maxillopoda (Dahl, 1956, 1963) fills this need.

The most primitive Thoracica, the Lepadomorpha—previously known from the Silurian, now appears to be represented in the Cambrian (Collins and Rudkin, 1981). This is important in establishing a very early radiation of the Maxillopoda, since the fossil record for all except the cirripeds and ostracodes is virtually nil. The Cambrian form adds to the apparent fact that the earliest thoracicans lacked calcareous plates.

It has generally been agreed that the Acrothoracica stemmed from the Lepadomorpha (Fig. 1), and since the discovery of a calcareous plate that is inferred to be homologus with the rostrum (Newman, 1974), derivation from a scalpelloid seems likely (Fig. 5). The scalpelloids also take a central position in the origin and radiation of the sessile barnacles: the Brachylepadomorpha, Verrucomorpha, and Balanomorpha (Fig. 3). The possibility that the superfamilies of the Balanomorpha are polyphyletic was explored by Newman and Ross (1976), and while the evidence was considered insufficient to demonstrate polyphyly, the possibility remains. If one or more of the balanomorph superfamilies could be demonstrated to have independently evolved from a scalpelloid ancestor, sessility among the Thoracica would have been achieved four or five rather than just three times.

The scalpelloids extend back with certainty to the Triassic, and possibly into the late Paleozoic (Whyte, 1976); but diversification, marked by complex increases in skeletal armament, awaited the Cretaceous—particularly the late Cretaceous when all three sessile orders (Brachylepadomorpha, Verrucomorpha, and Balanormorpha) had appeared. The impetus for the increase in armament in scalpelloids, and the relatively sudden appearance of sessile forms correlates with the so-called "late Mesozoic marine revolution" in which predation pressures on sedentary shelled forms apparently dramatically increased (Vermeij, 1977).

The Balanomorpha replaced the Brachylepadomorpha and Verrucomorpha in numerical abundance and diversity, beginning with the chthamaloids in the late Cretaceous, the coronuloids in the Paleocene, and the balanoides in the Eocene. The balanoids apparently replaced the chthamaloids and free-living coronuloids in shallow water, likely as early as the Oligocene, but certainly by the Miocene (Newman and Stanley, 1981). In addition to a more advanced feeding mechanism (Anderson, 1978), the balanoids developed a unique tubiferous wall which provided them with the competitive advantage of rapid growth (Stanley and Newman, 1980). By the Oligocene, higher balanoids (balanids) developed a complex interlocking system between the wall plates and tubiferous calcareous basis. This not only greatly increased the strength and rigidity of the wall, but apparently freed such forms from having to rely on internal contractile elements between the wall and basis in order to remain attached (Fig. 4C). This represents the most perfected wall structure in the sessile barnacles, and it undoubtedly contributes significantly to the current abundance and diversity of the highest benthic thoracican barnacles, the balanids.

ACKNOWLEDGMENTS

Thanks are due R. R. Hessler and F. R. Schram for discussions on the crustaceans in general and the Maxillopoda in particular; likewise, M. J. Grygier, especially concerning the Ascothoracica; and all three for helpful comments on the manuscript. I am especially grateful to the late L. E. Ritchie for demonstrations and discussions of many wonders of the Rhizocephala.

REFERENCES

Anderson, D. T. (1978). Cirral activity and feeding in the coral-inhabiting barnacle *Boscia anglicum* (Cirripedia). *J. Mar. Biol. U. K.* **58**(3), 607–626.

Anderson, D. T. (1980a). Cirral activity and feeding in the verrucomorph barnacles *Verruca recta* and *Verruca stroemia* (Cirripedia). *J. Mar. Biol. Assoc. U. K.* **60**(2), 349–366.

Anderson, D. T. (1980b). *Catomerus polymerus* and the evolution of the balanomorph form in barnacles. Memoirs of the Australian Museum.

Aurivillius, C. W. S. (1894). Studien über Cirripeden. *Könglelige Svenska Vetenskaps-Akademiens Handlingar* **26**(7), 1-107, 9 pls.

Baccetti, B. (1979). Ultrastructure of sperm and its bearing on arthropod phylogeny. *In* "Arthropod phylogeny" (A. P. Gupta, ed.), pp. 609-644. Van Nostrand Reinhold, New York.

Beklemischev, W. N. (1952). Osnovy sravnitel' noi anatomii bespozvonochnykh. *In* "Principles of comparative anatomy of invertebrates." (Translated by J. M. MacLennan, edited by Z. Kabata), pp. 430-478. Oliver & Boyd, Edinburgh. Nauka, Moscow.

Beurlen, K. (1930). *Vergleichende Stammesgeschichte und Fortschritte der Geologie Paleontologie* **8,** 317-586.

Bocquet-Védrine, J. (1961). Monographie de *Chthamalophilus delagei* J. Bocquet-Védrine, Rhizocéphale parasite de *Chthamalus stellatus* (Poli). *Cah. Biol. Mar.* **2**(5), 459-600, pls. I-XII.

Bocquet-Védrine, J. (1971). Redescription du Cirripède pédonculé *Calantica calyculus* (Aurivillius) et analyse de ses rapports avec *Scalpellum pilsbryi. Arch. Zool. Exp. Gén.* **112**(4), 761-770.

Bocquet-Védrine, J. (1972a). Suppression de l'ordre des Apodes (Crustacés Cirripèdes et rattachement de son unique représentant, *Proteolepas bivincta*, à la famille des Crinoniscidae (Crustacés Isopodes, Cryptonisciens). *C. R. Acad. Sci., Paris Sér. D* **275,** 2145-2148.

Bocquet-Védrine, J. (1972b). Les Rhizocéphales. *Cah. Biol. Mar.* **13,** 615-626.

Bocquet-Védrine, J. (1979). Interpretation actulle de la description de *Proteolepas bivincta* Darwin, 1854 (Représentant unique de l'ancien ordre des cirripèdes Apodes). *Crustaceana* **37**(2), 153-164.

Bourget, E. (1980). Barnacle shell growth and its relationship to environmental factors. *In* "Skeletal growth of aquatic organisms" (D. C. Rhoads and R. A. Lutz, eds.), pp. 469-491. Plenum, New York.

Bourget, E., and Crisp, D. J. (1975). An analysis of the growth bands and ridges of barnacles shell plates. *J. Mar. Biol. Assoc. U. K.* **55**(2), 439-461.

Broch, H. (1922). Studies on Pacific Cirripeds, papers from Dr. Th. Mortensen's Pacific Expedition. 1914-1916, no. X. *Vidensk. Meddel.* Dansk. naturh. Foren. **73,** 215-358.

Buckeridge, J. S. (1980). The fossil barnacles (Cirripdia:Thoracica) of New Zealand and Australia. Doctoral thesis, Geology Department, Univ. Auckland, New Zealand. pp. 1-431.

Burnett, B. (1975). Blood circulation in four species of barnacles (*Lepas, Conchoderma:* Lepadidae). *Trans. San Diego Soc. Nat. Hist.* **17**(21), 293-304.

Calman, W. T. (1909). Crustacea. *In* "A Treatise on Zoology" (Sir R. Lankester, ed.), pp. 1-346.

Collins, D., and Rudkin, D. M. (1981). *Priscansermarinus barnetti*, a probable lepadomorph barnacle from the Middle Cambrian Burgess Shale of British Columbia. *J. Paleontol.* **55**(5), 1006-1015.

Dahl, E. (1956). Some crustacean relationships. *In* "Hanström Festschrift" (K. G. Wingstrand, ed.), pp. 138-147, Lund Zool. Instit.

Dahl, E. (1963). Main evolutionary lines among recent Crustacea *In* "Phylogeny and Evolution of Crustacea" (H. B. Whittington, and W. D. I. Rolfe, eds.), pp. 1-15. Museum of Comparative Zoology, Special Publication.

Dando, P. R., and Southward, A. J. (1980). A new species of *Chthamalus* (Crustacea: Cirripedia) characterized by enzyme electrophoresis and shell morphology, with a revision of other species of *Chthamalus* from the western shores of the Atlantic Ocean. *J. Mar. Biol. Assoc. U.K.* **60**(3), 787-831.

Darwin, C. (1851). A Monograph on the subclass Cirripedia, with figures of all species. The Lepadidae; or pedunculated cirripedes. *Ray Soc., London,* v-xi, 1-400.

Darwin, C. (1854). A Monograph on the subclass Cirripedia, with figures of all the species. The Balanidae, the Verrucidae, etc. *Ray Soc., London,* vii-viii, 1-684.

Dayton, P. K., Newman, W. A., and Oliver, J. (1982). The vertical zonation of deep-sea Antarctic acorn barnacle, *Bathylasma corolliforme* (Hoek): Experimental transplants from the shelf into shallow water. *J. Biogeography* **9,** (in press).

Delage, Y. (1884). Evolution de la Sacculine (*Sacculina carcini* Thompson) Crustacé endoparasite de l'ordre nouveau des Kentrogonides. *Arch. Zool. Exp. Gén. Sér.* 2 **2,** 417-736.

Foster, B. A. (1978). The Marine Fauna of New Zealand: Barnacles (Cirripedia: Thoracica). *Memoir 69,* 1-160. New Zealand Oceanographic Institute, Wellington.

Ghiselin, M. T. (1974). The economy of nature and the evolution of sex. Univ. California Press, pp. 1-346.

Ghiselin, M. T., and Jaffe, L. (1973). Phylogenetic classification in Darwin's monograph on the sub-class Cirripedia. *Syst. Zool.* **22**(2), 132-140.

Grygier, M. J. (1982a). Sperm morphology in Ascothoracida (Crustacea: Maxillopoda): Confirmation of generalized nature and phylogenetic importance. *Int. F. Invertebr. Reprod.* **4,** (in press).

Grygier, M. J. (1982b). Ascothoracida and the unity of the Maxillopoda. *In* "Crustacean Phylogeny" (F. R. Schram, ed.). Balkema Pubs., Rotterdam (in press).

Gurney, R. (1942). Larvae of Decapod Crustacea. *Ray Soc. London,* 1-306.

Hattin, D. E. (1977). Articulated lepadomorph cirripeds from the Upper Cretaceous of Kansas: Family Stramentidae. *J. Paleontol.* **51(4),** 797-825.

Hedgecock, D. (1979). Biochemical genetic variation and evidence of spciation in *Chthamalus* barnacles of the tropical eastern Pacific. *Mar. Biol.* **54,** 207-214.

Krüger, P. (1940). Ascothoracida. *In* "Bronns Klassen und Ordnungen des Tierreichs *5,*" Abt. 1, Buch 3, Teil IV, 1-46.

Lacaze-Duthiers, H. (1880). Histoire de la *Laura gerardiae,* type nouveau de crustacé parasite. *Arch. Zool. Exp. Gén. Sér. 1* **8,** 537-581.

Newman, W. A. (1961). On the nature of the basis in certain species of the *Hembeli* section of *Chthamalus* (Cirripdia, Thoracica). *Crustaceana* **2(2),** 142-150.

Newman, W. A. (1974). Two new deep-sea Cirripedia (Ascothoracia and Acrothoracica) from the Atlantic. *J. Mar. Biol. Assoc. U.K.* **54**(2), 437-456.

Newman, W. A. (1979). A new scalpellid (Cirripdia); a Mesozoic relic living near an abyssal hydrothermal spring. *Trans. San Diego Soc. Nat. Hist.* **19(11),** 153-167.

Newman, W. A. (1980). A review of extant *Scillaelepas* (Cirripedia: Scalpellidae) including recognition of new species from the North Atlantic, Western Indian Ocean and New Zealand. *Tethys* **9(4),** 379-398.

Newman, W. A. (1982). Origin of the Maxillopoda; Urmalacostracan ontogeny and progenesis. *In* "Crustacean Phylogeny" (F. R. Schram, ed.). Balkema Pubs., Rotterdam (in press).

Newman, W. A., and Ross, A. (1971). Antarctic Cirripedia. *Am. Geophys. Union, Antarc. Res. Ser.* **14,** v-xiii, 1-257.

Newman, W. A., and Ross, A. (1976). Revision of the balanomorph barnacles; including a catalog of the species. *Mem. 9, San Diego Soc. Nat. Hist.* 1-108.

Newman, W. A., and Ross, A. (1977). Superfamilies of the Balanomorpha (Cirripedia, Thoracica). *Crustaceana* **32(1),** 102.

Newman, W. A., and Schram, F. R. (1982). Eocene barnacles (Cirripedia: Thoracica) of Tonga. *U.S. Geol. Surv. Professional Pap.* 640-H (in preparation).

Newman, W. A., and Stanley, S. (1981). Competition wins out overall: Reply to Paine. *Paleobiology* **7**(4), 561-569.

Newman, W. A., Zullo, V. A., and Wainwright, W. A. (1967). A critique on recent concepts of growth in Balanomorpha (Cirripdia, Thoracica). *Crustaceana* **12(2),** 167-178.

Newman, W. A., Zullo, V. A., and Withers, T. H. (1969). Cirripedia. *In* "Treatise on Invertebrate Paleontology" (R. C. Moore, ed.), Part R, Arthropoda 4, *1*:R206-R295. *Geol. Soc. Am. Inc.,* Univ. Kansas.

Paine, R. T. (1981). The forgotten roles of disturbance and predation. *Paleobiology* **7**(4), 553-560.

Ritchie, L. E., and Høeg, J. T. (1981). The life history of *Lernaeodiscus porcellanae* (Cirripedia: Rhizocephala) and coevolution with its procellanid host. *J. Crustacean Biol.* **1(3),** 334-347.

Rodriquez, J., and Gutschick, R. C. (1977). Barnacle borings in live and dead hosts from the Louisiana Limestone (Famennian) of Missouri. *J. Paleontol.* **51(4),** 718-724.

Sanders, H. L. (1963). The Cephalocarida: Functional morphology, larval development, comparative external anatomy. *Mem. Conn. Acad. Arts Sci.* **15,** 80 pp.

Schram, F. R. (1975). A Pennsylvanian lepadomorph barnacle from the Mazon Creek Area, Illinois. *J. Paleontol.* **49**(5): 928-930.

Schram, F. R., and Newman, W. A. (1980). *Verruca withersi* n.sp. (Crustacea: Cirripedia) from the middle of the Cretaceous of Columbia. *J. Paleontol.* **54(1),** 229-233.

Southward, A. J. (1976). On the taxonomic status and distribution of *Chthamalus stellatus* (Cirripedia) in the north-east Atlantic region: with a key to the common intertidal barnacles of Britain. *J. Mar. Biol. Assoc. U.K.* **56**(4), 1007-1028.

Spivey, H. (1981). Origins, distribution and zoogeographic affinities of the Cirripedia (Crustacea) of the Gulf of Mexico. *J. Biogeography* **8,** 153-176.

Stanley, S. M. (1974). What has happened to the articulate brachiopods? *Geol. Soc. Am., Abstr.* with Programs **6,** 966-967.

Stanley, S. M., and Newman, W. A. (1980). Competitive exclusion in evolutionary time: the case of the acorn barnacle. *Paleobiology* **6(2),** 173-183.

Tomlinson, J. T. (1969). The burrowing barnacles (Cirripedia: Order Acrothoracica). *U.S. Natl. Mus. Bull.* **296,** 1-162.

Turquier, Y. (1970). Recherches sur la biologie des Cirripèdes Acrothoraciques. III. La métamorphose des cypris femelles de *Trypetesa nassarioides* Turquier et de *T. lampas* (Hancock). *Arch. Zool. Exp. Gén.* **111**(4), 573-628.

Turquier, Y. (1978). Le tégument des Cirripèdes Acrothoraciques. *Arch. Zool. Exp. Gén.* **119(1),** 107-125.

Veillet, A. (1964). La metamorphose des cypris femellies des Rhizocéphales. *Zoöl. Meded. Leiden* **39,** 573-576.

Vermeij, G. J. (1977). The Mesozoic marine revolution: evidence from snails, predators and grazers. *Paleobiology* **3**(3), 245-258.

Wagin, W. L. (1976). "Meshkogrudie Raki: Ascothoracida." Kazan Univ. Press, Kazan, U.S.S.R. pp. 1-141.

Whyte, M. A. (1976). A Carboniferous pedunculate barnacle. *Proc. Yorks. Geol. Soc.* **41** Part 1(1), 1-12.

Wills, L. J. (1963). *Cyprilepas holmi* Wills 1962, a pedunclate cirripede from the Upper Silurian of Oesel, Esthonia. *Palaeontology* **6(1),** 161-165.

Yanagimachi, R. (1961). The life-cycle of *Peltogasterella* (Cirripedia, Rhizocephala). *Crustaceana* **2(3),** 183-186.

Zevina, G. B. (1978a). A new classification of the Scalpellidae (Cirripedia, Thoracica). Subfamilies Lithotryinae, Calanticinae, Pollicipinae, Scalpellinae, Brochiinae and Scalpellopsinae. *Zool. Zh. Akad. Nauk SSSR* **57(7),** 998-1007.

Zevina, G. B. (1978b). A new classification of the Scalpellidae (Cirripedia, Thoracica). 2. Subfamilies Arcoscalpellinae and Meroscalpellinae. *Zool. Zh. Akad. Nauk SSSR* **57(9),** 1343-1352.

Zevina, G. B. (1980). A new classification of Lepadomorpha (Cirripedia). *Zool. Zh. Akad. Nauk SSSR* **54(5),** 689-698.

Zevina, G. B. (1981). Barnacles of the suborder Lepadomorpha (Cirripedia, Thoracica) of the World Ocean. Part 1: Family Scalpellidae. Guides to the Fauna of the USSR published by the Zoological Institute of the Academy of Sciences of the USSR. **127,** 1-406. Leningrad.

Zullo, V. A., and Baum, G. R. (1979). Paleogene barnacles from the coastal plain of North Carolina (Cirripedia, Thoracica). *Southeast. Geol.* **20**(4), 229-246.

Part 4: Ostracoda (Maddocks)

I. ORIGIN AND AFFINITIES

The exact relationship of Ostracoda to other Crustacea remains uncertain, despite nearly a century of speculation. This isolation in understanding, if not in evolution, from other Crustacea is usually expressed taxonomically by ranking Ostracoda as a major subdivision; for example, as one of four subclasses (Dahl, 1956; Hartmann, 1963) or as one of nine classes (Manton, 1969). There is still no compelling morphological evidence for any of the several affinities that have been proposed, because the required supporting homologies of body segmentation and biramous limb structure can hardly be discerned in Ostracoda, except with the eye of faith. Many of the characters shared with other Crustacea are functional adaptations. Until more is known of the way of life and functional morphology of the oldest ostracodes, it may not be possible to recognize accurately any remaining plesiomorphic characters. We know that there has been frequent evolutionary convergence just within Ostracoda in structures of appendages and carapace.

The often-suggested derivation of ostracodes from "phyllopods" (Branchiopoda) (Grobben, 1892; Giesbrecht, 1912; Müller, 1894; Skogsberg, 1920; Pokorný, 1958) is based on similarities of the bivalved carapace, laterally compressed body, and conspicuous reduction of segments and appendages, none of which survive close scrutiny (Hartmann, 1963). The carapace of ostracodes is divided dorsally into two articulating halves: embryologically it arises from the antennal segment (III) and is already fully developed in the just-hatched nauplius larva (Müller, 1894; Weygoldt, 1960). The carapace of branchiopods, on the other hand, is undivided and has no articulated hinge; it is derived from the second maxillar segment (VI) of the embryo and develops very gradually through several metanaupliar larval stages (Hartmann, 1963). In Conchostraca it is multilamellar and shows growth lines. In Cladocera it does not cover the head, shows no attachment scars, lacks pores, and frequently lacks sensory bristles (Dahm, 1976). A carapace of some kind has been repeatedly and convergently invented by many lineages of Crustacea.

Most ostracodes have a basically rod-shaped limb (Hartmann, 1963), and the exceptions all occur in filter-feeding specialists. Conchostraca have the leaflike phyllopod limb. Adult cladocerans have totally lost the second maxilla, but, contrary to some authors (Müller, 1894; Kesling, 1961; van Morkhoven, 1962), it is probably present in ostracodes (Hartmann, 1966; Weygoldt, 1960). In many anatomical features, in their restriction to terrestrial habitats, and in their limited fossil record (Oligocene-Recent), it is plain that Cladocera are highly specialized and recently derived.

Conchostracans, on the other hand, include many primitive lines and occur much earlier (Silurian, according to Adamczak, 1961; Devonian, in Tasch, 1969). Adamczak (1961) argued that certain Ordovician ostracodes, for which he proposed the Order "Eridostraca," share numerous similarities of carapace structure with Conchostraca, especially molt retention, growth lines, layered structure, and prominent umbos. He concluded that the "Eridostraca" provide an intermediate link between primitive Conchostraca and some other ostracodes. Regretfully, however, he decided that ostracodes might be diphyletic, and he considered it necessary to derive the Leperditiida and Podocopida separately from a bradoriid ancestor. Hartmann (1963) affirmed a close connection of "Eridostraca" with Conchostraca, but suggested instead that most of these forms are not true ostracodes. Levinson (1951) and Scott (1961) interpreted them as true ostracodes with molt retention, and they are classified at present in the Superfamily Leperditellacea.

Hartmann (1963), following Remane (1957), Dahl (1956), and Siewing (1959), preferred instead to derive the Ostracoda from a point near the ancestral stock of the "maxillopoda" (Ascothoracida, Cirripedia, Branchiura, Mystacocarida, and Copepoda). Although ostracodes share the rod-shaped limb, the tendency to reduction of the posterior region, and the tendency to restrict feeding functions to the cephalic limbs, he warned that they have today diverged very far from the other stocks of maxillopods, and that detailed similarities are few. A more direct connection to maxillopods is suggested by the three-part nauplius eye (Elofsson, 1966), the inferred vestiges of up to ten postcephalic segments in *Cytherella* (Müller, 1894; Schulz, 1976), and very tentatively inferred homologies of the chitinous cephalic exoskeleton with that of calanoid copepods (Schulz, 1976).

Barnacles (Cirripedia and Ascothoracica) share the naupliar larval development, but without any carapace. The detailed anatomy of the so-called "cypris-larva," except for a few marginal valve characters (McKenzie, 1972), is not much like an ostracode, however, and the adult structure and ecologic strategies of these taxa are totally dissimilar.

In skeletal musculature, fairly successful though slightly contradictory detailed homologies can be drawn between Ostracoda and the Cephalocarida (Hessler, 1964, 1969; McKenzie, 1972), which may support claims for the near-ancestral role of the latter in models of crustacean phylogeny, as well as the very early divergence of Ostracoda from other Crustacea.

II. APPENDAGE HOMOLOGIES

The homologization of ostracode limbs with those of other Crustacea has long been a matter of controversy. Five to seven paired appendages are developed in ostracodes, with vestigial remnants of perhaps one more. These legs are basically rod-shaped, although flattened leaflike limbs, analogous to those of the Branchiopoda, have been developed in some filter-feeding taxa. In most of these appendages, the supposedly biramous form of the primitive crustacean limb can no longer be reliably discerned. Although the terms exopodite, endopodite, epipodite, and the like, are applied to ostracode limbs by various authors (not always harmoniously), in fact, it is very difficult to homologize the individual parts of ostracode limbs with those of other Crustacea.

It seems clear that the first four limbs belong to the cephalon and the last two to the thorax, but the fifth limb has been claimed for both. The arguments for homology of the fifth limb depend on four lines of reasoning:

(1) Should the vibratory plate be identified as epipodite or exopodite, and should the jointed leglike branch be considered the exopodite or endopodite? By one line of reasoning, this limb shows strong serial homologies to the maxilla (fourth limb), and by the other to the thoracic legs (Müller, 1894; Skogsberg, 1920).

(2) In some ostracodes, a rather strong line of demarcation in the main body separates the forehead and mouth region, with the first four limbs, from the following part, with the last three limbs. This division is probably functional rather than morphological (Hartmann, 1963; Weygoldt, 1960).

(3) The fact that in some (though not in all) Podocopida, no new appendage is added at the second molt was thought to indicate that ostracodes do not have the second maxilla of other crustaceans (Müller, 1894; Kesling, 1951; Howe *et al.*, 1961; van Morkhoven, 1962). It is now known that the order of appearance of appendages may vary slightly in different taxa (Hartmann, 1968; McKenzie, 1972; Weygoldt, 1960).

(4) In those ostracodes that have retained both segmental glands, they are situated in the segments of the antenna and of the fifth limb (Cannon, 1925; Hartmann, 1967; Hartmann, in Neale, 1977).

Hartmann's reasoning is accepted here, and the fifth limb is considered to be a cephalic appendage homologous with the second maxilla of other Crustacea. Because of these divergent opinions regarding homology, various terms have been used for the limbs. Here, they will be named as follows: (1) antennule (= first antenna of some authors), (2) antenna (= second antenna), (3) mandible, (4) maxilla (= maxillule or first maxilla), (5) fifth limb (= maxilla, second maxilla, maxilliped, or first thoracic leg), (6) sixth limb (= first or second thoracic leg), and (7) seventh limb (= second or third thoracic leg).

The antennule in all groups is a long jointed limb equipped with setae for a sensory or locomotory (swimming) function, with no trace of biramous structure.

The antennae are the main locomotory structures for walking, climbing, digging, or swimming. Biramous structure is conspicuous in some but not all. They carry a variety of sexually dimorphic features, such as sensory setae, glands, or clasping structures.

The mandibles have robust bases with strong teeth and a jointed setose palp for manipulating and ingesting food.

The maxillae, though variously constructed, always function as feeding structures. Some forms have a large vibratory plate that generates a continuous water current across the posterior region of the body for respiration.

The fifth limb, as may be inferred from the variability of its terminology, has different configurations and functions in different groups. In some it is clearly a mouthpart, similar to and assisting the maxilla. In others, the feeding function has been abandoned, and the jointed leg is expanded to serve as a sexually dimorphic clasping structure or even as a true walking leg. It frequently bears a vibratory plate to help circulate water for respiration. It is considered to be homologous with the second maxilla of other Crustacea (Hartmann, 1963, 1966; Hartmann, in Neale, 1977).

The sixth limb is the chief walking leg in crawling forms, but in many swimming forms it is a much reduced accessory mouthpart with a large vibratory plate.

The seventh limb may be a walking leg or may function as a recurved cleaning limb, to protect the posterior body region and eggs.

The brush-shaped sensory (?) organ of certain male Podocopida is thought to be the vestige of an eighth pair of appendages (Weygoldt, 1960; Hartmann, 1966).

The furcae are generally considered to be homologous with the telson of other Crustacea and hence not true paired appendages. However, Bowman (1971) and Kornicker (1975) have shown that the furca of Podocopida and Platycopida is located ventral to the anus, whereas in Myodocopida it is located dorsal to the anus. This would make the furca of Podocopida and Platycopida equivalent to a uropod. Schulz (1976) identified a chitinous remnant that he labeled as a telson on *Saipanetta* and *Cytherella,* the most primitive living members of these two orders. In the majority of swimming and burrowing ostracodes, where the furcae are a major locomotory structure, they consist of two broad lamellae that are never jointed, are fused at their base, and are armed with sturdy claws. In the Podocopida, the thoracic legs have taken over the walking function, and the furcae correspondingly are much reduced or even lost.

III. THE PRIMITIVE OSTRACODE

By detailed comparative morphological analysis of the living orders, Müller (1894) deduced the following composite profile of the characteristics of the protostracode: A free-swimming animal living near the bottom; with a calcified bivalved carapace enclosing the whole body and shut by a closing muscle, with a rostral incisure and a convex ventral margin; body externally segmented, with at least eleven postcephalic segments, not all of which bore limbs; with a heart, simple digestive system, compound lateral eyes, and a three-part median nauplius-eye; with eight pairs of limbs, and perhaps a ninth already modified as the male copulatory organ; an eight-jointed sensory/locomotory antennule like that of the Cypridinacea; a biramous antenna like that of the Platycopida, with well-developed exopodite and endopodite; a mandible with enlarged basal chewing part and four-jointed palp, like that of the Cypridinacea; a maxilla like that of the Platycopida, with a three-jointed protopodite, small exopodite, and three-jointed endopodite; a fifth limb like that of Cypridacea (*Macrocypris*), but with a vibratory plate like that of the Halocypridacea; a sixth limb like that of the Halocypridacea females; a seventh limb with cleaning function, as in the Halocypridacea; an eighth appendage like the brush-shaped organ of Bairdiacea and Cypridacea, but in both sexes; and a pair of furcae similar to those of Halocypridacea.

This interpretation was analyzed in exhaustive detail by Skogsberg (1920), who agreed with nearly all of the general conclusions but almost none of the specific ones. He considered the functional adaptations of each of the limbs and warned, indeed, he almost despaired, of the difficulties of drawing accurate detailed homologies between orders. These judgments, rating Cypridinacea and Polycopacea as most primitive overall, have been largely accepted by later zoologists (Hartmann, 1963, 1966) and paleontologists (Pokorný, 1958). More recently, the primitive quality of certain features of Platycopida (*Cytherella*) and Podocopida (*Saipanetta*) has been appreciated (McKenzie, 1967b, 1970; Maddocks, 1972, 1973, 1976; Schornikov and Gramm, 1974; Schulz, 1976). Nevertheless, none of the surviving orders of ostracodes can be considered truly primitive (generalized). All are evolutionary mosaics, and many supposedly primitive forms are now ecological specialists. Many of the most conspicuous trends in ostracodes have been reductions, which are notoriously subject to convergence. Since the search for a least-common-denominator ancestral "type" appears unlikely to yield unambiguous results, attention has turned to the fossil record.

The oldest known ostracodes are Cambrian Bradoriida (= Archaeocopida of some authors) with phosphatic carapaces (Sylvester-Bradley, 1961; Müller, 1964; Andres, 1969; Jones and McKenzie, 1980), though this identification has not been universally accepted. Ulrich and Bassler (1931) believed them to be the ancestors of Ostracoda but assigned them to the Conchostraca. Raymond (1935) also considered them to be the ancestors of Ostracoda but transferred them to the Archaeostraca. Adamczak (1968) compared the structure of certain Bradoriida (Phosphaticopina) with the egg cases (ephippia) of modern Cladocera. Because convergence in carapace form is so common among crustaceans, Öpik (1968) warned that "without the body, empty shells (moults) of *Canadaspis* would be placed in the Bradoriida." The phyllocarid affinities of *Canadaspis* (Middle Cambrian, Burgess Shale) were reaffirmed by Briggs (1978), but he speculated, "It seems unlikely that the Bradoriids were phyllocarids but without further evidence of their morphology this possibility can not be ruled out."

Sylvester-Bradley (1961), Kozur (1974), Müller (1964, 1979a), Andres (1969), and McKenzie and Jones (1979) have argued convincingly for the classification of Bradoriida as Ostracoda on the basis of the shape, hingement, and inferred sexual dimorphism of the carapace. Jones and McKenzie (1980) considered the Bradoriida to be heterogeneous, including the ancestral ostracodes, other aberrant ostracode-like animals, and phyllocarid-like or branchiopod-like crustaceans. The phosphatic composition may be regarded as a primitive character. This change from dominantly phosphatic to calcareous mineralization of skeletons occurred also in Cambro-Ordovician Brachiopoda and other invertebrates, which has been considered to reflect the evolution of biochemical mechanisms of calcification or changes in ocean chemistry (Simkiss, 1964; Lebedev and Lebedeva, 1965; Rudwick, 1970; Rhodes and Bloxam, 1971). Bradoriida are diverse, including more than 100 species and two, three, or more separate lines, among which Kozur (1974) speculatively identified the ancestors of three later orders of ostracodes. They occur in the Lower Cambrian and thus may be one of the very first lineages derived in the radiation of the protocrustaceans.

Fragmentary appendages preserved by secondary phosphatization within Upper Cambrian Bradoriida (Phosphaticopina) from Sweden (Müller, 1979a) show us a peculiar animal that in some ways fulfills the predictions for the primitive ostracode: curved ventral margin, many-jointed exopodites, biramous (!) antennule in post-oral position, and a well-developed carapace within the unhatched egg. Other features of these specimens are more unexpected: laminated carapace, unsegmented body, only six or seven limbs, limbs not differentiated by function, feeding functions of antennules and antennae, absence of respiratory plates and furcae, and up to 16 podomeres on the supposed exopodite. To eyes accustomed to the modern ostracodes, these Cambrian candidates are strange indeed! The partially preserved body (without appendages) of a single specimen of Middle Cambrian bradoriid from Australia appears to show at least four distinct thoracic segments (McKenzie and Jones, 1979).

Otherwise, few fossils with preserved appendages are known: a cypridacean with modern anatomy from the Lower Cretaceous of Brazil (Bate, 1972), a myodocopid in stomach contents of a Jurassic plesiosaur from the U.S.S.R. (Dzik, 1978), and a cypridacean from the Carboniferous of France

(Brongniart (1876). Tantalizing fragments of Paleocopida were illustrated by Müller (1979b) and of Bairdiacea by Gocht and Goerlich (1957). Thus, what we know of phyletic affinities and evolutionary trends in Ostracoda has been interpreted almost entirely from carapace remains.

IV. PHYLOGENY

If the conchostracan aspects of certain Leperditellacea are attributed to convergence, the Ostracoda may be considered to be monophyletic, with an ancestral stem within the Bradoriida (Scott, 1961; Sylvester-Bradley, 1961; Hartmann, 1963; Müller, 1964; Andres, 1969; Kozur, 1972, 1974; Jones and McKenzie, 1980). The other orders, Leperditicopida, Paleocopida, Platycopina, Podocopida, and possibly Myodocopida, all appear in the Ordovician, with rare and poorly known representatives of Leperditiida and Paleocopida reported to occur in the Upper Cambrian (Scott, 1961; Jones and McKenzie, 1980) (see Table I and Fig. 1). This Ordovician burst of adaptive radiation of calcareous shelled types, near-simultaneous origin of the major post-Cambrian orders, and rapid disappearance of Cambrian stocks is a common evolutionary pattern in many invertebrate phyla.

The Leperditicopida may not be closely related to other ostracodes (Levinson, 1951; Adamczak, 1961), although Abushik (1979) placed them near the Paleocopida. Their radial carapace markings are internal structures within the lamellae and thus do not represent evidence for a heart and blood vessels (Sohn, 1974), which removes the chief reason for linking them with Myodocopida (Kozur, 1972). Nothing is known of their soft parts. These large grazing herbivores flourished in intertidal and shallow subtidal carbonate environments in the Ordovician and Silurian, declined in the Devonian, and then vanished without known descendants (Abushik, 1979).

In their relatively large size, occasional presence of a heart and paired compound eyes, biramous antennae, and large dorsal furca (telson), the Myodocopida are very distinct from and possibly more primitive than the other living ostracodes (Skogsberg, 1920; Hartmann, 1963). It is not certain whether the ridged spinose posterodorsal body of Polycopacea and Thaumatocypridacea represents vestigial segmentation, or whether it suggests a close connection of these taxa (Kornicker and Sohn, 1976b). The prominent rostral incisure of most Cypridinacea and Halocypridacea, which functions as an oarlock for the swimming branch of the antenna, is constructed differently and thought to be of independent origin in the two groups (Kornicker and Sohn, 1976b). Of the three suborders, the Myodocopina (Cypridinacea) are considered to be the most primitive in appendage anatomy (Müller, 1894; Skogsberg, 1920). Early separation

TABLE I

Classification of Ostracoda

Order Bradoriida Raymond, 1935, Lower Cambrian-Lower Ordovician
Order Leperditicopida Scott, 1961, Upper Cambrian?, Lower Ordovician-Upper Devonian
Order Myodocopida Sars, 1866, Ordovician?, Silurian-Recent
  Suborder Myodocopina Sars, 1866, Silurian-Recent
    Superfamily Cypridinacea Baird, 1850, Silurian-Recent
  Suborder Halocypridina Dana, 1852, Silurian-Recent
    Superfamily Halocypridacea Dana 1852, Cretaceous-Recent
    Superfamily Thaumatocypridacea G. W. Müller, 1906, Permian-Recent
    Superfamily Entomoconchacea Brady, 1868, Silurian-Lower Carboniferous
  Suborder Cladocopina Sars, 1866, Ordovician?, Silurian-Recent
    Superfamily Polycopacea Sars, 1866, Upper Devonian?, Carboniferous-Recent
    Superfamily Entomozoacea Přibyl, 1951, Ordovician?, Silurian-Carboniferous
Order Paleocopida Henningsmoen, 1953, Lower Ordovician-Triassic
  Suborder Beyrichicopina Scott, 1961, Silurian-Lower Carboniferous
    Superfamily Beyrichiacea Matthew, 1886, Silurian-Lower Carboniferous
  Suborder Hollinomorpha Henningsmoen, 1965, Lower Ordovician-Triassic, Neogene-Triassic?
    Superfamily Hollinacea Swartz, 1936, Lower Ordovician-Triassic
    Superfamily Eurychilinacea Ulrich and Bassler, 1923, Lower Ordovician-Permian
    Superfamily Primitiopsacea Swartz, 1936, Middle Ordovician-Middle Devonian
    Superfamily Punciacea Hornibrook, 1949, Neogene-Recent
  Suborder Binodicopina Schallreuter, 1972, Lower Ordovician-Middle Permian
    Superfamily Drepanellacea Ulrich and Bassler, 1923, Middle Ordovician-Middle Permian
    Superfamily Aechminacea Bouček, 1936, Middle Ordovician-Lower Carboniferous
    Superfamily Aparchitacea T. R. Jones, 1901, Lower Ordovician-Upper Carboniferous
  Suborder Kirkbyocopina Gründel, 1969, Ordovician-Triassic
    Superfamily Kirkbyacea Ulrich and Bassler, 1906, Ordovician-Triassic
Order Platycopida Sars, 1866, Lower Ordovician-Recent
  Suborder Platycopina Sars, 1866, Lower Ordovician-Recent
    Superfamily Leperditellacea Ulrich and Bassler, 1906, Lower Ordovician-Permian, Triassic-Jurassic?
    Superfamily Kloedenellacea Ulrich and Bassler, 1908, Lower Ordovician-Triassic
    Superfamily Cytherellacea Sars, 1866, Silurian-Recent
Order Podocopida Sars, 1866, Middle Ordovician-Recent
  Suborder unknown
    Superfamily Paraparchitacea Scott, 1959, Devonian-Permian
  Suborder Metacopina Sylvester-Bradley, 1961, Middle Ordovician-Cretaceous
    Superfamily Healdiacea Harlton, 1933, Silurian-Cretaceous
    Superfamily Quasillitacea Coryell and Malkin, 1936, Devonian-Carboniferous
    Superfamily Thlipsuracea Ulrich, 1894, Middle Ordovician-Devonian
  Suborder Podocopina Sars, 1866, Middle Ordovician-Recent
    Superfamily Sigilliacea Mandelstam, 1960, Silurian-Recent
    Superfamily Darwinulacea Brady and Norman, 1889, Carboniferous-Recent
    Superfamily Cypridacea Baird, 1845, Devonian-Recent
    Superfamily Bairdiacea Sars, 1888, Middle Ordovician-Recent
    Superfamily Cytheracea Baird, 1850, Middle Ordovician-Recent

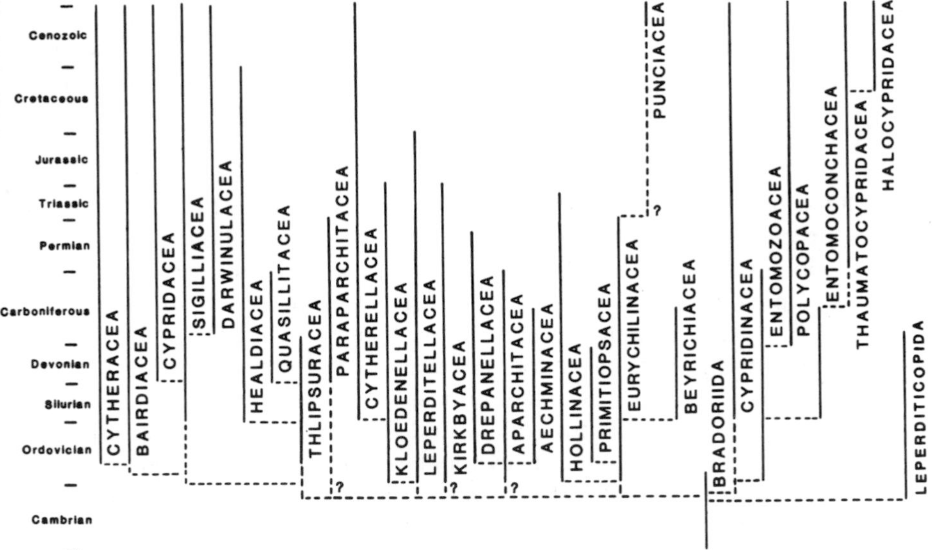

Fig. 1. Inferred phylogeny and chronostratigraphic ranges of major taxa of Ostracoda.

(pre-Silurian) of Cypridinacea from other Myodocopida was also postulated by Kornicker and Sohn (1976b) in a Hennigian analysis of living representatives.

The unpaired genitalia and several appendage features unite the Cladocopina and Halocypridina, which may also have diverged before the Silurian (Kornicker and Sohn, 1976b). Polycopacea show several quite primitive features, including a biramous maxilla, but the total absence of the sixth and seventh limbs is clearly a derived character related to miniaturization for their interstitial life. The Entomozoacea are transitional into and may represent the ancestors of Polycopacea (Kozur, 1972). The Entomoconchacea are ancestral to the Thaumatocypridacea and Halocypridacea, and similarities of the carapace suggest a more distant connection with the Cladocopina (Kornicker and Sohn, 1976a). The planktonic Halocypridacea date only from the Cretaceous (Pokorný, 1964), perhaps because the uncalcified carapace is not commonly preserved. Although Kornicker and Sohn (1976b) speculate that Halocypridacea diverged from Thaumatocypridacea before the Permian, a middle to late Mesozoic origin would be conformable with the known history of most other components of the modern planktonic ecosystem.

As defined here, the Paleocopida comprise a rather heterogeneous collection of largely Paleozoic lineages, whose mutual relationships are not yet clear. Both dimorphic and nondimorphic, as well as lobate and nonlobate, forms are already present in the Lower Ordovician, so that the nature of the presumed ancestral stock(s) has not been delineated, and it is also not agreed whether the nondimorphic stocks represent the primitive condition (Schallreuter, 1973) or regressive offshoots (Henningsmoen, 1965).

Within the Hollinomorpha, with conspicuous antral dimorphism, the Eurychilinacea seem to provide an appropriate ancestral stem for both the Hollinacea and Primitiopsacea. The Beyrichiacea were also derived from the Eurychilinacea by transformation of the antrum into the crumina (Henningsmoen, 1954; Kesling, 1957; Jaanusson, 1957; Martinsson, 1962). The Binodicopina are more perplexing, because they are essentially nondimorphic. The Aparchitacea may furnish appropriate ancestry for both the Drepanellacea and Aechminacea, but the origin of Aparchitacea is not known. The nondimorphic Kirkbyacea also appear without known ancestor, and, for this reason, both the Kirkbyocopina and Binodicopina are frequently removed as separate orders. Platycopid affinity was proposed for the Kirkbyacea by Schallreuter (1968) and podocopid affinity by Gramm and Pozner (1972) and Kozur (1972). The mysterious "Paleozoic-looking" Punciacea are probably descended from Eurychilinacea (Hornibrook, 1949, 1963), although Schallreuter (1968, 1972) suggested an origin from the Kirkbyacea. Although punciacean carapaces have been found in modern sediments in the western Pacific, the living animal has not yet been found.

The Podocopida and Platycopida have frequently been considered to share enough carapace features to be classified as suborders Podocopina and Platycopina within one order Podocopida (Müller, 1894; Pokorný, 1958, 1978; Hartmann, 1963, 1964; Kozur, 1972; Hartmann and Puri, 1974; Schulz, 1976). Among these characters are the strongly calcified carapace, the straight to concave ventral margin and more or less convex dorsal margin, the domiciliar sexual dimorphism, and the evolutionary trend of the adductor muscle-scar pattern from aggregate to discrete. But the body and appendages of *Cytherella,* the only living platycopid, are so very different from those of all other ostracodes that almost no detailed homologies can be traced (Maddocks, 1976). Very early on, taxonomic separation of these two orders was proposed on anatomical grounds (Sars, 1866, 1922–1928; Skogsberg, 1920), and it has found additional support in the Lower Paleozoic fossil record (Schallreuter, 1968; Gründel, 1967, 1968; Adamczak, 1966, 1967, 1968, 1969, 1976b). Adamczak (1969) analyzed marginal valve construction and concluded that the filter-feeding habit of modern Cytherellacea was shared by Paleozoic Cytherellacea, Kloedenellacea, Primitiopsacea, and Beyrichiacea (Platycopida and Paleocopida), whereas Lower Paleozoic Podocopina and Metacopina (Podocopida) appear to have been detritus-feeders like the modern representatives. On the other hand, there is much likelihood that these two orders share a common origin (Skogsberg, 1920; Hartmann, 1963; Gründel, 1967; Kristan-Tollmann, 1977a,b). Older representatives converge morphologically, and additional extinct lineages help to bridge the gap between modern representatives. The various taxonomic schemes are merely different compromises between vertical and horizontal classification (Simpson, 1961; Sylvester-Bradley, 1962).

Schallreuter (1968) and Henningsmoen (1965) proposed that the ancestors of Platycopida probably were nondimorphic Paleocopida, either Aparchitacea or Drepanellacea. Guber and Jaanusson (1964) speculated that the roots of the Kloedenellacea lie within either the Leperditellacea or the velate Paleocopida. The Leperditellacea remain taxonomically heterogeneous and of uncertain phyletic position. The Cytherellacea are obviously descended from the Kloedenellacea (Adamczak, 1966, 1976b; Schallreuter, 1968; Gründel, 1967; Kozur, 1972). They combine an elaborate but primitive soft-part anatomy with a highly specialized filter-feeding habit and a conservative carapace; Gramm (1973b) proposed that neoteny has been an important agent in their evolution.

The Paraparchitacea have been classified in the Paleocopida (Scott, 1961), Platycopida (Gründel, 1967; Schallreuter, 1968), and Podocopida (Sohn, 1971; Adamczak, 1976b; Gründel, 1978), and their phyletic relationships remain uncertain.

The definition and classification of Podocopida has been controversial because the supposedly diagnostic characters (the attachment scars of the adductor muscles, hingement, and development of the duplicature) are evolving within this group and may be absent or only weakly developed in early representatives. These difficulties are enhanced by the frequency of convergent, parallel, and iterative evolution of these characters in different podocopid lineages and with contemporary Platycopida, which only very careful morphological analysis can discriminate.

This confusion is aptly symbolized by the nominal suborder Metacopina. It was originally proposed as a horizontal taxon to include the primitive representatives of both Platycopina and Podocopina plus presumed intermediate lineages (Sylvester-Bradley, 1961, 1962; Scott, 1961), and the unifying morphologic character was considered to be the healdiacean aggregate muscle-scar pattern. To this concept van den Bold (1974) added the Darwinulacea. In later schemes, ancestors have been reunited with descendants in an expanded Platycopina and/or Podocopina (Sohn, 1961; van Morkhoven, 1962; Gründel, 1967, 1969; Hartmann, 1964; Adamczak, 1966, 1968; Kozur, 1972; Gramm, 1968, 1972), leaving a reduced Metacopina to be interpreted as an evolutionary grade preceding Podocopina, or sometimes to be deleted. Szczechura and Blasyk (1968) emphasized the vertical continuity and separateness of Podocopina and

Metacopina, leaving Platycopina as an offshoot. Kristan-Tollmann (1977a,b) discarded the Platycopina altogether. Sohn (1965) removed the Healdiacea to Podocopina, while Schornikov and Gramm (1974) restricted the Metacopina to include only the Healdiacea.

Meanwhile, detailed morphological and evolutionary analyses of these aggregate muscle-scar patterns (Gramm, 1967, 1968, 1969, 1970a,b, 1972, 1973a,b, 1976, 1977; Gründel, 1964a,b; Kristan-Tollmann, 1971, 1973a,b, 1977a,b, 1979) and of valve marginal characters (Adamczak, 1966, 1967, 1976a) have shown that many of these similarities are only superficial, and that important differences both in arrangement and evolutionary tendencies characterize each lineage. In this view, the Metacopina were redefined as a separate evolutionary lineage in their own right, descended from primitive Podocopina and showing parallel evolution with contemporary lineages (Adamczak, 1967, 1976a,b; Gründel, 1967, 1968, 1969).

Because of its primitive soft parts, the "living fossil" *Saipanetta* was first interpreted as a living healdiacean (McKenzie, 1967a,b, 1970, 1972, 1975; Maddocks, 1972, 1973), and thus the chronostratigraphic and taxonomic range of the Metacopina was greatly expanded (Szechura and Blaszyk, 1968; Hartmann and Puri, 1974). Schulz (1976) considered *Saipanetta* to represent a phylogenetic intermediate between ancestral Platycopina (*Cytherella*) and descendant Podocopina. Several characters of *Saipanetta* indicate affinity with Darwinulacea and, to a lesser extent, with Cypridacea (McKenzie, 1967a, 1970, 1975; Maddocks, 1972, 1973). For this reason, several workers (McKenzie, 1972, 1975; Gründel, 1978; Kozur, 1972) have transferred the Darwinulacea and Cypridacea to the Metacopina (= Cypridocopina), transforming it into an enormous vertical taxon that has engulfed all Podocopida except the rather specialized Bairdiacea and Cytheracea. On the other hand, it has been determined (Gramm, 1977; Schornikov and Gramm, 1974) that the similarities of *Saipanetta* to Healdiacea are only superficial, and that it appears to be much more closely related to a different podocopine lineage (Sigilliacea). Danielopol (1972) and Maddocks (1976) emphasized the podocopine aspects of *Saipanetta,* pointing out its separate resemblances to each of the other branches, and Maddocks (1976) considered it to be a plausible model for certain characters of the ancestral podocopine.

Of the Metacopina as here defined, the Healdiacea are the most closely related to the Podocopina and probably are descended from a primitive Ordovician podocopine (Adamczak, 1967; Kozur, 1972). Although many authors have postulated the reverse derivation of Podocopina from Healdiacea, it is not the phylogeny that is in question, but the taxonomic placement of the presumed ancestral Family Bairdiocyprididae. Henningsmoen (1965) speculated that the Metacopina (Quasillitacea) arose from Kloedenellacea, but Thlipsuracea and Quasillitacea are fairly closely connected branches without known descendants (Gründel, 1967, 1978; Adamczak, 1967).

Within the Podocopina, there is considerable evidence for recognizing two main branches, with Sigilliacea, Darwinulacea, and Cypridacea as one, and Bairdiacea and Cytheracea as the other (Müller, 1894; Skogsberg, 1920; Alm, 1915; Hartmann, 1963; Gründel, 1967; McKenzie, 1972; Schulz, 1976; Maddocks, 1976). The Sigilliacea parallel the Healdiacea in many characters, but they are considered (Schornikov and Gramm, 1974; Gramm, 1977) to be a separate podocopine lineage. They have been extremely conservative, and living *Saipanetta* retains a very primitive anatomy. The Darwinulacea are closely allied to Sigilliacea and also have retained many primitive features. The adductor muscle-scar pattern has been stable since the Carboniferous (Sohn, 1976), and they have changed hardly at all since the Jurassic, perhaps because all known living species are parthenogenetic. The Cypridacea share several important characters of appendages and genitalia, notably the Zenker's organ, with Sigilliacea and Darwinulacea, but otherwise they can hardly be considered primitive. Since their origin, they have radiated vigorously in marine and freshwater environments. A few have adapted to the supralittoral zone and other water-logged terrestrial habitats. Although the latter have bizarre setae for retaining water, phyletically they are not a separate lineage of podocopines (the "Terrestricytheracea" of Schornikov, 1969, 1980), but are acutely specialized Cypridacea.

The oldest podocopines known may belong to the Bairdiacea, and this ancient stock probably stands near the ancestral line of the order. The carapace has been very conservative in its evolution, and today the soft-part anatomy appears to be equally conservative. It must be presumed that this bairdiacean anatomy has been derived by simplification and loss of many primitive (sigilliacean?) features (Maddocks, 1976). The Cytheracea are obviously descended from Bairdiacea (Müller, 1894; Skogsberg, 1920; Gründel, 1967, 1978; Gründel and Kozur, 1971, 1973), although Scott (1961) and Sylvester-Bradley (1961) postulated affinity with Quasillitacea. They have continued the reductional evolutionary trends of the Bairdiacea and are today very simple and conservative in soft-part anatomy. In contrast, the carapace has become ecologically plastic and a site of much evolutionary novelty. After explosive adaptive radiation in the Mesozoic, they have become the dominant ostracodes in marine environments and comprise more than 90% of the present-day fauna.

REFERENCES

Abushik, A. F. (1979). Otrad Leperditicopida—ctroenie, klassifikatsiya rasprostranenie. *In* "Taxonomy, Biostratigraphy and Distribution of Ostracodes" (N. Krstić, ed.), pp. 29-34. Serb. Geol. Soc., Belgrade.

Adamczak, F. (1961). Eridostraca—a new suborder of ostracods and its phylogenetic significance. *Acta Palaeontol. Pol.* **6,** 29-102.

Adamczak, F. (1966). On Kloedenellids and Cytherellids (Ostracoda Platycopa) from the Silurian of Gotland. *Stockholm Contrib. Geol.* **15,** 7-21.

Adamczak, F. (1967). Morphology of two Silurian metacope ostracodes from Gotland. *Geol. Foeren. Stockholm Foerh.* **88,** 462-475.

Adamczak, F. (1968). Palaeocopa and Platycopa (Ostracoda) from Middle Devonian rocks in the Holy Cross Mountains, Poland. *Stockholm Contrib. Geol.* **17,** 1-109.

Adamczak, F. (1969). On the question of whether the palaeocope ostracodes were filter-feeders. *In* "Taxonomy, Morphology and Ecology of Recent Ostracoda" (J. Neale, ed.), pp. 93-98. Oliver & Boyd, Edinburgh and London.

Adamczak, F. (1976a). Morphology and carapace ultrastructure of some Healdiidae (Ostracoda). *Abh. Verh. Naturwiss. Ver. Hamburg* [N.S.] **18/19,** Suppl., 315-318.

Adamczak, F. (1976b). Middle Devonian Podocopida (Ostracoda) from Poland; their morphology, systematics and occurrence. *Senckenbergiana Lethaea* **57,** 265-467.

Alm, G. (1915). Monographie der Schwedischen Süsswasser-Ostracoden nebst systematischen Besprechungen der Tribus Podocopa. *Zool. Bidr. Uppsala* **4,** 1-247.

Andres, D. (1969). Ostracoden aus dem mittleren Kambrium von Öland. *Lethaia* **2,** 165-180.

Bate, R. H. (1972). Phosphatized ostracods with appendages from the Lower Cretaceous of Brazil. *Palaeontology* **15,** 379A-393.

Bowman, T. E. (1971). The case of the nonubiquitous telson and the fraudulent furca. *Crustaceana* **21,** 165-175.

Briggs, D. E. W. (1978). The morphology, mode of life, and affinities of *Canadaspis perfecta* (Crustacea: Phyllocarida), Middle Cambrian, Burgess Shale, British Columbia. *Geol. Surv. Can., Bull.* **264,** 439-487.

Brongniart, C. (1876). Note sur un nouveau genre d'entomostracé fossile provenant du Terrain Carbonifère des environs de Saint-Étienne (Paleocypris edwardsii). *Ann. Sci. Geol.* **7,** 1-6.

Cannon, H. G. (1925). On the segmental excretory organs of certain freshwater ostracods. *Philos. Trans. R. Soc. London* **214,** 1-27.

Dahl, E. (1956). Some crustacean relationships. *In* "Bertil Hanström, Zoological Papers in Honour of His Sixty-Fifth Birthday, November 20, 1956" (K. G. Wingstrand, ed.), pp. 138-147. Lund Zool. Inst., Lund, Sweden.

Dahm, E. (1976). The carapace of Cladocera—a morphological comparison of Cladocera and Ostracoda. *Abh. Verh. Naturwiss. Ver. Hamburg* [N.S.] **18/19,** Suppl, 331-336.

Danielopol, D. (1972). Supplementary data on the morphology of *Neonesidea* and remarks on the systematic position of the Family Bairdiidae (Ostracoda: Podocopida). *Proc. Biol. Soc. Wash.* **85,** 39-48.

Dzik, J. (1978). A myodocopid ostracode with preserved appendages from the Upper Jurassic of the Volga River region (USSR). *Neues Jahrb. Geol. Palaeontol., Monatsh.* **7,** 393-399.

Elofsson, R. (1966). The nauplius eye and frontal organs of the non-malacostraca (Crustacea). *Sarsia* **25,** 1-128.

Giesbrecht, W. (1912). Crustacea. *Handb. Morphol. Wirb. Tiere* **4,** 9-21.

Gocht, H., and Goerlich, F. (1957). Reste des Chitin-Skelettes in fossilen Ostracoden-Gehausen. *Geol. Jahrb.* **73,** 205-214.

Gramm, M. N. (1967). Rudimentarnye muskulnyepjatna u triasovych *Cytherelloidea* (Ostracoda). *Dokl. Akad. Nauk SSSR* **173,** 931-934.

Gramm, M. N. (1968). Morskie triasovye ostrakody iz juzhnogo Primorya. Ob evoliutsii adduktoria kavellinid. *Probl. Paleontol. Mezhdunar. Geol. Kongr., 23rd Sess., 1968* pp. 109-122.

Gramm, M. N. (1969). Stroenie gomologichnykh organov i ikh èvolutsia kak sposob ustanovlenia rodstvennykh otnosheniĭ (na primere adduktora iskopaemykh ostrakod). Problemy Filogeniĭ i Sistematiki (M. N. Gramm and V. A. Krassilov, eds.). Vsesouznoe Paleontologicheskoe Obshchestvo, Vladivostokoe Otdelenie, Dal'nevostochnyĭ Geologicheskiĭ Institut, Akademia Nauk SSSR, pp. 72-91.

Gramm, M. N. (1970a). Otpechatki adduktora triasovykh citerellid (Ostracoda) Primorya i nekotoryie teorii filembryogeneza. *Paleontol. Zh.* pp. 88-103.

Gramm, M. N. (1970b). Ostrakody semeyctva Healdiidae iz Triasovykh otlozheniy yuzhnopo primorya. *In* "Triasovye Bespozvonochnye i Rasteniia Vostoka SSSR, Vladivostok 1970" (E. V. Krasnov, ed.), pp. 41-93. Dal'nevostochnyĭ Nauchnyĭ Tsentr, Dal'nevostochnyĭ Geologicheskiĭ Institut, Akademiiâ Nauk, SSSR.

Gramm, M. N. (1972). Marine Triassic ostracodes from South Primorye. On the evolution of the adductors of Cavellinids. *Proc. Int. Paleontol. Union, Int. Geol. Congr., 23rd, 1968* pp. 135-148.

Gramm, M. N., ed. (1973a). Neoteniya i napravlennost evolyutsii otpechatka adduktora ostrakod. *Mater. Evol. Semin., Vladivostok* **1,** 31-41. Dal'nevostochnyĭ Nauchnyĭ Tsentr, Biologo-Pochvennyĭ Institute, Institut Biologii Moriâ.

Gramm, M. N. (1973b). Neotenicheskie yable niya u iskogaemykh ostrakod. *Paleontol. Zh.* pp. 1-12.

Gramm, M. N., and Pozner, V. M. (1972). Morfologiiâ i ontogenez otnechatka adduktora paleozoĭskikh ostrakod *Scrobicula scrobiculata*. *Paleontologicheskiĭ Zhurnal,* 99-105.

Gramm, M. N. (1976). On two tendencies on the evolution of ostracod adductor muscle scar. *Abh. Verh. Naturwiss. Ver. Hamburg* [N.S.] **18/19,** Suppl., 287-294.

Gramm, M. N. (1977). The position of the families Microcheilinellidae and Sigilliidae (Ostracoda) within the Suborder Podocopa. *Palaeontographica A* **155,** 193-199.

Grobben, K. (1892). Zur Kenntnis des Stammbaumes und des Systems der Crustaceen. *Sitzungsber. Kais. Akad. Wiss. Wien, Math.-Naturwiss. K.* **150,** 237-274.

Gründel, J. (1964a). Zur Ausbildung und taxionomische Bedeutung der Narben der zentralen Muskelgruppe in der Unterklasse Ostracoda. *Neues Jahrb. Geol. Palaeontol., Monatsh.* **10,** 577-597.

Gründel, J. (1964b). Zur Gattung *Healdia* (Ostracoda) und zu einigen verwandten Formen aus dem unteren Jura. *Geologie* **13,** 456-477.

Gründel, J. (1967). Zur Grossgliederung der Ordnung Podocopida G. W. Müller, 1894 (Ostracoda). *Neues Jahrb. Geol. Palaeontol., Monatsh.* **6,** 321-332.

Gründel, J. (1968). Zur Gliederung der Familie Healdiidad (Ostracoda) und zu ihrer Stellung innerhalb der Ordnung Podocopida. *Ber. Dtsch. Ges. Geol. Wiss., Reihe A* **13,** 225-232.

Gründel, J. (1969). Neue taxionomische Einheiten der Unterklasse Ostracoda (Crustacea). *Neues Jahrb. Geol. Palaeontol., Monatsh.* **6,** 353-361.

Gründel, J. (1978). Die Ordnung Podocopida SARS, 1866 (Ostracoda). Stand und Probleme der Taxonomie und Phylogenie. *Freiberg. Forschungsh. C* **334,** 49-68.

Gründel, J., and Kozur, H. (1971). Zur Taxonomie der Bythocytheridae und Tricorninidae (Podocopida, Ostracoda). *Monatsber. Dtsch. Akad. Wiss. Berlin* **13,** 907-937.

Gründel, J., and Kozur, H. (1973). Zur Phylogenie der Tricorninidae und Bythocytheridae (Podocopida, Ostracoda). *Freiberg. Forschungsh. C* **282,** 99-111.

Guber, A. L., and Jaanusson, V. (1964). Ordovician ostracodes with posterior domiciliar dimorphism. *Bull. Geol. Inst. Univ. Uppsala* **42,** 1-43.

Hartmann, G. (1963). Zur Phylogenie und Systematik der Ostracoden. *Z. Zool. Syst. Evolutionsforsch.* **1,** 1-154.

Hartmann, G. (1964). Neontological and paleontological classification of Ostracoda. In, "Ostracods as Ecological and Palaeoecological Indicators" (H. S. Puri, ed.), *Pubbl. Stn. Zool. Napoli* **33,** Suppl., 550-587.

Hartmann, G. (1966). Ostracoda. *Bronn's Klassen* **5,** Part 4, No. 1, 1-216.

Hartmann, G. (1967). Ostracoda. *Bronn's Klassen* **5,** Part 4, Nos. 1, 2, 217-408.

Hartmann, G. (1968). Ostracoda. *Bronn's Klassen* **5,** Part 4, No. 3, 409-568.

Hartmann, G., and Puri, H. S. (1974). Summary of neontological and paleontological classification of Ostracoda. *Mitt. Hamb. Zool. Mus. Inst.* **70,** 7-73.

Henningsmoen, G. (1954). Lower Ordovician ostracods from the Oslo region, Norway. *Nor. Geol. Tidsskr.* **33,** 41-68.

Henningsmoen, G. (1965). On certain features of palaeocope ostracodes. *Geol. Foeren. Sockholm foerh.* **86,** 329-334.

Hessler, R. R. (1964). The Cephalocarida: Comparative skeleto-musculature. *Mem. Conn. Acad. Arts Sci.* **16,** 1-97.

Hessler, R. R. (1969). Cephalocarida. *In* "Treatise on Invertebrate Paleontology" (R. C. Moore, ed.), Part R, Arthropoda 4, Crustacea (except Ostracoda) Myriapoda-Hexapoda, Vol. 1, pp. R120-R128. Geol. Soc. Am., Boulder, Colorado, and the Univ. of Kansas Press, Lawrence.

Hornibrook, N. de B. (1949). A new family of living Ostracoda with striking resemblances to some Palaeozoic Beyrichiidae. *Trans. R. Soc. N. Z.* **77,** 469-471.

Hornibrook, N. de B. (1963). The New Zealand ostracode family Punciidae. *Micropaleontology* **9,** 318-320.

Howe, H. V., Kesling, R. V., and Scott, H. W. (1961). Morphology of living Ostracoda. *In* "Treatise on Invertebrate Paleontology" (R. C. Moore, ed.), Part Q, Arthropoda 3, pp. Q3-Q17. Geol. Soc. Am., Boulder, Colorado, and the Univ. of Kansas Press, Lawrence.

Jaanusson, V. (1957). Middle Ordovician ostracodes of central and southern Sweden. *Uppsala Universitet, Mineralogisk-geologiska Institut, Bulletin* **37,** 173-442.

Jones, P. J., and McKenzie, K. G. (1980). Queensland Middle Cambrian Bradoriida (Crustacea): New taxa, palaeobiogeography and biological affinities. *Alcheringa* **4,** 203-225.

Kesling, R. V. (1951). Terminology of ostracod carapaces. *Contrib. Mus. Paleontol., Univ. Mich.* **9,** 93-171.

Kesling, R. V. (1957). Origin of Beyrichiid ostracods. *Contrib. Mus. Paleontol., Univ. Mich.* **14,** 57-80.

Kesling, R. V. (1961). Ontogeny of Ostracoda. *In* "Treatise on Invertebrate Paleontology" (R. C. Moore, ed.), Part Q, Arthropoda 3, pp. Q19-Q21. Geol. Soc. Am., Boulder, Colorado, and the Univ. of Kansas Press, Lawrence.

Kornicker, L. S. (1975). Antarctic Ostracoda (Myodocopina). *Smithson. Contrib. Zool.* **163,** 1-720.

Kornicker, L. S., and Sohn, I. G. (1976a). Evolution of the Entomoconchacea. *Abh. Verh. Naturwiss. Ver. Hamburg* [N.S.] **18/19,** Suppl., 55-61.

Kornicker, L. S., and Sohn, I. G. (1976b). Phylogeny, ontogeny and morphology of living and fossil Thaumatocypridacea (Myodocopa; Ostracoda). *Smithson. Contrib. Zool.* **219,** 1-124.

Kozur, H. (1972). Einige Bemerkungen zur Systematik der Ostracoden und Beschreibung neuer Platycopida aus der Trias Ungarns und der Slowakei. *Geol.-Palaeontol. Mitt. Innsbruck* **2,** 1-27.

Kozur, H. (1974). Die Bedeutung der Bradoriida als Vorlaeufer der post-Kambrischen Ostracoden. *Z. Geol. Wiss.* **2,** 823-830.

Kristan-Tollmann, E. (1971). Aur phylogenetischen und stratigraphischen Stellung der triadischen Healdiiden (Ostracoda). *Erdoel-Erdgas-Z.* **87,** 428-438.

Kristan-Tollmann, E. (1973a). Zur phylogenetischen und stratigraphischen Stellung der triadischen Healdiiden (Ostracoda). II. *Erdoel-Erdgas-Z.* **89,** 150-155.

Kristan-Tollmann, E. (1973b). Zur Ausbildung des Schliessmuskelfeldes bei triadischen Cytherellidae (Ostracoda). *Neues Jahrbuch für Geologie und Paläontologie, Monatshefte* **6,** 351-373.

Kristan-Tollmann, E. (1977a). Zur Evolution des Schliessmuskelfeldes bei Healdiidae und Cytherellidae (Ostracoda). *Neues Jahrb. Geol. Palaeontol., Monatsh.* **10,** 621-639.

Kristan-Tollmann, E. (1977b). On the development of the muscle-scar pattern in Triassic Ostracoda. *In* "Aspects of Ecology and Zoogeography of Recent and Fossil Ostracoda" (H. Löffler and D. Danielopol, eds.), pp. 133-143. Junk, The Hague.

Kristan-Tollmann, E. (1979). Taxonomie der Mesozoischen Healdiidae. *In* "Taxonomy, Biostratigraphy and Distribution of Ostracodes" (N. Krstić, ed.), pp. 41-45. Serb. Geol. Soc., Belgrade.

Lebedev, V. I., and Lebedeva, A. I. (1965). On the reasons for the relatively early appearance and the single period of flourishing of organisms which build their shells of fluorapatite. *Geokhimiya* **12,** 1404-1409.

Levinson, S. A. (1951). Thin sections of Palaeozoic Ostracoda and their bearing on taxonomy and morphology. *J. Paleontol.* **25,** 553-560.

McKenzie, K. G. (1967a). Ostracod "living fossils": New finds in the Pacific. *Science* **155,** 1005.

McKenzie, K. G. (1967b). Saipanellidae: A new family of podocopid Ostracoda. *Crustaceana* **13,** 103-113.

McKenzie, K. G. (1970). The 'soft parts' of *Saipanetta tumida* (Brady, 1890) (Ostracoda, Metacopina), with an expansion of the diagnosis of Saipanettidae. *Crustaceana* **19,** 104-105.

McKenzie, K. G. (1972). Contribution to the ontogeny and phylogeny of Ostracoda. *Proc. Int. Paleontol. Union, Int. Geol. Congr., 23rd, 1968*, pp. 165-188.

McKenzie, K. G. (1975). *Saipanetta* and the classification of podocopid Ostracoda: A reply to Schornikov and Gramm (1974). *Crustaceana* **29,** 221-224.

McKenzie, K. G., and Jones, P. J. (1979). Partially preserved soft anatomy of a Middle Cambrian Bradoriid (Ostracoda) from Queensland. *Search* **10,** 444-445.

Maddocks, R. F. (1972). Two new living species of *Saipanetta* (Ostracoda, Podocopida). *Crustaceana* **23,** 28-42.

Maddocks, R. F. (1973). Zenker's organ and a new species of *Saipanetta* (Ostracoda). *Micropaleontology* **19,** 193-208.

Maddocks, R. F. (1976). Quest for the ancestral podocopid: Numerical cladistic analysis of ostracode appendages, a preliminary report. *Abh. Verh. Naturwiss. Ver. Hamburg* [N.S.] **18/19,** Suppl, 39-53.

Manton, S. M. (1969). Introduction to classification of Arthropoda. *In* "Treatise on Invertebrate Paleontology" (R. C. Moore, ed.), Part R, Arthropoda 4, Crustacea (Except Ostracoda) Myriapoda-Hexapoda, Vol. 1, pp. 3-15. Geol. Soc. Am., Boulder, Colorado, and the Univ. of Kansas Press, Lawrence.

Martinsson, A. (1962). Ostracodes of the family Beyrichiidae from the Silurian of Gotland. *Bull. Geol. Inst. Univ. Uppsala* **41,** 1-369.

Müller, G. W. (1894). Die Ostracoden des Golfes von Neapel und der angrenzenden Meeres-Abschnitte. *Fauna Flora Golfes Neapel* **21,** 1-404.

Müller, K. J. (1964). Ostracoda (Bradoriina) mit phosphatischen Gehäusen aus dem Oberkambrium von Schweden. *Neues Jahrb. Geol. Palaeontol., Abh.* **121,** 1-46.

Müller, K. J. (1979a). Phosphatocopine ostracodes with preserved appendages from the Upper Cambrian of Sweden. *Lethaia* **12,** 1-27.

Müller, K. J. (1979b). Body appendages of Paleozoic ostracods. *In* "Taxonomy, Biostratigraphy and Distribution of Ostracodes" (N. Krstić, ed.), pp. 5-8. Serb. Geol. Soc., Belgrade.

Neale, J. W. (1977). Discussion on ostracod terminology. *In* "Aspects of Ecology and Zoogeography of Recent and Fossil Ostracoda" (H. Löffler and D. Danielopol, eds.), pp. 495-497. The Hague.

Öpik, A. A. (1968). Ordian (Cambrian) Crustacea Bradoriida of Australia. *Bull.—Bur. Miner. Resour., Geol. Geophys.* (*Aust.*) **103,** 1-37.

Pokorný, V. (1958). "Grundzüge der Zoologischen Mikropaläontologie," Vol. 2, pp. 1-453. VEB Dtsch. Verlag Wiss., Berlin.

Pokorný, V. (1964). *Conchoecia? cretacea* n. sp., first fossil species of the family Halocyprididae (Ostracoda, Crustacea). *Acta Universitatis Carolinae, Geologica No. 2,* 175-180.

Pokorný, V. (1978). Ostracodes. *In* "Introduction to Marine Micropaleontology" (B. U. Haq and A. Boersma, eds.), pp. 109-149. Am. Elsevier, New York.

Raymond, P. E. (1935). *Leanchoilia* and other mid-Cambrian Arthropoda. *Bull. Mus. Comp. Zool.* **76,** 205-230.

Remane, A. (1957). Die Geschichte der Tiere. *In* "Die Evolution der Organismen" (G. Heberer, ed.), pp. 340-422. Verlag G. Fischer.

Rhodes, F. H. T., and Bloxam, T. W. (1971). Phosphatic organisms in the Paleozoic and their evolutionary significance. *In* "Proceedings of the North American Paleontological Convention" (E. L. Yochelson, ed.), Part K, pp. 1486-1513. Allen Press, Lawrence, Kansas.

Rudwick, M. J. S. (1970). "Living and Fossil Brachiopods." Hutchinson, London.

Sars, G. O. (1866). Oversigt af Norges marine Ostracoden. *Nor. Vidensk.-Akad., Forh.* pp. 1-130.

Sars, G. O. (1922-1928). "An Account of the Crustacea of Norway," Vol. 9, Ostracoda, pp. 1-277. Bergen Museum, Bergen, Norway.

Schallreuter, R. (1968). Ordovizische Ostracoden mit geradem Schlossrand und konkaven Ventralrand. *Wiss. Z. Ernst-Moritz-Arndt-Univ. Greifsw., Math.-Naturwiss. Reihe* **17,** No. 1/2, 127-152.

Schallreuter, R. (1972). Drepanellacea (Ostracoda, Beyrichicopida) aus mittelordovizischen Bachsteinkalk geschieben. *Ber. Dtsch. Ges. Geol. Wiss., Reihe A* **17,** 139-145.

Schallreuter, R. (1973). Die Ostracodengattung *Hyperchilarina* und das *Aparchites*-Problem. *Geol. Foeren. Stockholm Foerh.* **95,** 37-49.

Schornikov, E. I. (1969). Novoe semeystvo rakushkovykh rachkov (Ostracoda) iz supralitorali kurilskikh ostrovov. *Zool. Z.* **48,** 494-498.

Schornikov, E. I. (1980). Ostrakody b nazemnykh biotorakh. *Zool. Zh.* **59,** 1306-1319.

Schornikov, E. I., and Gramm, M. N. (1974). *Saipanetta* McKenzie, 1967 (Ostracoda) from the northern Pacific and some problems of classification. *Crustaceana* **27,** 92-102.

Schulz, K. (1976). Das Chitinskelett der Podocopida (Ostracoda, Crustacea) und die Frage der Metamerie dieser Gruppe. Doctoral Dissertation, Universität Hamburg.

Scott, H. W. (1961). Classification of Ostracoda. *In* "Treatise on Invertebrate Paleontology" (R. C. Moore, ed.), Part Q, Arthropoda 3, Crustacea Ostracoda, pp. 74-92. Geological Society of America, Boulder, Colorado, and the University of Kansas Press, Lawrence.

Siewing, R. (1959). Neue Ergebnisse der Verwandtschaftsforschung bei Crustaceen. *Wiss. Z. Univ. Rostock, Math.-Naturwiss. Reihe* **3,** 343-348.

Simkiss, K. (1964). Phosphates as crystal poisons. *Biol. Rev. Cambridge Philos. Soc.* **39,** 487-505.

Simpson, G. G. (1961). "Principles of Animal Taxonomy," pp. 1-247. Columbia Univ. Press, New York.

Skogsberg, T. (1920). Studies on marine ostracods, part I (Cypridinids, Halocyprids and Polycopids). *Zool. Bidr. Uppsala, Suppl.* **1,** 1-784.

Sohn, E. G. (1961). *Aechminella, Amphissites, Kirkbyella* and related genera. Review of some Paleozoic genera. *Geol. Surv. Prof. Pap.* (*U.S.*) **330-B,** B107-B160.

Sohn, I. G. (1965). Classification of the Superfamily Healdiacea and the genus *Pseudophanasymmetria* Sohn and Berdan, 1952 (Ostracoda). *Geol. Surv. Prof. Pap.* (*U.S.*) **525-B,** B69-B72.

Sohn, I. G. (1971). New Late Mississippian ostracode genera and species from northern Alaska: A revision of the Paraparchitacea. *Geol. Surv. Prof. Pap.* (*U.S.*) **711-A,** A1-A22.

Sohn, I. G. (1974). Evidence for the presence of a heart in Paleozoic ostracodes inconclusive. *J. Res. U. S. Geol. Surv.* **2,** 723-726.

Sohn, I. G. (1976). Antiquity of the adductor muscle attachment scar in *Darwinula* Brady & Robertson, 1885. *Abh. Verh. Naturwiss. Ver. Hamburg* [N.S.] **18/19,** Suppl., 305-308.

Sylvester-Bradley, P. C. (1961). Order Archaeocopida. *In* "Treatise on Invertebrate Paleontology" (R. C. Moore, ed.), Part Q, Arthropoda 3, pp. Q100-Q103, Q358-Q359. Geol. Soc. Am., Boulder, Colorado, and the Univ. of Kansas Press, Lawrence.

Sylvester-Bradley, P. C. (1962). The taxonomic treatment of pbylogenetic patterns in time and space, with examples from the Ostracoda. *Syst. Assoc. Publ.* **4,** 119-133.

Szczechura, J., and Blaszyk, J. (1968). *Cardobairdia inflata* n. sp. from the Middle Jurassic of Poland and its taxonomic position within Ostracoda. *Acta Palaeontol. Pol.* **23,** 185-197.

Tasch, P. (1969). Branchiopoda. *In* "Treatise on Invertebrate Paleontology" (R. C. Moore, ed.), Part R, Arthropoda 4, Crustacea (except Ostracoda) Myriapoda-Hexapoda, Vol. 1, pp. R128-R191. Geol. Soc. Am., Boulder, Colorado, and the Univ. of Kansas Press, Lawrence.

Ulrich, R. O., and Bassler, R. S. (1931). Cambrian bivalved Crustacea of the Order Conchostraca. *Proc. U. S. Natl. Mus.* **78,** 1-130.

van den Bold, W. A. (1974). Taxonomic status of *Cardobairdia* (van den Bold, 1960) and *Abyssocypris* n. gen.: Two deepwater ostracode genera of the Caribbean Tertiary. *Geosci. Man* **6,** 65-79.

van Morkhoven, F. P. C. M. (1962). "Post-Paleozoic Ostracodes," Vol. 1. Elsevier, Amsterdam.

Weygoldt, P. (1960). Embryologische Untersuchungen an Ostrakoden: Die Entwicklung von *Cyprideis littoralis* (G. S. Brady) (Ostracoda, Podocopa, Cytheridae). *Zool. Jahrb., Abt. Anat. Ontog. Tiere* **78,** 369-496.

6

Biogeography

LAWRENCE G. ABELE

THE BIOLOGY OF CRUSTACEA, VOL. 1

ISBN 0-12-106401-8

I. INTRODUCTION

In 1852 Dana listed the distribution of all 2689 known species of Crustacea and suggested that the major patterns were correlated with *isocrymal lines* that indicated the coldest 30 consecutive days of the year. Dana's provinces do not differ greatly from those currently recognized (Ekman, 1953; Briggs, 1974). There are probably more than 40,000 Recent species of Crustacea recognized today. They are distributed throughout the world in virtually all habitats, though by far the greatest number are marine. The fossil record is extensive, beginning in the Cambrian (Schram, Chapter 4 of this volume), and any consideration of distribution patterns must deal with this diversity. To extend Dana's analysis would require knowledge of systematic relationships, the fossil record, geography, paleogeography, and the present day distribution of species. For the Crustacea, little information concerning the above areas is available, and here I can only summarize what appear to be distribution patterns and suggest ideas that might be testable. While other approaches to biogeography might be more rigorous (see Ball, 1975; Rosen, 1978; Simberloff *et al.*, 1981), I do not believe that the available data for Crustacea warrant further treatment.

II. METHODS AND MATERIALS

The crustacean literature is vast, and almost every faunal list, taxonomic note, systematic monograph, and expedition report contains distributional data. There is no way to represent adequately or fairly the efforts of these individual authors, especially authors of reports on expeditions such as the *Siboga, Dana,* and *Galathea,* which are a gold mine of information. I have been restricted by the available library resources and my own limited knowledge of most languages other than English.

In compiling data for the latitudinal gradient figures, I considered a species range to include those regions between the northern and southern records unless there was information suggesting otherwise. For example, a species known to occur in North Carolina and Florida was considered to have a range from North Carolina to Florida. The data sources are cited in the legend of each figure.

III. PALEOZOOGEOGRAPHY

The fossil record of crustaceans is summarized by Schram (Chapter 4 of this volume). Branchiopods, ostracodes, cirripeds, and phyllocarid malacostracans were present during the early Paleozoic. By the late Paleozoic a diversity of malacostracans were present (Schram, 1974b, 1977), and an analysis of their distribution is provided by Schram (1977).

The late Paleozoic crustaceans were predominately Phyllocarida, Hoplocarida, and primitive Eumalacostraca. Among the latter were eocarids, syncarids, pygocephalomorph mysids, spelaeogriphaceans, and phreatoicid isopods. Schram (1977) suggests that the higher malacostracans (hoplocarids and eumalacostracids) originated during the early Devonian in the tropical waters of Laurentia (Fig. 1). The fossil record suggests that these groups were restricted to tropical marine waters during the late Devonian and Carboniferous periods. The formation of Pangea during the Permian resulted in dispersal through shallow-water seas. By the Permian, then, hoplocarid and eumalacostracans had spread into the temperate marine and freshwater

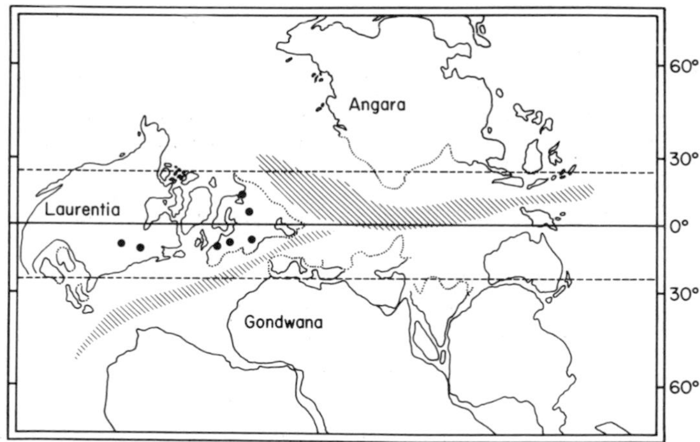

Fig. 1. Records of fossil Malacostraca from the Late Devonian. Hatched areas denote seaways. (From Schram, 1977.)

habitats of Gondwana and Laurentia. This fauna was related to and formed a taxonomic continuum with the Carboniferous fauna. However, the late Permian marked a faunal change (the first decapods appeared) that was to continue into the Mesozoic.

At the beginning of the Triassic, the taxonomic structure of the malacostracan fauna had changed. The only taxa with Paleozoic affinities were a phreatoicidean isopod and an anaspidacean syncarid found in fresh to brackish deposits in and around Sydney, Australia. Both of these groups are extant today and are Gondwanan in distribution: phreactoicidean isopods occur in freshwater habitats of Australia, New Zealand, southern Africa, and India, and anaspidacean syncarids occur in freshwater in Australia, Tasmania, New Zealand, and South America.

The remainder of the Triassic malacostracan fauna is modern in form and consists of marine mysid mysidaceans, flabelliferan isopods, and dendrobranchiate and pleocyemate decapods. These are widely distributed in marine deposits of Pangea. Thus the Recent families of malacostracans (or their immediate ancestors) have had a long time during which they could disperse to all continents through shallow-water seas.

IV. PATTERNS OF SPECIES RICHNESS

A. Latitudinal Patterns

Although there is some variation within taxonomic groups, two major latitudinal patterns emerge. The first is the well known increase in species numbers with decreasing latitude that is shown by ostracodes, copepods (to a certain extent given the limited data), somewhat by euphausiids, and clearly by cirripeds, stomatopods, and decapods (see Figs. 12-15, 19-20). Possible explanations for this have been reviewed by Fischer (1960), Pianka (1966), and Ricklefs (1979), but the available data do not permit any real choice from among them.

The second major latitudinal pattern is shown by amphipods and isopods (and possibly other peracarid groups). These groups are very rich in species, probably as rich in total as the Decapoda, yet the greatest number of species is found in the temperate zone about 30° latitude. It may not be a coincidence that taxonomic groups that are species rich in the temperate regions have direct development, whereas those that are species rich in the tropics have some type of planktonic larval development. Thorson (1950; also Mileikovsky, 1971) showed that, in general, the percentage of prosobranch mollusk species with pelagic larval development increased with decreasing latitude. He suggested that the proportion of species with pelagic development should increase toward the tropics because the production of phytoplankton occurs there throughout the year. Direct development should predominate in the polar regions because of the relatively short period when planktonic food is available coupled with the low water temperature, which might slow development.

B. Longitudinal Patterns

The data available on numbers of species in the four main tropical regions of the world are shown in Table I. For all groups of crustaceans, the greatest number of species is found in the extensive Indo-West Pacific region. Cypridiniform ostracodes, thoracic barnacles, stomatopods, and decapods all have the least number of species in the Eastern Atlantic. The two peracarid groups for which there are data and acrothoracic barnacles have the second greatest number of species in the Eastern Atlantic. Taxonomic groups associated with coral reefs, such as stomatopods, are more diverse in the Western Atlantic, while groups associated with soft bottoms or the rocky intertidal are more diverse in the Eastern Pacific, where these habitats are abundant. Two observations are apparent from the data in Table I. The first is that a very large part of the species richness of the Indo-West Pacific is due to the large number of congeneric species found there compared to other regions. For example, there are 128 species of the snapping shrimp genus

TABLE I

Distribution of Numbers of Species of Crustaceans in the Four Main Tropical Regions of the World

Group	Eastern Atlantic	Eastern Pacific	Western Atlantic	Indo-West Pacific
Ostracoda Myodocopa[a]	4	6	34	92
Cirripedia Thoracica	61	114	90	247
Cirripedia Acrothoracica	7	4	3	34
Mysidacea	109	73	101	382
Amphipoda	185	125	70	575
Stomatopoda	29	46	71	196
Decapoda (estimate)	1000	1500	1300	5000
Portunidae	22	17	25	175
Parthenopidae	9	16	16	114
Sesarma	6	8	15	80
Alpheus	19	22	25	128
Area (km^2)	400,000	380,000	1,280,000	6,570,000

[a] Includes only Cypridiniformes from *Dana* stations.

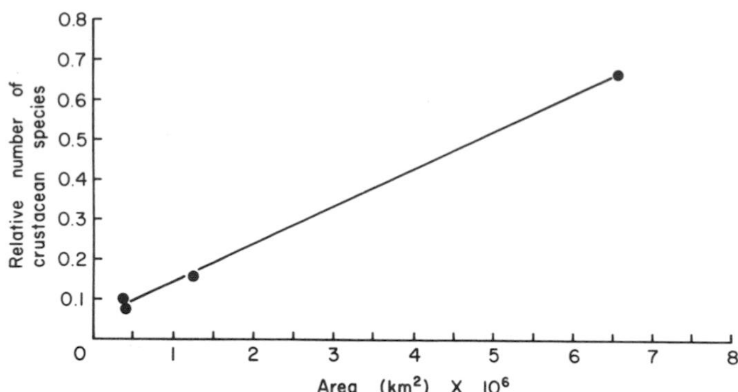

Fig. 2. The relative number of crustacean species in the Eastern Atlantic, Eastern Pacific, Western Atlantic, and Indo-West Pacific plotted against the area of continental shelf in each of the four regions ($r = 0.99$, $p < 0.01$). Relative number of crustaceans was estimated from data in text and Table I, and area data were provided by John C. Briggs.

Alpheus in the Indo-West Pacific and 66 in the rest of the world. The grapsid crab genus *Sesarma* and many other crustacean genera show a similar pattern. The second observation, I believe, is related to the first. In Fig. 2, the relative number of crustacean species is plotted against the amount of shelf area in the regions. The relative number of crustaceans was estimated from Table I and data in the text. The areal extent of shallow water explains about 98% of the variation in numbers of species. Obviously, the relationship is influenced very much by the huge area and large numbers of species in the Indo-West Pacific region. However, additional data presented below suggest that area is important in affecting numbers of species. The Indo-West Pacific region is not only large but consists of many archipelagoes. The large area and many island groups would seem ideal for geographic isolation and speciation. Under these conditions one would expect a large number of congeneric species. This explanation for the diversity of the Indo-West Pacific does not depend on competition (see Heck and McCoy, 1978) and differs markedly from that of Briggs (1966, 1974) and Stehli and Wells (1971).

C. Influence of Area

Area may influence species richness in at least three ways (see Connor and McCoy, 1979, for a review): (1) an increase in area may be correlated with an increase in habitat heterogeneity; (2) an increase in area may permit larger population sizes and therefore reduce the probability of extinction; and (3) an increase in area may simply increase the sample size resulting in

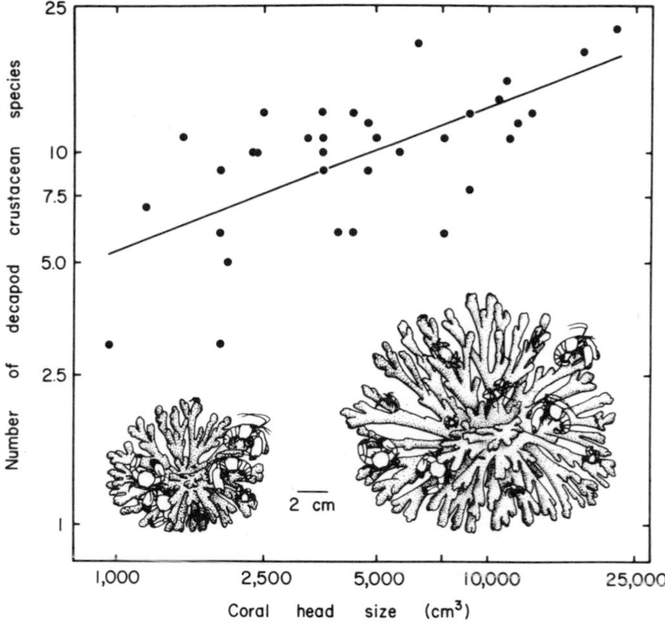

Fig. 3. Number of decapod crustacean species plotted against size ($l \times w \times h$) of coral heads of *Pocillopora damicornis* coral heads. The best fit is the log-log relationship, log S = 0.356 log size − 0.316, where $r = 0.64$ and $p < 0.001$. (Data from Abele and Patton, 1976.)

more species. The relative importance of each mechanism may differ under different ecological conditions.

Area can be shown to influence crustacean species numbers on scales from habitats to major geographic regions. Abele and Patton (1976) demonstrated that there is a significant correlation between the number of decapod crustacean species and the size of coral head (Fig. 3). It is important to consider this when comparing samples from different regions (Abele, 1976a). Abele and Blum (1977) examined the freshwater decapod faunas of the Perlas Islands, Panama, and compared species numbers to size of island and elevation of island. Although species numbers were correlated with both, island elevation was the best predicator of number of freshwater decapod species. A similar result was found here for the freshwater decapods of the West Indies (Fig. 4). However, when the number of West Indian marine shrimps was compared to island area, elevation, and perimeter, perimeter was found to be the best predicator (Fig. 5). This is probably because it reflects the shallow-water habitats of the island. The relationship between the area of shallow-water and the number of tropical crustacean species has already been discussed (Fig. 2).

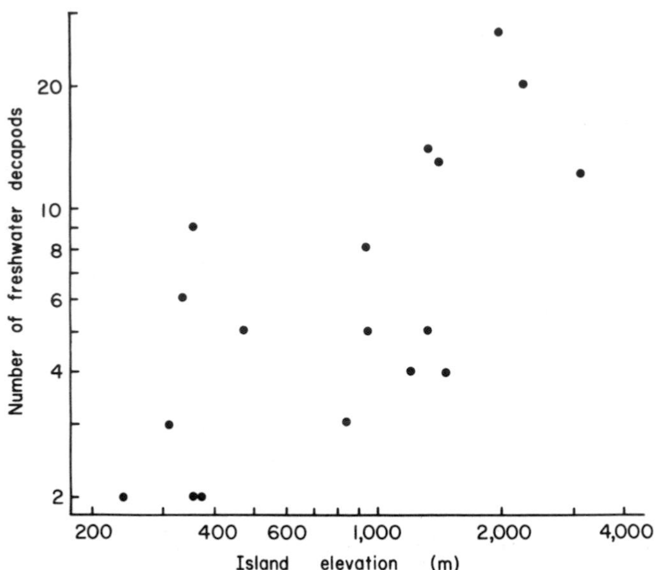

Fig. 4. The number of freshwater decapod crustaceans plotted against island elevation for 18 West Indian islands. The best fit is the log-log relationship, $\log S = 0.694 \log \text{elevation} - 4.05$, where $r = 0.694$ and $p < 0.01$. (Data from Chace and Hobbs, 1969.)

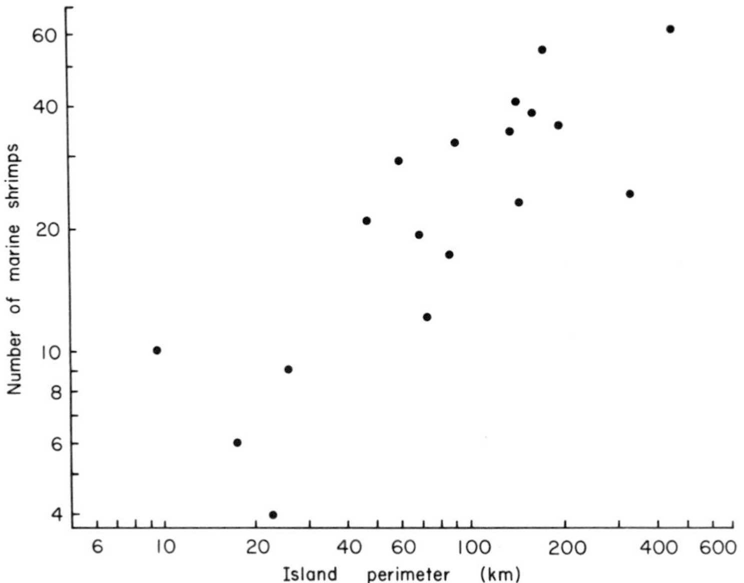

Fig. 5. The number of marine shrimp species plotted against island perimeter for 18 West Indian islands. The best fit is a log-log relationship, $\log S = 0.612 \log \text{perimeter} - 1.875$, where $r = 0.831$ and $p < 0.01$. (Data from Chace, 1972.)

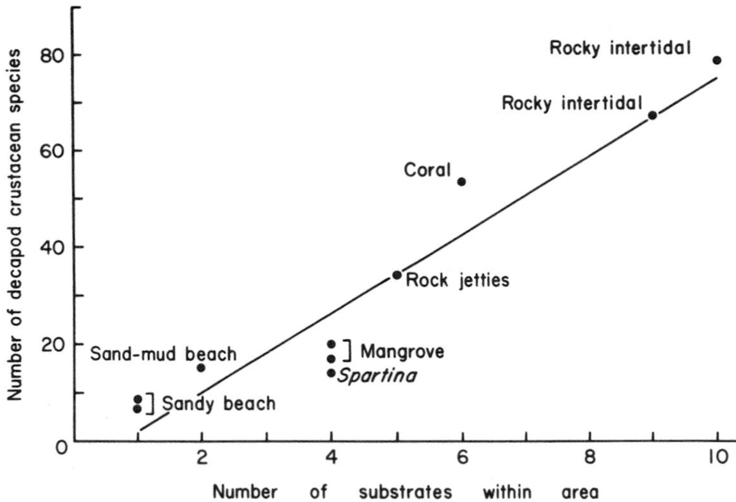

Fig. 6. The number of marine decapod crustacean species plotted against substrate diversity of some habitats. (Data from Abele, 1974.)

D. Habitat Heterogeneity

The relationship between habitat complexity and species richness is usually positive (Ricklefs, 1979). For crustaceans, Abele (1974) found that the number of decapods in a habitat was a function of the structural complexity of the habitat (Fig. 6). Habitat subdivisions are arbitrary to a certain extent, but the relationship would probably hold for other possible subdivisions. Whiteside and Harmsworth (1967) found a positive relationship between habitat complexity and species diversity of chydorid cladocerans. Within the grassbed habitat, the number of decapod species is positvely correlated with the biomass of the plants rather than with the number of plant species (Heck and Wetstone, 1977).

Abele (1974) suggested that if structural complexity was important in determining species richness, similar habitats might have similar numbers of species regardless of the size of the species pool. This hypothesis was based on his own results and those of Dahl (1953) and Dexter (1972), who both found a similar number of crustaceans on sandy beaches of different latitude and longitude. The hypothesis was tested by Heck (1979), who compared the number of invertebrates, primarily decapod crustaceans, found in tropical and temperate grassbeds. Although the total number of decapods was greater in the tropics, Heck suggested that this was due to transients from adjacent habitats, and once these were excluded the number of resident species was similar in temperate and tropical regions. Nelson (1980) similarly found no relationship between latitude and gammarid amphipod species richness in seagrasses from Nova Scotia to Florida. Another test of the hypothesis was performed by Abele (1981), who compared the number of decapods found on the coral *Pocillopora damicornis* in the Eastern Pacific (Panama), Central Pacific (Palau), and the Great Barrier Reef (Lizard Island). In these three areas, the number of species occurring on *P. damicornis* was similar despite the fact that there are about five times as many potential colonists in the Central Pacific and tropical Australia. These results suggest that the increase in species numbers across the Pacific is not the result of an increase in within-habitat diversity.

E. Species Swarms

There are several regions in the world where a relatively large number of closely related species are confined to a narrowly circumscribed area. The most famous species swarm among the Crustacea is the gammarid amphipod fauna of Lake Baikal, from which more than 240 species and 34 genera have been reported. All but one of the species are endemic to Lake Baikal and are ecologically and morphologically diverse. Other crustacean species swarms reported from the lake include isopods (*Asellus*), ostracodes (*Candona, Pseudocandona* and *Cytherissa*), and the harpacticoid copepod genus *Moraria*. In addition to being very old (it probably originated some 30 million years ago), the lake is the deepest in the world (1741 m) and among the largest, extending about 636 km in length and up to 80 km in width (Kozhov, 1963). The Baikal amphipods have been cited as support for the notion that long periods of time are required to evolve a highly diverse fauna (Sanders, 1968). Other factors, however, have certainly been important (see Barnard, 1969b; Brooks, 1950). For example, Lake Baikal in the past consisted of a ramified system of shallow lakes, and geographic isolation has probably been a major factor in the formation of species. We also do not know the rate of species formation over time and an asymptotic model should be considered (Strong, 1974).

Another species swarm of amphipods is that in the genus *Hyalella,* involving 11 species from Lake Titicaca high in the Andes (Barnard, 1969b).

The African rift lake, Lake Tanganyika, contains a diverse crustacean fauna (Brooks, 1950). Calman (1906) recognized a species swarm of 11 species in 3 genera of atyid shrimp endemic to the lake. Two ostracode genera, *Paracypris* (11 species) and *Cypridopsis* (8 species), the harpacticoid copepod genus *Schizopera,* and two cyclopoid copepod genera *Eucyclops* and *Microcyclops* are also considered to form species swarms in Lake Tanganyika (Brooks, 1950).

Brooks lists the ostracode genus *Candona* from Ohrid Lake (Albania area) and the atyid genus *Cardina* and the copepod *Eodiaptomus* from the Malili River (Celebes) as other crustacean species swarms.

Mention should be made of the crustacean fauna of Artesian Well, San Marcos, Texas, which contains ten amphipod species as well as other crustacean species, making it probably the richest cave crustacean fauna in the world (Holsinger and Longley, 1980).

F. Patterns with Depth

For the majority of crustaceans there is a negative relationship between number of species and depth. This is shown for the Cumacea and Ostracoda in Fig. 7, based on the data in Jones (1969) and Kornicker (1975). Similar plots would be obtained for amphipods (Barnard, 1962) and other groups but, as with the cumaceans, these would include species from a wide variety of habitats. Although the samples were not from exactly the same microhabitat, Coull (1972) has shown that harpacticoid copepods increase in number of species with increasing depth. Although samples were taken down to 5165 m, examination of his figures suggests that samples below 1000 m may not differ in species richness. However, there are some taxa, within larger groups, that increase in species richness with depth. The Desmosomatidae, a family of isopods, shows an increase in number of species with increasing depth (Hessler, 1970). Groups within the Tanaidacea show similar patterns (Gardiner, 1975).

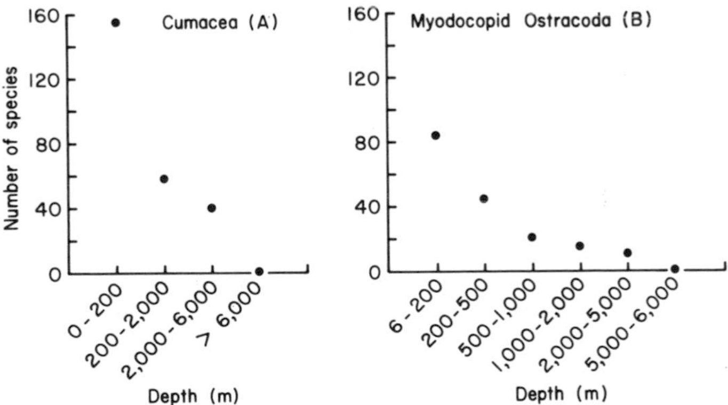

Fig. 7. The number of cumacean (A) and myodocopid ostracode species (B) plotted as a function of depth. (Cumacean data from Jones, 1969; ostracode data from Kornicker, 1975.)

V. BIOGEOGRAPHIC MIGRATIONS

A. Lessepsian Migration

Lessepsian migration refers to a unidirectional movement of species from the Red Sea through the Suez Canal into the Mediterranean (Por, 1971). Decapod crustaceans are thought to be particularly important in this migration, constituting 21.3% (30 species) of the putative migrants (Por, 1973). The process is thought to be unidirectional into the Mediterranean because the Levant Basin of the Mediterranean has high salinities and temperature and is undersaturated with a temperate fauna. The Red Sea fauna is then considered to be preadapted to the habitat and dominant in competitive abilities (Topp, 1969; Briggs, 1974; see Vermeij, 1978, for a review).

The role of competition in the migration cannot be evaluated with the present data. Caution is suggested for at least four reasons. First, some species considered to be Red Sea migrants are circumtropical and therefore may have immigrated from the Atlantic (John and Lawson, 1974). Second, Kornfield and Nevo (1976) have presented electrophoretic data on *Aphanius dispar* which suggest that the fish is not a recent Red Sea migrant, as previously believed. Third, our knowledge of the Mediterranean biota prior to the construction of the Suez Canal is scanty, and therefore any species may have been present in the Mediterranean before the canal opened. Knowledge of the Red Sea is also weak: in 1958, Holthuis reported on 58 species of macruran decapods from the Red Sea, 15 of which were recorded from the area from the first time. Finally, I would suggest that biotic

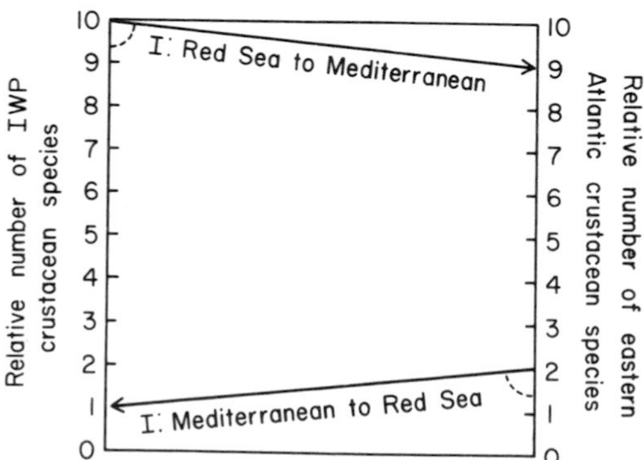

Fig. 8. Proposed relationship between number of crustacean species and Lessepsian migration. The number of Indo-West Pacific crustacean species is about five times that of the Eastern Atlantic-Mediterranean. If the immigration rate for the two regions is about equal, then we should expect migration to be predominately from the Red Sea through the Suez Canal to the Mediterranean independent of any competitive interactions or preadaptations.

exchange should be predominately from the Red Sea on the basis of diffusion alone, in the absence of any factor other than the difference in species richness between the two regions. This is diagrammed in Fig. 8. Considering the difference between the Eastern Atlantic and Indo-West Pacific, the Red Sea region probably contains five times the number of crustacean species found in the Mediterranean. Thus, even if biotic exchange were random the result would be largely unidirectional from the Red Sea to the Mediterranean.

B. Indo-West Pacific to East Pacific

Another almost exclusively unidirectional migration has been suggested, involving crustaceans and other organisms. This is movement from the Indo-West Pacific to the Eastern Pacific (Garth, 1974). Garth (1974) points out that, with few exceptions, the species common to these two regions are obligate symbionts of corals, particularly corals of the genus *Pocillopora*. Again, I would suggest caution in concluding that these species recently migrated from the Indo-West Pacific. First, there is the distinct possibility that the species have always been in the Eastern Pacific as part of the Tethyian fauna (see a discussion dealing with corals by Heck and McCoy, 1978). Second, there are a number of taxonomic questions concerning the species involved. For example, Castro (1981) recently showed that *Trapezia ferruginea,* considered to be an Indo-West Pacific migrant, actually consists of three species, two of which may be endemic to the Eastern Pacific. Additional studies may reveal similar problems.

C. Migration and Salinities

As salinity declines from about 10 to 5 ppt, the number of marine species declines markedly. Similarly, the number of freshwater species drops markedly as salinities increase from about 3 to 5 ppt (Remane, 1934). Salinities between 3 and 5 ppt seem to be a barrier between marine and freshwater species and represent a species minimum for both groups. This relationship appears to be true for decapods. In Fig. 9, data on relative species numbers and salinities are plotted for the decapods of the Panama Canal. The data are based on collections made (L. G. Abele, unpublished) from the freshwater canal through Gatun locks into the Caribbean. Gainey and Greenberg (1977) reviewed this phenomenon for mollusks: they related the species minimum to physiochemical discontinuities in the ionic composition of seawater in this region, which are reflected in the physiological mechanisms of mollusks. It is possible that decapods are responding similarly to these changes.

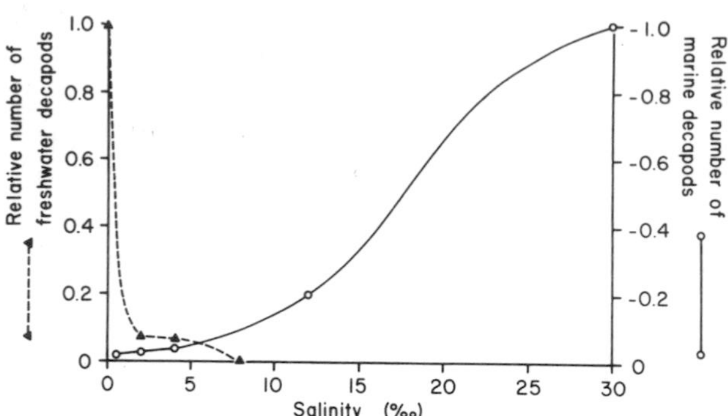

Fig. 9. Relationship between relative species numbers of marine and freshwater decapod crustaceans of the Panama Canal as a function of salinity. (Data from L. G. Abele, unpublished.)

D. Introduction by Man

Man, both purposefully and accidentally, has infleunced the distributions of some crustacean species. Kaestner (1970) notes that many terrestrial species of isopods are cosmopolitan in greenhouses. Abele (1972) reported on the probable introductions of two decapods. One, *Neorhynchoplax kempi,* collected in the Panama Canal, was previously known from a single locality in Iraq. Lachner *et al.* (1970) summarized some additional introductions of decapods. Additional introductions have been reported by Barnard (1970), Fulton and Grant (1900), Stubbings (1967), Newman (1963), Chilton (1911), Marchand (1946), Wolff (1954), Hurley (1968), Naylor (1957, 1960), Edmondson (1962), McKenzie (1971b), Holthuis and Provenzano (1970), and Carlton (1975). Additional data are reviewed by Jazdzewski (1980), with special reference to gammarid amphipods. This list of references is far from complete, and a review of introductions would be very desirable.

VI. MORPHOLOGY AND DISTRIBUTION

A. Size

Barnard (1962) examined the striking relationship between latitude and size among gammaridean amphipods. Arctic and Antarctic amphipods have average lengths of 17 and 13 mm, respectively, while tropical amphipods are less than half this size (5–6 mm). Poulsen (1965) examined this relationship for a number of myodocopid ostracodes (Fig. 10). He found that species from higher latitudes are generally larger (two to four times larger) than those

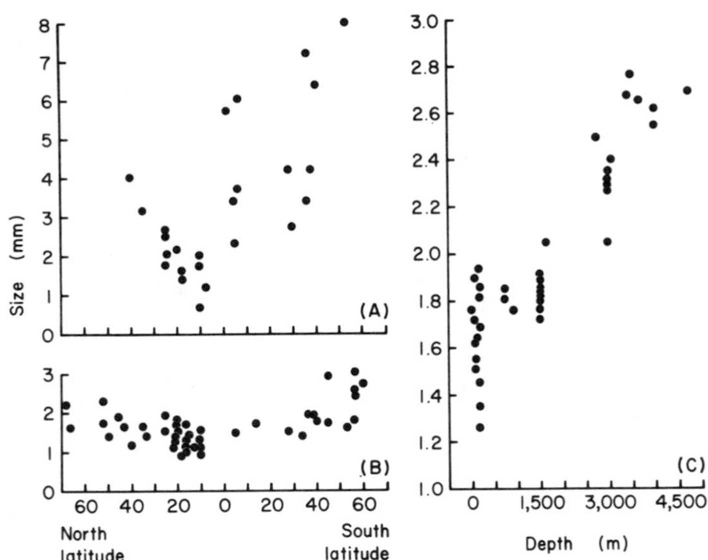

Fig. 10. Relationship between body size and latitude for (A) Cyclasteropinae and (B) Asteropinae ostracodes. (C) Relationship between body size and depth in the myodocopid ostracodes, *Synasterope* and *Spinacopia*. (Latitude data from Poulsen, 1965; depth data from Kornicker, 1975.)

from the tropics. In the Cyclasteropinae (Fig. 10A) the mean length of species from 0° to 20° latitude is 1.7 mm; from 21° to 44° it is 3.6 mm; and from 45° to 70° it is 8.0 mm. Populations of *Parasterope muelleri* show a slight increase from the Virgin Islands (1.1 mm) to Falkland Island (1.6 mm). Hall and Hessler (1971) found a correlation between body size and latitude for the mystacocarid *Derocheilocaris typica*. Except for individuals from the extreme southern portion of the range, individual size increased with increasing latitude. The crab *Sesarma reticulatum* has a carapace breadth of 13-16.5 mm in Florida and 26-27 mm in Massachusetts (Abele, 1973). Similarly, males of *Cyclograpsus cinereus* increase in carapace length from a range of 4.2-7.5 mm at 20°S to 5.7-12.7 mm at 41°S (Garth, 1957).

Among a number of crustaceans there appears to be an increase in body size with increasing depth (Belyaev, 1972). Wolff (1956a; see also Gardiner, 1975) found a trend of increasing body size in tanaids with increasing depth. He also suggested a similar trend for isopods (Wolff, 1956b). However, Menzies and George (1967) examined size and depth for 37 species of the isopod genus *Storthyngura* from about 1000 to 7000 m and found no obvious relationship. Barnard (1962) found no relationship between body size and depth for gammaridean amphipods. An increase in body size with increasing depth has been noted for some podocopid ostracodes (van Morkhoven, 1972) and some genera of myodocopid ostracodes (Kornicker, 1975). Among myodocopids, the relationship is particularly striking (Fig. 10C) in the genera *Synasterope* and *Spinacopia,* where shallow-water species (ca. 100 m) are 1.4–1.9 mm in length and deep-water species (ca. 3000 m) are 2.6–2.7 mm. However, as with other crustacean taxa, the relationship does not hold for all ostracodes.

An increase in depth and an increase in latitude both result in lower water temperatures, so that the relationship may be between body size and water temperature. Sewell (1948) examined body size and temperature in copepods and found that species that occur in regions of 1°–10°C increased more in body size with decreasing temperature than did species that occurred in warmer waters. However, the relationship is not straightforward for many taxa (Belyaev, 1972).

Recently, Vermeij (1977, 1978) proposed that crabs which prey on mollusks have relatively larger crushing claws in tropical regions than they do in temperate regions and that within the tropics, relative size increases in the order: Eastern Atlantic, Western Atlantic, Eastern Pacific, and Indo-West Pacific. Relative size of the crushing claw was measured by the ratio of claw height or thickness to carapace breadth. A re-examination of these and additional data by Abele *et al*. (1982) demonstrates that the relationship is more complex, and additional data, specifically actual measurements of strength, are required to test the hypothesis. At present, no unequivocal relationship between crusher claw size and geography is known. Similarly, Kent (1979) examined relative mandible size in spring lobsters and found no clear interoceanic or latitudinal patterns.

Within many lineages of coral-dwelling stomatopods there is a positive, significant relationship between body size and geographic range (Reaka, 1980). Reaka (1980) suggests that this is consistent with the hypothesis that large species disperse more widely and are relatively better opportunistic colonizers than are small species.

B. Eye Development

Menzies *et al.* (1968) examined the relationship between isopod eye development, latitude, depth, and light penetration. Briefly, they found that the percentages of species with eyes decreased markedly with depth from 100% in the intertidal region to 0% in the abyss. There were two latitudinal patterns. At high latitudes, the percentage of eye-bearing species drops off immediately from the shelf to the deep sea. The second pattern is a gradual decrease in this percentage at low latitudes. Poulsen (1965) and especially Kornicker (1975) found that the percentages of blind myodocopid ostracodes increase with depth. In a general survey, Clarkson (1967) concluded that almost all blind species of crustaceans are benthic in depths exceeding 600 m.

VII. DISTRIBUTION OF TAXONOMIC GROUPS

A. Cephalocarida

Cephalocarids are small (~ 2-4 mm), benthic crustaceans that burrow in the top layers of fine sediment. *Hutchinsoniella macracantha* carries only two eggs at a time and development begins with a benthic metanauplius stage. There is no fossil record.

Cephalocarids consist of four genera and nine species and are cosmopolitan in temperate and tropical regions (Fig. 11; Hessler and Sanders, 1973). Of the nine species, *Hutchinsoniella macracantha* has the widest distribution, with disjunct populations known in depths of 1-69 m from Long Island Sound, Buzzards Bay, Massachusetts, off New England, in Chesapeake Bay, off North Carolina, and off Angra dos Reis, Brazil. There are small but consistent differences among the populations, but most authors recognize only a single species (Hessler and Sanders, 1973). The genus *Lightiella* is widely distributed in shallow water, being known from the southeastern Pacific in Saint Vincent's Bay, New Caledonia (*L. monniotae*), the northeastern Pacific in San Francisco Bay (*L. serendipita*), the northern Gulf of Mexico and off the west coast of Florida (*L. floridana:* McLaughlin, 1976; Stoner, 1981), and the West Indies (Barbados, Puerto Rico) and Florida (*L. incisa*). Species of *Sandersiella* are similarly widespread: *S. bathyalis* is known from southwest Africa off Walvis Bay and Brazil (Wakabara and Mizoguchi, 1976), *S. calmani* from Peru, and *S. acuminata* from Tomioka Bay, Japan. There is a single species in the genus *Chiltoniella* (*C. elongata*) known from Hawke Bay, New Zealand (Knox and Fenwick, 1977). In addition to a wide horizontal distribution, cephalocarids occur from the intertidal to approximately 1550 m depth.

Hutchinsoniella macracantha is a true hermaphrodite. The ova and sperm develop separately, but the oviducts and vasa deferentia join into a common genital duct so that the potential for self-fertilization exists (Hessler *et al.*, 1970). In addition, the first larval stage is benthonic, and thus there is no planktonic dispersal stage. It would seem that hermaphroditism, the potential for self-fertilization, low fecundity, and absence of a planktonic dispersal stage would lead to much local differentiation—yet little is observed. Hessler and Sanders (1973) interpreted this zoogeographic pattern to be that of a relict group that was more broadly distributed in the past. They further

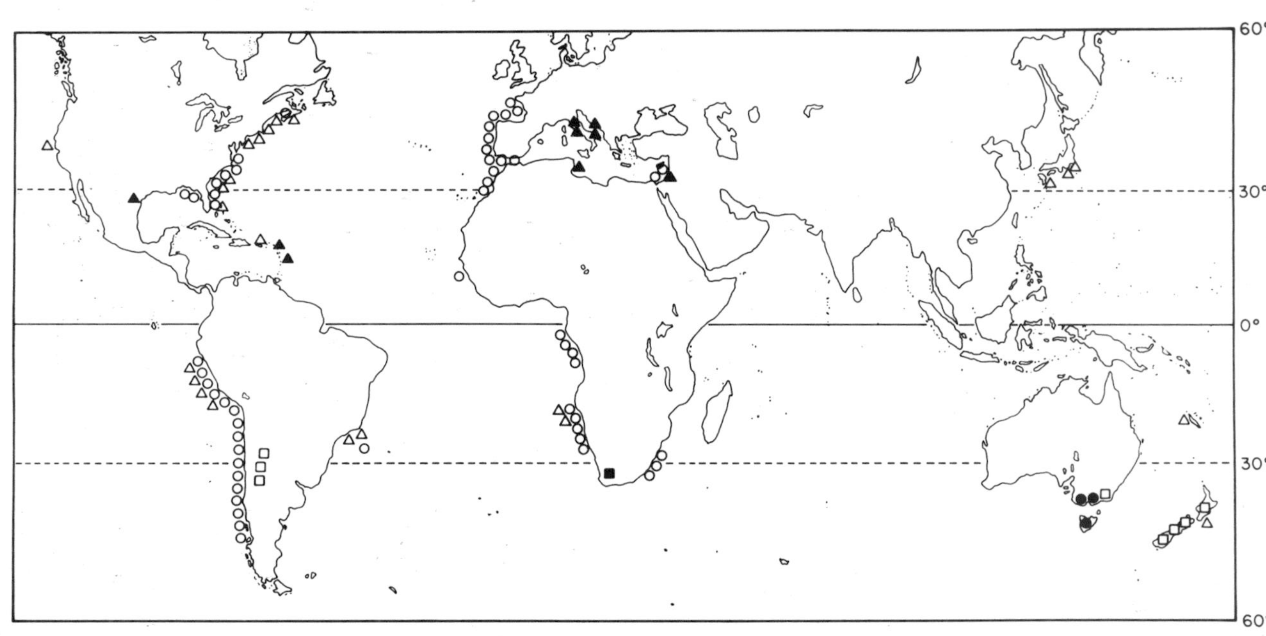

Fig. 11. World distribution of Cephalocarida (△), Anaspidacea (excluding Stygocarididae) (●), Stygocarididae (□), Spelaeogriphacea (■), Mystacocarida (○), and Thermosbaenacea (▲). All localities are approximate.

argued that the cephalocarids are generalists (restricted to flocculent sediments), which are prevented from evolving through specialization by the presence of morphologically more highly evolved forms. While the latter argument may be true, it is not testable and does not answer the question: Why has there been little local differentiation? It could just as easily be argued that cephalocarids are highly specialized (e.g., Schram, Chapter 4 of this volume) and as such are restricted to flocculent sediments by their own morphology, not by competition.

B. Branchiopoda

Branchiopods are mostly small (3–30 mm, though anostracans may reach 100 mm) crustaceans that occur in continental waters ranging from temporary freshwater ponds to hypersaline lakes. Reproduction usually involves both resistant eggs and fast hatching eggs with a free swimming metanauplius larva. Parthenogenetic and hermaphroditic species are known. The fossil record is extensive, beginning in the Devonian. Four Recent orders are currently recognized: Notostraca, Conchostraca, Cladocera, and Anostraca.

The Notostraca (reviewed by Longhurst, 1955b; Linder, 1952, with additions by Lynch, 1966, 1972) consist of a single family, Triopidae, with two genera: *Triops* and *Lepidurus*. *Triops* is known from the Late Carboniferous and *Lepidurus* from the Triassic. Species of both genera are usually found in fresh or brackish water of temporary pools in dry areas of the world (Longhurst, 1955b). The eggs resist desiccation and can withstand temperatures to within 1°C of boiling (Carlisle, 1968). At least some populations of *Triops* species are selfing hermaphrodites (Longhurst, 1954). The following distributional ranges are from Longhurst (1955b), Linder (1952), and Williams (1968), unless otherwise stated. The taxonomy of species in both genera are confused, but at present four wide-ranging species of *Triops* and at least seven of *Lepidurus* are recognized. Subspecies or races have been named for species in both genera. *Triops cancriformis* occurs in Western Europe from Spain to Sweden and east to Russia, North Africa, Asia Minor, and the Middle East to India. *Triops granarius* occurs from South Africa across India and Asia to China. *Triops longicaudatus* occurs in Western North America, Central and South America, the West Indies, Galapagos Islands, Hawaii, Japan, and New Caledonia. *Triops australiensis* occurs in the drier regions of Australia (see Williams, 1968), with a subspecies known from Madagascar. *Lepidurus apus* may be one of the most widely distributed of the freshwater crustaceans, being reported from Europe (Fox, 1949), North Africa, Israel, Asia Minor, Russia, North and South America, New Zealand, and Australia. However, *L. apus* is considered a senior synonym of at least six species, and there are geographic races recognized so perhaps *L.*

apus is not a single species. *Lepidurus arcticus* is circumpolar in the Arctic regions and occurs in the Aleutians, North America (Alaska to Labrador), Greenland, Iceland, Bear Island, Spitzbergen, and from Scandinavia to Siberia. *Lepidurus batesoni* is known only from Chilik Kul in the Kazak region of Russia. In addition to *L. apus* and *L. arcticus,* four species occur in North America: *L. packardii* in California; *L. bilobatus* in Utah, Colorado, and probably Arizona; *L. couesii* in Idaho, Utah, Oregon, and Saskatchewan, Canada; and *L. lemmoni* (= *lynchi*) in California, Oregon, Washington, Montana, Wyoming, and Nevada. The large number of species known from North America probably reflects the large number of workers in North America compared to other regions.

Most authors cite the drought-resistant eggs of notostracans as important in passive dispersal. However, several species have quite narrow ranges which would suggest that they do not disperse well. Two species, *T. granarius* and *T. longicaudatus,* occur in rice fields (Crossland, 1965; Grigarick *et al.,* 1961), where they are serious pests, and some distributional records may be due to transport by man. Longhurst (1955a) suggests that the resistant eggs and hermaphroditism (which may be associated with range extension) have resulted in their generally wide distribution. This in turn has prevented geographic isolation and accounts for the relatively few species in the group.

Conchostracans, or clam shrimp, have a fossil record extending back to the Lower Devonian. Conchostracans can produce resistant eggs and some species are parthenogenetic. There are 5 Recent families and approximately 180 species in about 10 genera. Some of the genera known from the Recent have extensive fossil records, such as that of *Lynceus,* which extends back to the Lower Cretaceous, or of *Cyzicus,* which extends back to the Lower Devonian (Tasch, 1969). As with the Notostraca, there are species with very wide distributions, but others have quite restricted ranges: e.g., *Eulimnadia diversa* or *E. stoningtonensis,* which are known from single ponds in North America. The basic distribution pattern of conchostracans appears to be widespread genera with very few cosmopolitan species (Tasch, 1969). These distribution patterns appear to be very old, since fossils congeneric with Recent conchostracan species occur over a 200-million-year time gap at the same locality (Tasch and Zimmerman, 1961). Tasch (1979) suggests that the distribution of genera is due to continental drift.

The largest group of branchiopods is the Cladocera, or water fleas, with more than 420 species recognized in 8 families. Chydorid cladocerans have been reported from the Permian (Smirnov, 1970). Cladocerans can produce resistant eggs and the majority of species occur in fresh to brackish water, although a few marine genera (e.g., *Penilia, Evadne,* and *Podon*) are known.

Genera and many species tend to be widely distributed: e.g., *Daphnia,*

Chydorus, Alona, Alonella, Ilyocryptus, and *Macrothrix* are cosmopolitan, while *Percantha, Pleuroxus,* and *Leptodora* have representatives in Europe, North America, and Asia. Within the family Macrothricidae there are 16 genera distributed as follows (Fryer, 1974): 6 Holarctic, 2 cosmopolitan, 2 Australia–New Zealand, 2 New World tropics, 1 circumtropical, and 1 each in the Nearctic, Palearctic, and Old World tropics. Overall, approximately one-third of cladoceran genera occur only in North America and Europe, suggesting a Laurentian distribution. A number of species are also widespread: *Leptodora kindtii* occurs in North America and Europe, *Bosmina longirostris* is cosmopolitan, and *Polyphemus pediculus* occurs in North America and Eurasia. Crawford (1974), while arguing for a greater Gondwanaland, cites the distribution of the cladoceran genus *Daphiniopsis* (and that of the fossil reptile group *Lystrosaurus*) as evidence for the attachment of parts of western and central China to India during the Permian. *Daphiniopsis* today is known only from Kerguelen, Antarctica, Australia, Tibet, and Inner Mongolia, a distribution presumably achieved during the mid-Mesozoic. Another interesting distribution is that of the terrestrial genus *Bryospilus,* represented by two species: *B. repens,* from Puerto Rico and Venezuela and questionably from New Zealand, and *B. bifidus,* known only from New Zealand. Both species occur in forest litter, one in a cloud forest and the other in a temperate rain forest (Frey, 1980).

The Anostraca consists of about 175 species in 7 families (Linder, 1941). The fossil record begins in the Lower Devonian. The vast majority of fairy shrimp occur in small, temporary alkaline pools and have drought-resistant eggs.

The Branchinectidae is widely distributed, with representatives in the Arctic, Europe, northern and western North America, North Africa, southern South America, South Georgia, and West Antarctica. The Artemiidae is cosmopolitan, excluding the Arctic and Antarctic. Members of the Branchipodidae are known from Europe, Africa, South Australia, and Asia and are absent from the New World, Arctic, and Antarctic regions. The family Streptocephalidae is cosmopolitan except for South America and Australia. The Thamnocephalidae consists of several genera which are widely distributed in the warmer parts of the world. Representatives of the Chirocephalidae occur in the Arctic, Europe, North Africa, Asia, and North America. The Polyartemiidae consists of two genera known from the Arctic region.

Although the fossil record of the Anostraca extends back to the Lower Devonian, none of the Recent genera is known as a fossil and of the Recent families only the Artemiidae has a fossil record, being known from the Pleistocene (Tasch, 1969).

Branchiopods may be widely distributed for two reason. The first is their extensive fossil record. Genera, and possibly a species of *Triops* (Longhurst, 1954), are known from the Lower Devonian, Carboniferous, and Triassic (Tasch, 1969), i.e., when the continents formed a single land mass and dispersal was probably easier. Second, many species have resistant eggs which facilitate dispersal and some species have hermaphroditic (e.g., *Triops*) or parthenogenetic populations. Thus, species can and do occur over wide geographic regions, even in relatively inhospitable habitats. Rzoska (1961), for example, reported ten species of branchiopods from temporary pools in scrub country of the Northern Sudan (Khartoum).

C. Remipedia

This recently discovered class (Yager and Turner, 1980; Yager, 1981) is represented by the single species *Speleonectes lucayensis,* known from one locality in a marine cave in the Bahamas. The species is fairly large (ca. 24 mm), lacks eyes, and is pelagic. There is no fossil record and nothing is known of its development.

D. Ostracoda

There are approximately 30,000 species of ostracodes of which about 2700 are Recent (Benson, 1966; McKenzie, 1971b). The fossil record extends from the Cambrian to the Recent. Modern ostracodes are divided into two major groups: the entirely marine Myodocopa, with approximately 625 species (Kornicker, 1977; Cohen, 1982), and the Podocopa, with about 1200 marine species (Benson, 1966), 1000 freshwater species (McKenzie, 1971b), and a few terrestrial species (Chapman, 1961; Harding, 1953). The dispersal stage for most marine ostracodes is the nauplius larva which hatches from eggs that are either deposited on the bottom or held in the valves prior to hatching. The dispersal of freshwater species is discussed later. There is a vast literature on the distribution of ostracodes, and I can summarize only some of the data here.

The number of myodocopid ostracode species increases with decreasing latitude for benthic species (Fig. 12A) (Poulsen, 1962, 1965) and the primarily pelagic halocypridids (Fig. 12B) (McKenzie, 1967; Poulsen, 1969). Similarly, Kornicker (1977) has shown that the number of species in coastal embayments increases with decreasing latitude on both coasts of North America. If the southern ocean is considered on a global basis (Kornicker, 1975) between Antarctica and 35°S, the number of species of myodocopids increases greatly with decreasing latitude from 15 species on the Antarctic shelf to 55 species on the shelf at 35°S. There is no obvious change in species numbers with latitude at greater depths (2000-6000 m). Within the

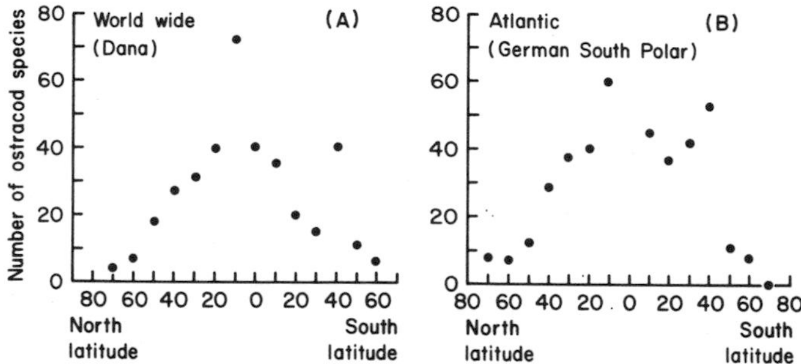

Fig. 12. Relationship between myodocopid ostracode species richness and latitude. Left (A) is for primarily benthic species collected by the Dana expedition (Poulsen, 1962, 1965). Right (B) is for the primarily pelagic halocypridids collected by the German South Polar expedition in the Atlantic (McKenzie, 1967).

tropics, the number of myodocopid ostracodes increases in the order Eastern Atlantic, Eastern Pacific, Western Atlantic, and Indo-West Pacific (Table I). This pattern of species richness is influenced especially by the genera *Codonocera, Cypridinodes,* and *Cypridina,* which contain many species in the Indo-West Pacific (Poulsen, 1962, 1965).

The number of marine podocopid ostracodes known from various localities is summarized by Puri (1967) and Benson (1964, 1966). At first glance there is no obvious relationship between latitude and species number (see Benson, 1966, Fig. 5), but the comparison is complicated by the relationship between the number of samples from a locality and the number of species collected there. For example, the number of species known from various geographic localities in the Atlantic is significantly and positively correlated with the number of samples taken from that locality ($r = 0.81$, $n = 12$, data from Benson, 1966, Figs. 4 and 5). If only reasonably well-studied localities are compared, there does seem to be an increase in species number with decreasing latitude, although the available data include only a narrow range of latitude. For example, 50 species of podocopids are known from the northern Arctic coast of Europe at about 80°N latitude, about 160 species are known from 60°N, and 310 species from the Mediterranean at about 35°N latitude (Benson, 1966, Fig. 5).

Biogeographic regions can be recognized for benthic ostracodes (Benson, 1964, 1966; Puri, 1967), but oceanic pelagic ostracodes tend to be widely distributed (Angel, 1972; Poulsen, 1977). Poulsen (1977) has summarized the zoogeography of 100 pelagic species with the following results: of 74 species found in at least three samples, 91% are found in all three major oceans; the greatest number of species are found in the tropics and subtropics; and regional groups of species may be found in the Arctic, Antarctic, temperate, and tropical regions, with perhaps the northwest Pacific being a special group.

There are more than 1000 species of freshwater ostracodes, primarily in the family Cyprididae, although the Darwinulidae and Cytheridae are also represented. McKenzie (1971b) has reviewed the paleozoogeography of these groups which arose during the Devonian but were not well developed until the Permian. The Darwinulidae were dominant during the Paleozoic, with the Cyprididae diversifying during the Mesozoic. The distribution of genera has probably been influenced by continental drift (McKenzie and Hussainy, 1968), and Krömmelbein (1966), noting the high proportion of identical species, including cytherids and a darwinulid, in the Brazil and West African Wealden assemblages, concluded that this is best explained by direct continental connections between Africa and South America during the early Cretaceous.

A few comments are necessary on the dispersal of freshwater ostracodes. Cypridids have resistant eggs, entocytherids deposit eggs on their crayfish host (Walton and Hobbs, 1971; Hart and Hart, 1974), and cytherids and darwinulids brood the early larval stages. Cypridid eggs may persist in mud for a number of years (e.g., Sars, 1895). Viable eggs have also been recovered experimentally from the lower digestive tracts of birds (Proctor, 1964; Proctor and Malone, 1965), and viable eggs have passed through the digestive tracts of fish (Kornicker and Sohn, 1971). Other comments on dispersal may be found in Löffler (1964), Klie (1939), Sandberg (1964), Sandberg and Plusquellec (1974), Fryer (1953), Delomore and Donald (1969), McKenzie (1971a), and De Deckker (1977). The general conclusion (McKenzie, 1971b) is that freshwater ostracodes may be dispersed by several mechanisms involving passive dispersal and/or other organisms: thus, ovigerous females and/or eggs may be carried (1) in mud on birds, (2) in the alimentary tracts of birds and fishes, (3) in dust-laden storm winds, and (4) by man transplanting rice. While such passive dispersal is possible, the idea has not been subjected to a rigorous test. For example, while occasional sympatry of Australian freshwater ostracodes may be the result of dispersal via birds, this dispersal route is unlikely to have influenced the major distribution patterns (see remarks by Kornicker following De Deckker, 1977).

E. Mystacocarida

The adults of this subclass are very small (~ 0.5 mm), benthic, and live in the interstitial spaces between sand grains, which would seem to limit dispersal. Development apparently begins with a nauplius larva in some species and with a metanauplius in others. All of the larvae appear to be benthic, living between sand grains. There is no fossil record.

This subclass contains the following species (Friauf and Bennett, 1974;

Hessler, 1972): *Derocheilocaris remanei,* under various subspecific names (Hessler, 1972, has suggested that some of these subspecies should be given specific rank), occurs from the Atlantic coast of France, in the Mediterranean, and along the coast of Africa around the Cape of Good Hope to Durban; *D. typica* occurs from about Cape Cod, Massachusetts, to Miami, Florida; *D. ingens* is known only from Reid State Park, Maine; *D. hessleri* occurs from northwest Florida to Skip Island, Mississippi, in the northern Gulf of Mexico; *D. tehiyae* is known only from the coast of Israel; *D. angolensis* is known from Angola, Africa; and *D. delamarei* is known from Swakopmund, southwest Africa. The genus *Ctenocheilocaris* contains two species: *C. galvarini,* which occurs along the coast of Peru and Chile, and *C. claudiae,* which is known from off Rio de Janeiro, Brazil (Renaud-Mornant, 1976).

Schram (Chapter 4 of this volume) points out that the group is essentially circum-Atlantic in distribution (Fig. 11) with outliers (Chile and Durban) adjacent to the Atlantic. Mystacocarids have not been found on beaches of the northeastern Pacific or of the Indo-West Pacific, although some areas have been intensively sampled (this, of course, is subject to change at any moment). Schram speculates that the group may have evolved in the Atlantic region and spread as the Atlantic Ocean formed. This would mean that, although no fossils are known, the group is relatively young (late Cretaceous), which would tend to support Hessler's (1971) interpretation of mystacocarids as neotenous (or of progenetic origin, Gould, 1977, p. 336) rather than primitive. In addition Renaud-Mornant and Delamare-Deboutteville (1977) suggest that the group is not necessarily primitive and that advanced character states can be recognized in mystacocarids.

F. Branchiura

Branchiurans are parasitic on cold-blooded vertebrates, especially fishes. Females leave the host and deposit eggs on a solid substrate. The larvae are initially free-swimming before attaching to a host, which in the juveniles of some species is an intermediate host (Fryer, 1968). The degree of host specificity ranges from slight (*Argulus africanus* has been recorded from 11 families of fish) to host specific at the genus level (Wilson, 1903; Fryer, 1968; Cressey, 1971, 1972). There is no fossil record. Branchiurans are represented by one family, the Argulidae, four genera (*Argulus, Dolops, Chonopeltis,* and *Dipteropeltis*), and approximately 150 species.

The genus *Argulus* contains about 130 species and all occur on inshore, coastal marine, and freshwater fishes. The genus is virtually cosmopolitan in distribution, but detailed distribution data are unavailable for most of the species. Yamaguti (1963) listed the distribution of the known species of

Argulus, but this work contains a number of errors (Fryer, 1969). There are 21 species recorded from African freshwater fishes (Fryer, 1968). One species, *A. africanus,* is widespread in Africa, occurring in the Upper Congo, in the Zambezi system, in Lakes Rukwa, Kitangiri, Victoria, Kioga, and Edward. Four other species occur in more than one watershed, while the remaining species, including a swarm of seven species in Lake Tanganyika, are known from one river system or lake. Cressey (1971, 1972) reviewed the 23 American species of *Argulus*. Excluding *A. japonicus*, which occurs on goldfish and has been introduced world-wide, there are 3 species known from western North America and 19 from eastern North America, 2 of which are known from only a single state each. The largest numbers of species are found along the Gulf coast, but apparently none of the species ranges into Central America. There are approximately 16 species known from tropical and subtropical South America (Rinquelet, 1943). According to the list in Yamaguti (1963), some approximate species numbers for large geographic regions are: Central America and Cuba, 4 species; Europe, 6 species; the region of Java-Sumatra-Ceylon, 5 species; India, 4 species; Japan, 6 species; and China, 7 species. There are obviously more species to be discovered in all of these regions.

The genus *Dipteropeltis* probably consists of a single species (see Monod, 1928; Fryer, 1969), *D. hirundo,* which is known from the interior of Brazil (Mato Grosso) and Argentina on freshwater fish.

Chonopeltis consists of seven species all known from freshwater fish in Africa (Fryer, 1968). Three species occur in the Congo system, one (unnamed) in West Africa, one from Lake Nyasa, one from the Limpopo system, and one from the Upper Nile, Tana, and Mugunbuzi Rivers (see Fryer, 1968, Fig. 8).

The most interesting genus of branchiurans in terms of distribution is *Dolops:* eleven species are known from South America, one from Africa, and one from Tasmania (Fryer, 1969). All occur on freshwater fish. Fryer (1969) reviewed the distribution of the genus and concluded that continental drift was the most plausible explanation for the present pattern. Continental drift may also have played a role in the distribution of *Chonopeltis* and *Dipteropeltis,* which may be sister genera derived from a common *Argulus*-like ancestor (Fryer, 1956).

G. Copepoda

There are probably 8000 Recent species in the class Copepoda. Currently these are distributed into seven orders (Bowman and Abele, Chapter 1 of this volume). A beautifully preserved fossil parasitic copepod is known from the Lower Cretaceous, suggesting that the history of the group extends much further back in time (Cressey and Patterson, 1973) than the Cretaceous. Copepods occur in marine and freshwater habitats, and a few harpacticoids occur in moist terrestrial habitats. Development occurs in the plankton with a nauplius as the first stage in most copepods, but those of harpacticoids are benthic.

There are three major ecological radiations within the copepods: a free-living planktonic group primarily in marine but with freshwater representatives; a parasitic group with representatives parasitizing sponges, cnidarians, mollusks, polychaetes, crustaceans, echinoderms, ascidians, fishes, and whales; and a smaller group inhabiting marine benthic environments consisting primarily of harpacticoids.

Sewell (1948) has summarized a massive amount of information on the distribution of free-swimming planktonic forms and, unless otherwise noted, the following data (Table II) are drawn from that work. Prior to Sewell (1948), Giesbrecht (1892) recognized three major zoogeographic regions based on planktonic copepods, and these were the Arctic, the Tropical and subtropical, and the Antarctic. Steuer (1933) subdivided these three regions into seven provinces with some subprovinces. More recently, McGowan (1974) analyzed distribution patterns of zooplankton in the Pacific Ocean and was able to recognize eight distinct provinces: Subarctic, Transition, North Central, Eastern Tropical, Equatorial, South Central, Transition, and Subantarctic. The warm water cosmopolitan species were not included in the analysis. The data suggest that additional regional differentiation (see below) may be recognized (Table II; see also Fleminger and Hulsemann, 1973).

There are 184 species reported in the Eastern Pacific region. In the northern portion (23°-50°N) 115 species occur: 4 appear to be endemic and 19 have probably been swept into the area from the Arctic region. In the southern portion (10°-50°S) 111 species have been recorded, and 35 of these do not occur in the northern region. If we exclude the 19 Arctic species and 2

TABLE II

Number of Epipelagic Copepod Species Known from Various Regions of the World

Region	Number of species	% Endemic
Eastern Pacific	184	12.5
Western Pacific	204	13.7
Indian Ocean	270	21.0
Red Sea	111	1.8
Atlantic Ocean	186	20.4-25.8
Mediterranean	120	8.3

more from the West Wind Drift, there are 167 Eastern Pacific species, and 112 or 67% of them occur in the Atlantic. At least 94 species or about 56% occur in tropical and southwestern Pacific waters. Approximately 87 species or 52% occur in the Indian Ocean. Only a maximum of 21 species or 12.5% appear to be endemic, and a few of these are questionable.

Of 204 species recorded from the West Pacific (including the southwestern Pacific), 94 occur in the East Pacific, 110 occur throughout the Indo-West Pacific, and 28 are endemic.

There are 270 species recorded from the Indian Ocean region of which 56 or 21% are endemic. In this region the number of calanoid copepods decreases moving west: the Malay Archipelago (117 species), South Burma region (85 species), Ceylon region (71 species), Araluan Sea (67 species), and Red Sea (52 species).

In the Red Sea, 111 species have been recorded, and of these, 2 are endemic, 78 occur in the East Pacific, and the remainder in the Indo-West Pacific.

There are probably 186 species in the temperate and tropical regions of the Atlantic, though including species that occasionally stray from the South or North Atlantic would increase the number to about 235. Of the 186 species, 29 in the northern Atlantic, 6 along the west coast of Africa, and 13 in the southern Atlantic (9 of which are brackish) appear to be endemic. The remaining species occur in the Eastern Pacific and Indo-West Pacific region.

In the Mediterranean, 120 species have been recorded; 103 occur also in the Indo-Pacific. Including the Black Sea there may be as many as 10 species endemic to the region.

The above data deal primarily with marine species. Recently, Turner (in press) examined estuarine copepods along the coast of North America. His results (Fig. 13) show that combined calanoid and cyclopoid species richness and number of families increase toward the equator. He points out that a closer look shows that cyclopoid species numbers are highest in tropics and calanoid species richness is higher in the temperate zone.

To summarize epipelagic copepods: the greatest number of species is found in the tropics, and many species occur in all tropical and temperate oceans. There is a general increase in species numbers seaward from coastal regions (Bowman, 1971). The following minimum percentages of the fauna appear to be endemic: East Pacific (12.5%), West Pacific (13.7%), Indian Ocean (20.7%), and Atlantic Ocean (20.9% without brackish species, 25.8% with them). The Atlantic figures are probably high because of the exclusion of questionable records in the northern and southern regions. There appears to be a higher percentage of endemic species among the brackish water faunas. Fleminger and Hulsemann (1973) analyzed several calanoid genera in detail and showed that some species previously consid-

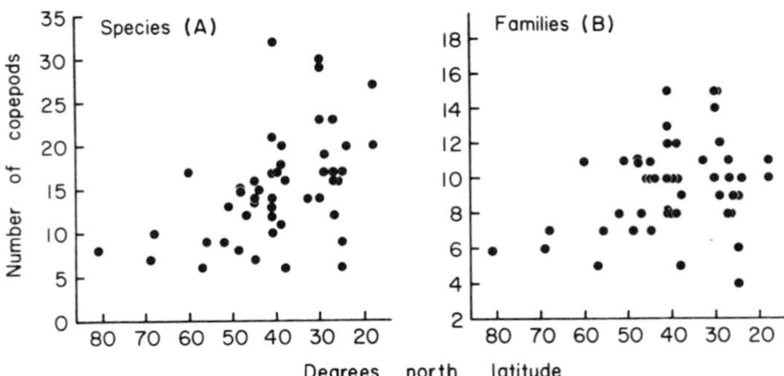

Fig. 13. Relationship between latitude and species number for estuarine calanoid and cyclopoid species (left) and families (right) for the east coast of the United States. (Redrawn from Turner, 1981.)

ered to be circumtropical actually consist of two or more similar species. They concluded that warm water epiplanktonic copepods, bathypelagic fish, and euphausiids share two main geographic patterns. The first pattern is that circumglobal species tend to occur in subtropical waters up to the subtropical convergences, often penetrating well into temperate waters, whereas species occurring in low latitudes tend to show regional provincialism.

The previous discussions have dealt with the copepods occurring between the surface and about 100 m. Sewell (1948) provides data on the deep-water copepods, some of which are summarized in Table III. There are two apparent patterns in the distribution of deep-water copepods. The first is that the number of species at any one site increases from 100 m to a peak at about 900–1200 m, the exact depth depending on the region. The numbers of individuals roughly follows this same trend. The second pattern is that the percentages of species endemic to deep-water masses is very much larger than for shallow-water species. For example, about 94% of the 357 species known from the water masses of the North Atlantic Basin are endemic to that region.

Sewell (1940) summarized the distribution of harpacticoid copepods, most of which were benthic in habits. Knowledge of harpacticoids is uneven so caution must be used. With that caveat in mind we find that pelagic harpacticoids tend to be cosmopolitan. Sewell found also that of 41 cosmopolitan benthic species, the majority (56%) whose habits were known were associated with littoral algae. In addition to cosmopolitan forms, many harpacticoids tend to be widely distributed. Among 350 species reported from Norway are 53 species known from India. Similarly about 50% of the West Pacific fauna is known from the Indian Ocean. More recently Wells

TABLE III

Distribution by Depth of Deep-Sea Free-Swimming Planktonic Copepoda[a]

	Number of species			
Depth	Bay of Biscay	Malay Archipelago	Ireland	Indian Ocean
100	20	—	4	9
200	17	8	8	9
300	47	—	8	7
400	43	—	2	11
500	44	—	17	14
600	50	—	11	16
700	57	69	19	—
800	—	—	21	—
900	—	43	32	24
1000	66	80	21	—
1500	65	79	25	24
2000	22	39	—	2
2500	14	—	—	1
3000	11	—	—	1

[a] Excluding species that occur between 0 and 100 m. The depth regions are rounded off to facilitate comparisons. Data from sources in Sewell (1948).

(1967) examined the copepods (primarily harpacticoids) of Inhaca Island, Mozambique. Of 89 species, 24 to 26 are cosmopolitan, another 20 have been found in all oceans but the Pacific, 6 are pantropical, 13 are confined to warm temperate and tropical regions, 17 occur only in the Indo-Pacific, and the remaining species are local or poorly known. Thus, approximately 80% of the species are widespread in distribution. Similar results were found by Coull and Herman (1970) in Bermuda.

A treatment of the parasitic forms is beyond the scope of this paper. However, Cressey and Collette (1970) presented a very interesting analysis of the relationship between parasitic copepods and their needlefish hosts. (See also Cressey and Cressey, 1980, for data on scombrid fishes and their copepod parasites.) Humes and Stock (1973) present data on cyclopoid parasites associated primarily with invertebrates.

H. Cirripedia

The cirriped crustaceans consist of approximately 1000 species in four orders. The fossil record extends from the Middle Cambrian Burgess Shale (Collins and Rudkin, in Newman, 1979) to the Recent. All species are marine, although some species occur in almost freshwater, and all are sedentary, symbiotic, or parasitic. Dispersal usually involves planktonic naupliar and cypris larvae. There are a number of papers dealing with distributional aspects of cirripeds. Nilsson-Cantell (1938) summarized information on the Indian fauna and (1978) dealt with part of the Scandinavian fauna. Stubbings (1967) summarized data on the West African fauna, Foster (1978) dealt with the thoracic barnacles of New Zealand, and Newman and Ross (1971) dealt with the Antarctic fauna. Spivey (1981) summarized a great deal of information on the cirripeds of the Gulf of Mexico and Western Atlantic. Tomlinson (1969) reviewed the burrowing acrothoracican barnacles.

There are about 30 species of the small, parasitic Ascothoracica, which are found on hexacorals and echinoderms. The fossil record begins in the Cretaceous and there are four Recent families. None of the Recent families or genera has a fossil record. The Synagogidae is known from the Mediterranean, Japan, Okhotsk Sea, the northwest Pacific, Arctic, and Antarctic regions. The Lauridae occur in the Mediterranean, and in the Central and Indo-Pacific. The Petrarcidae consists of a single genus known from deep water off Japan. The Dendrogastridae has representatives in the White Sea, northeastern Atlantic, California, and the Indo-Pacific region.

Acrothoracic barnacles are known from the Upper Devonian to the Recent. There are approximately 50 Recent species in 10 genera and 3 families, and all bore into calcareous substrates (Tomlinson, 1969, 1973). The order is cosmopolitan in distribution with the largest number of species found in the Indo-West Pacific. Approximately 4 species are known from the Eastern Pacific, three from the Western Atlantic, 7 from the Eastern Atlantic, and 32 from the Indo-West Pacific region. Worldwide, the greatest numbers of species occur between 20°N and 5°S latitude (Fig. 14B). A few species are known from widely scattered localities: *Lithoglyptes spinatus* occurs in the Pacific Ocean, Red Sea, Indian Ocean, and Carribbean Sea; *Kochlorine*

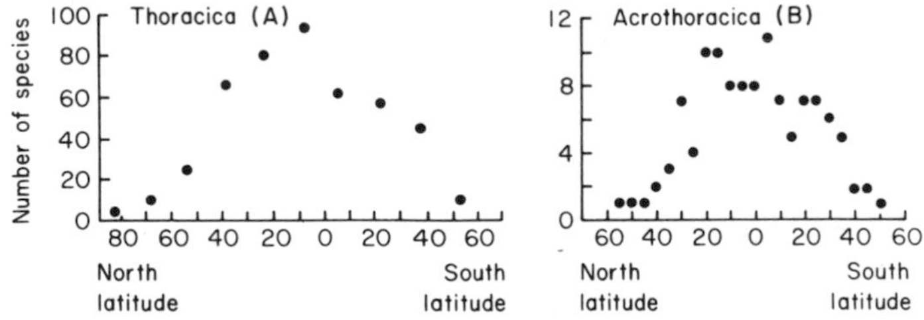

Fig. 14. Relationship between latitude and species number for littoral Thoracica (A) and Acrothoracica (B) barnacles of the world. Thoracica data from H. R. Spivey (personal communication), Nilsson-Cantell (1938, 1978), Stubbings (1967), Newman and Ross (1971), and Spivey (1981); Acrothoracica data from Tomlinson (1969, 1973).

hamata is known from the Eastern and Western Pacific, Indian Ocean, and Mediterranean Sea; and *K. floridana* is known from the Gulf of Mexico, Western Atlantic, and Madagascar. In contrast, *Australophialus melampygos* occurs only in New Zealand. Four of the genera (*Kochlorine, Lithoglyptes, Trypetesa,* and *Cryptophialus*) are nearly cosmopolitan in distribution. *Weltneria* occurs in the Indo-West Pacific, *Berndtia* in the Western Pacific, *Kochlorinopsis* in the Eastern Atlantic, and *Balanodytes* in the Marshall Islands and Taiwan. *Australophialus* has two species in South Africa and one in New Zealand.

The Thoracica include three Recent groups, the Lepadomorpha, the Verrucomorpha, and the Balanomorpha. There are approximately 700 species, and the fossil record extends back to the Upper Silurian. The number of species of littoral thoracicans as a function of latitude is shown in Fig. 14A, where it can be seen that there is an inverse relationship between latitude and species numbers. There is also a longitudinal pattern with the number of species increasing in the order Eastern Atlantic (61), Western Atlantic (90), Eastern Pacific (114), and Indo-West Pacific (247). As with other groups of crustaceans, the longitudinal pattern is largely a result of an increase in the number of species per genus. For example, in the *Balanus* "amphitrite" complex there are 6 species known from the Eastern Atlantic, 9 from the Western Atlantic, 10 from the Eastern Pacific, and 13 from the Indo-West Pacific (Henry and McLaughlin, 1975)

There is an interesting relationship between lepadomorph and balanomorph barnacles; i.e., the species ratio changes with latitude (Newman and Ross, 1971). Spivey (1981) examined this ratio for ten geographic regions and showed that the lepadomorph:balanomorph ratio decreases markedly toward the equator. It appears that the decrease is the result of an increase in the number of balanomorphs rather than a decrease in the number of lepadomorph species.

Rhizocephalan barnacles are an entirely parasitic group found exclusively on crustaceans, primarily decapods. There are approximately 226 Recent species in 31 genera, and no fossils are known. Rhizocephalans are probably the poorest known of the cirripeds. For example, 5 of 14 rhizocephalans known from the Gulf of Mexico are thought to be endemic, yet none of their hosts is endemic (Spivey, 1981).

I. Phyllocarida, Leptostraca

The nebaliids are marine and benthic with the exception of the nektonic *Nebaliopsis,* although Calman (1917) reports *Nebalia longicornis* from a plankton tow off Three King Island, New Zealand. Development is direct in *Nebalia bipes* and probably in other species and genera as well. The fossil record of the phyllocarids extends back to the Cambrian, and that of the nebaliids extends to the Upper Permian, represented by a species assigned to the Recent genus *Nebalia* (Rolfe, 1969).

The Nebaliidae, with four recognized genera (Cannon, 1960), is the only extant family. The species are distributed as follows: *Paranebalia longipes* occurs from 1 to 9 m in the regions of Bermuda, Virgin Islands, Japan, Gulf of Siam, and the Torres Straits. *Nebaliopsis typica* is a widely distributed nektonic species known from 3500 m to possibly the surface waters, although closing nets have not been regularly used to determine the exact depth range. The species has been collected from the southern and southwestern Indian Ocean, the southern Pacific, and the southern, southeastern, and southwestern Atlantic (Cannon, 1931). *Nebaliella caboti* has been taken from 378 m between Newfoundland and Cape Breton and from 2085 m off New Jersey (Hessler and Sanders, 1965); *N. antarctica* has been taken from 9 to 30 m off Kerguelen and from Akaroa Harbor, New Zealand; *N. extrema* from 380 to 385 m off Kaiser Wilhelm II Land (in western Antarctica) and from 160 to 335 m in the Palmer Archipelago (off the Antarctic Peninsula). The status of species of *Nebalia* is unsettled (Thiele, 1904; Hansen, 1920; Cannon, 1960). *Nebalia typhlops* is known from the west coast of northern and central Norway at 270–360 m, the west coast of Ireland at 210–369 m, and the Mediterranean Sea along the coast of Italy at 115–1100 m, and a subspecies, *occidentalis,* from 467 to 509 m off the coast of New Jersey in the United States (Hessler and Sanders, 1965). *Nebalia pugettensis* (= *Epinebalia pugettensis*) is known from the Sea of Okhotsk, Puget Sound, Washington, Tomales Bay, California, and Morro Bay, California (Clark, 1932; Menzies and Mohr, 1952; Hessler, 1964). It is with the remaining species that taxonomic problems exist. Cannon (1960) listed *N. bipes, N. geoffroyi,* and *N. lonigcornis* without commenting on subspecies or synonymies, while Thiele (1904) considered *N. bipes* to be a northern species and *N. longicornis* to be a southern species. Regardless of taxonomic problems, specimens of *Nebalia* referred to *bipes* are known from 7 to 15 m in the Arctic-Boreal region, including the North Pacific, Barents Sea, Western Atlantic, Novaya Zemlya, throughout both sides of the North Atlantic down to the Bay of Fundy on the U.S. coast. Other localities include the Suez Canal (Lake Timsah), Ceylon, Japan, and the Red Sea (Calman, 1927; Tattersall, 1905). Specimens referred to *N. geoffroyi* have been reported from the Mediterranean, Madeira, Red Sea, and Puerto Rico. Subspecies of *N. longicornis* have been reported from Cuba, New Zealand, Simon's Town, South Africa, South Georgia, Falkland Islands, and Chile (Cannon, 1931; Calman, 1917, Hansen, 1920). Additional unidentified specimens of *Nebalia* have been reported from the Bahamas (Brattegard, 1970) and Brazil (Wakabara, 1965).

The distribution of the Recent species is probably the result of continental movements and speciation events. Any further comments would require information on the phylogenetic relationships of both fossil and Recent groups which is not yet available.

J. Hoplocarida

Stomatopods are predominately tropical, shallow-water crustaceans. Larval development occurs in the plankton. There are more than 342 species in 12 Recent families (Manning, 1980). The fossil record begins in the Carboniferous (Schram, Chapter 4 of this volume). The following data are drawn from the excellent zoogeographic analysis provided by Manning (1977; see also Reaka and Manning, 1980). Two major distribution patterns are apparent. First, the number of species in the four main tropical regions of the world increase in the order Eastern Atlantic (29 species), Eastern Pacific (46 species), Western Atlantic (71 species), and Indo-West Pacific (196 species) (Table I). Second, within each region the number of species increases with decreasing latitude. For example, in the Eastern Atlantic (Fig. 15A) there are no species north of 50°N latitude, 4 species at 25°N, 8 species at 20°N, and 22 species at 5°N. The latitudinal range of the various species is shown in Fig. 15B. As with some other crustacean taxa (Fig. 14A), the greatest number of species is found just north of the equator.

The tropical distribution of stomatopods is shown also by the distribution of genera. Of 68 known genera, only 2, *Hemisquilla* and *Pterygosquilla*, with 2 species each, are restricted to warm temperate regions. Of the remaining 66 genera, 6 are pantropical in distribution (*Eurysquilla, Acanthosquilla, Lysiosquilla, Parasquilla, Pseudosquilla,* and *Meiosquilla*), representing 61 species. The genus *Pterosquilla* is represented by subspecies or species circumglobally in the southern hemisphere. A single species, *Heterosquilla mccullochae,* is probably circumtropical in distribution.

There is high endemism at the species level in each of the regions. In the Eastern Atlantic 23 (82%) are endemic and 8 of these occur in the Mediterranean which has no endemic species. Similarly 83% of the 46 Eastern Pacific species are endemic and 8 (17%) occur in the Western Atlantic. Again, 82% of the 71 Western Atlantic species are endemic: 7 species (10%) occur in the Indo-West Pacific, 8 (11%) occur in the Eastern Pacific, and 4 (6%) occur in the Eastern Atlantic and Indo-West Pacific. From the data in Manning (1977), about 96% of the Indo-West Pacific are endemic to that region.

As with other crustacean groups, there are some odd distribution patterns. *Bathysquilla microps* is known from the Gulf of Mexico and the Caribbean-Western Atlantic in depths between 604 and 1281 m and off

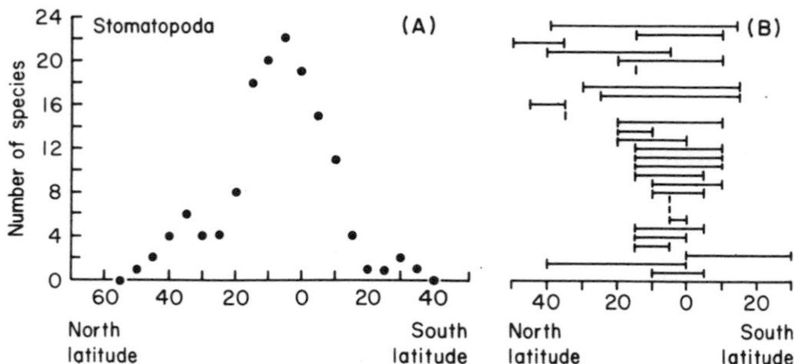

Fig. 15. Relationship between latitude and species number for stomatopods of the Eastern Atlantic (A) and the latitudinal range of the species (B). (Data from Manning, 1977.)

Maui, Hawaii, in 731–786 m depth. Manning and Struhsaker (1976) discuss this and point out that several taxa of fish exhibit similar distributions.

K. Syncarida

There are three orders of syncarid crustaceans (Noodt, 1965; Schminke, 1974a; Knott and Lake, 1980): the fossil Palaeocaridacea known from the Carboniferous to the Permian, the Anaspidacea known from the Permian to the Recent, and the Recent Bathynellacea (Schram, Chapter 4 of this volume).

Four Recent families of Anaspidacea are currently recognized (Anaspididae, Koonungidae, Stygocarididae, and Psammaspididae) and all occur in freshwater (Fig. 11). The eggs are deposited by the female, and development occurs within the egg to a stage resembling the adult. The family Anaspididae occurs only in Tasmania; one species of Koonungidae is found in Tasmania and another in southern Australia; psammaspidid syncarids were described in 1974 (Schminke, 1974a) from southern Australia and a second species was described in 1980 from Tasmania (Knott and Lake, 1980); stygocaridids occur at high elevations in southern South America (Noodt, 1965), Australia (Schminke, 1980) and New Zealand (Schminke and Noodt, 1968). Thus, the Recent Anaspidacea represent a freshwater Gondwana distribution.

The order Bathynellacea is Recent only and consists of three families: the Bathynellidae, the Leptobathynellidae, and the Parabathynellidae. The dispersal capacity of the Bathynellacea is extremely limited (Schminke, 1974b). The eggs are shed freely and the young resemble the adults. There is no resting stage. Schminke (1974b) believes that this is a primarily freshwater group and that the few marine species have secondarily invaded the habitat. Although representatives of the group are found on all major continents, an analysis of the systematic relationships reveals patterns linked to continental movement during the Mesozoic (Schminke, 1974b).

L. Thermosbaenacea

Thermosbaenaceans are small (ca. 2–4 mm in length) crustaceans with a dorsal brood pouch that the young leave in an advanced state of development. The four genera and probably eight or nine species are usually interstitial in habits and have been recorded from hot springs, ground water, caves, wells, pools adjacent to the sea, and lagoons. The salinity of these habitats ranges from 0 to 64 ‰, while the hot springs may reach 48°C. There is no fossil record. The following data are from Stock (1976) unless otherwise noted.

Five (and possibly six) species occur around the Mediterranean: *Thermosbaena mirabilis* inhabits hot springs (ca. 3.4 ‰ salinity, temperature to 48°C) at the oasis of El Hamma near Gabes, Tunisia (Monod, 1924); *Monodella argentarii* (? = *M. halophila,* see Rouch, 1965) is known from a freshwater cave on Mount Argentario near the sea in Tuscany north of Rome and from salty interstitial water near Dubrovnik, Yugoslavia, and areas adjacent to the Aegean sea; *M. stygicola* is known from slightly brackish water in a cave about 1 km southwest of Castromania, Italy; *M. relicta* is known from thermomineral springs (31°C, ca. 56 ‰ salinity) at Hamei Zohor, Israel, on the shores of the Dead Sea; and *Limnosbaena finki* is known from interstitial freshwater of the Bosnia region of Yugoslavia. Some unidentified thermosbaenaceans have been recorded from other localities around the Mediterranean.

Two species occur in the West Indies: *Halosbaena acanthura* from hypersaline (ca. 64 ‰ salinity) to slightly less than marine conditions (ca. 32 ‰ salinity) in pools and lagoons along the beaches of Curaçao, and *Monodella sanctaecrucis* from deep wells (salinity ca. 1–1.8 ‰) on Saint Croix, Virgin Islands.

The fifth species of *Monodella, M. texana,* was collected from fresh, cool water in a cave (Ezell's Cave, San Marcos, Hays County) in Texas, of the United States (Maguire, 1964, 1965).

The distribution of thermosbaenaceans can be summarized as follows: species of *Monodella* occur in freshwater in the United States, in almost freshwater in the West Indies, and in fresh to hypersaline waters around the Mediterranean; *L. finki* also occurs around the Mediterranean, whereas *H. acanthrua* occurs in marine and hypersaline waters in the West Indies.

Biogeographic hypotheses concerning this group have been proposed by Barker (1959), Fryer (1965), and Maguire (1964, 1965). Stock (1976) summarizes these in light of his own discoveries of thermosbaenaceans in the

West Indies. The hypothesis of Stock (1976) and Fryer (1965) is essentially a vicarious explanation. They propose that the group was previously marine and widespread in the ancient Tethys Sea. Their present distribution is then the result of geographic isolation caused by their being "stranded" inland by changes in sea level during the Miocene. This hypothesis accounts for the present distribution of *Monodella* and *Limnosbaena* and suggests that *Halosbaena acanthura* represents only one of additional marine forms yet to be discovered.

M. Mysidacea

Mauchline (1980) recently reviewed the biology of mysids and unless otherwise stated the data given are from that work. There are 780 species of mysids in 120 genera and 6 families (Mauchline and Murano, 1977). The majority of species occur in shallow-water marine habitats; about 200 are oceanic, 25 are freshwater, and another 18 occur in freshwater caves and wells. Many species occur a few centimeters to a few meters off the bottom. There is direct development within a brood pouch. The fossil record begins in the Carboniferous with the Pygocephalomorpha, an extinct group that underwent an extensive radiation during the early Mesozoic (Schram, 1974a, 1977). The fossil record of the modern forms begins in the Mesozoic (Schram, Chapter 4 of this volume).

The numbers of mysid species known from various regions of the world are summarized in Table IV, based on the data in Mauchline and Murano

TABLE IV

Number of Mysidacea Known from Various Regions of the World

Region	Number of species
Arctic Ocean	8-23[a]
North Atlantic	110
Western Atlantic	101
Eastern Atlantic	109
Mediterranean Sea	60
Southern Atlantic	43
North Pacific	56
Eastern Pacific	73
Western Pacific	367
Southern Pacific	36
Indian Ocean	100
Red Sea	31

[a] Includes species occasionally taken in the Arctic.

(1977) and Mauchline (1980). All species, including deep-sea species, known from each region are included.

There are 8 species that regularly occur in the Arctic Ocean, although another 15 species have been recorded from north of 60°N latitude. About 110 species have been recorded from the North Atlantic south to about 40°N latitude. Only 8 species appear to be restricted to the Western Atlantic, while almost 50 are apparently restricted to the Eastern Atlantic. One of the species recorded from the Western Atlantic, *Praunus flexuosus,* appears to be a recent immigrant from European waters (Wigley and Burns, 1971). It was probably not present in American waters during the late 1800s, but is relatively common today. How the species arrived is unknown. Sixty species, including more than 20 endemics, have been reported from the Mediterranean Sea. There is faunal division within the Mediterranean: approximately 30 species occur in the Atlantic and Western Mediterranean, but do not occur in the easter portion. Twenty-two species have been reported from the Black Sea; 9 occur in the Mediterranean, 6 are endemic, and 7 are shared with the Caspian Sea. The mysid fauna of the Caspian consists of the 7 species shared with the Black Sea plus 15 endemic species. The western Atlantic fauna south of 40°N latitude consists of 101 species about 70 of which are endemic to the area. There is some faunal division between the *Carolinian* region (see Briggs, 1974) and the Caribbean. About 109 species have been recorded from the Eastern Atlantic, and 40 of these appear to be endemic.

Approximately 56 species have been recorded north of 40°N latitude in the Pacific Ocean, but this number includes species (about 20) from the Kurile-Kamchatka Trench and the adjacent regions. Seventy-three species have been recorded from the Eastern Pacific, but the shallow-water fauna of this region is poorly known. The huge area of the Western Pacific includes about 367 mysid species, but this number includes some truly oceanic species. Within this region, the area of the Malay Archipelago includes the greatest number of species, with 216 being reported. The Indian Ocean mysid fauna includes 100 species, plus 15 additional species endemic to the Red Sea.

Mauchline and Murano (1977) grouped the known species of mysids into 10° regions of latitude, plotting number of species against latitude. They found that the greatest number of species occurs at about 45°N latitude and that there are more species reported from the northern than from the southern hemisphere.

The above data and those from W. M. Tattersall (1951), Brattegard (1969, 1973, 1980), Wigley and Burns (1971), and Williams (1972) for the number of species in the Western Atlantic are plotted in Fig. 16A, and those for the Eastern Pacific are plotted in Fig. 16B. There appears to be no strong gradient

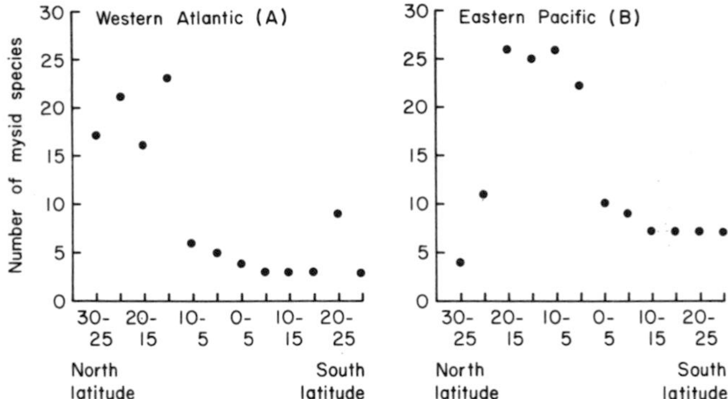

Fig. 16. Relationship between latitude and species number for Mysidacea from the Western Atlantic (A) and Eastern Pacific (B). Data from Tattersall (1951), Brattegard (1969, 1973), and Wigley and Burns (1971).

of increasing species richness with decreasing latitude in the Western Atlantic. If any gradient does occur, it is not strong, since at least 15 species occur off the coast of New England, while about 24 have been recorded from the tropical Caribbean. In the Eastern Pacific the greatest number of species occurs from 0° to 20°N latitude, though as pointed out earlier, the fauna there is poorly known.

N. Cumacea

There are about 700 species of cumaceans in 80 genera and 8 families (Jones, 1963, 1969; Zimmer, 1980). The fossil record extends from the Upper Permian to the Recent although fossils are rare (Schram, Chapter 4 of this volume). All species are marine, though some may be found a considerable distance up rivers. Development is direct. Jones (1963) suggests that the greatest number of species occurs in tropical coastal waters and this may prove to be true when more data are available. However, the data summarized by Jones (1969), who lists all known species, reveals a pattern similar to isopods and amphipods: i.e., the largest number of species occur in temperate regions (Table V).

The number of species known from various depths is plotted in Fig. 7A from the data of Jones (1969; see also Jones and Sanders, 1972). Although this includes species from a variety of habitats, the within-habitat data of Jones and Sanders (1972, p. 738) shows that species richness is fairly constant from about 1000 to 4000 m, at which point it drops to only two species below 5000 m.

TABLE V

Number of Cumacean Species from Various Localities

Region	Number of species
Norway	48
Rumania (Black Sea)	20
Mediterranean	54
Arctic-north boreal (total)	168
Japan	68
New Zealand	33
Temperate Australia	116
Tropical Australia	53
Antarctic-Subantarctic	24
West Africa	10
South Africa	26
Deep sea (>200 m)	232

O. Spelaeogriphacea

The single Recent species of this group, *Spelaeogriphus lepidops,* has direct development and is known only from a freshwater stream in Bats Cave (Fig. 11) on Table Mountain near Cape Town, South Africa (Gordon, 1957). A second species, *Acadiocaris novascotica,* has been reported from the Carboniferous of Canada (Schram, 1974a).

P. Amphipoda

Amphipods are a widely distributed, ecologically diverse group of crustaceans. All species have direct development. The fossil record begins in the Upper Eocene (Schram, Chapter 4 of this volume), and Recent species are

TABLE VI

Geographic Distribution of Marine Gammaridean Amphipoda

Region	Number of species
Arctic-boreal	1471
North warm temperate	361
Tropical	955
South warm temperate	384
Antarctic-antiboreal	301
Deep sea (> 2000 m)	239

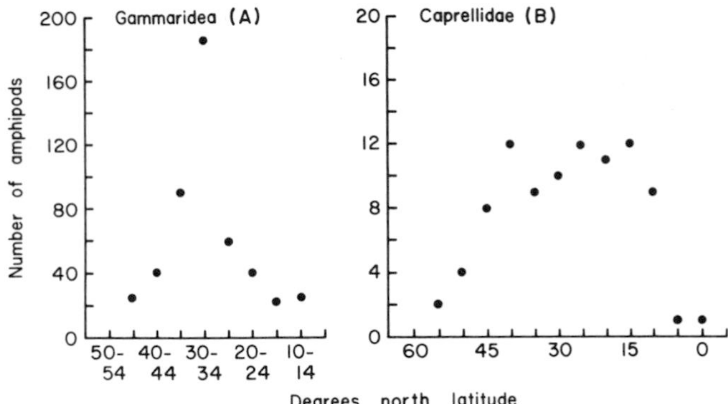

Fig. 17. Relationship between latitude and species number for gammarid amphipods in the Eastern Pacific (A) and caprellid amphipods of the Western Atlantic (B). (Gammarid data from Barnard, 1954, 1969a,c, 1979; Caprellid data from McCain, 1968.)

found in the deepest depths of the oceans and high in mountains. There are probably 6000 species divided into three major groups: the Gammaridea, Caprellidea, and Hyperiidea.

Barnard (1962, 1969a,b, 1970, 1976) has summarized the major distribution patterns of gammarid isopods and his data were used to construct Table VI. The greatest diversity is in the Arctic-Boreal region, where almost 1500 species have been found. Some data (Barnard, 1954, 1969a,c, 1979) from the Pacific northwest are plotted in Fig. 17A where the same pattern can be seen—most species and families occur in the cold-temperate region. Similarly, Watling (1979) found the greatest number of species in the cold-temperate region of the northeastern Atlantic.

A consideration of generic relationships is summarized by Barnard (1969b) and reveals a relationship similar to that of species. Only 53 of 526 genera are tropical in distribution, while 114 genera occur only in the Arctic-boreal region and 94 genera in the Antarctic-antiboreal region. Another 104 genera occur in the deep sea, 52 are cosmopolitan, and the remainder occur in temperate regions. Similarly only 6 of 53 families are tropical, while 21 are cosmopolitan. The remainder, approximately 50%, are primarily cold water in distribution.

There are approximately 800 species of freshwater gammarids and these, too, have their greatest diversity in cold-temperate regions (Barnard, 1969a; Table VII). The majority of freshwater gammarids are in 2 families, the Gammaridae in cold-temperate regions, and the tropical Hyalellidae. Barnard (1976) suggested that the present distribution of freshwater gammarids may be related to continental movements during the Mesozoic and to glaciation in the Nearctic during the late Cenozoic. During the Mesozoic

TABLE VII

Geographic Distribution of Freshwater Gammarid and Hyalellid Amphipoda

Family and locality	Genera	Species
Gammaridae		
Lake Baikal	37	240
Palearctic	57	475
Nearctic	9	118
South Africa	2	11
Australia-New Zealand	5	25
Hyalellidae		
Neotropical	1	28
African	1	1
Australia-New Zealand	2	5

the Gammaridae would have been able to disperse through Laurasia and reach Australia and South Africa. Glaciation in the Nearctic resulted in the present impoverished fauna (Table VII). Interestingly, South America lacks true freshwater Gammaridae, and the rift lakes of East Africa are apparently devoid of amphipods. This may complicate an explanation based on continental dispersal but, as Barnard (1969b, 1976) emphasizes, any statement is tentative until phylogenetic relationships are known.

There are approximately 220 species of semiterrestrial and terrestrial talitroid amphipods (Barnard, 1969b; Hurley, 1959, 1968). Of these, probably 50 are completely terrestrial, living in leafmold of tropical and southern cold-temperate forests (Hurley, 1968). Terrestrial amphipods occur in the tropics (especially on oceanic islands), and are known from Japan, Africa, Western Australia, the Philippines, and the Indo-Malayan region. They are apparently absent, except for introductions by man, from Europe, North America, and South America (but are present in Central America, Jamaica, and Haiti). Hurley (1968) suggests several mechanisms for this distribution but, with the data available, cannot select any one over the others.

The Caprellidea are shallow-water marine crustaceans which include approximately 140 species in four families of which the Caprellidae and especially *Caprella* is the largest. This group also includes the Cyamidae, which are parasitic on the skin of whales. Mayer (1903) lists the species of Caprellidae from localities around the world, but given the general knowledge and taxonomy of the time (McCain, 1968) these can only be considered the crudest estimates. More recently, McCain (1968) reviewed the caprellids of the Western North Atlantic (approximately Nova Scotia to the Equator) and reported on 26 species. In Fig. 17B, the number of species is plotted against latitude for this fauna based on the distributional maps in McCain's volume.

The numbers of species increase from 55° to 40°N and then are approximately constant to about 10°N, where there is a dramatic decrease in species number. Although the lower latitudes are not well surveyed, it is unlikely that there is an increase in species number with decreasing latitude. This same pattern is apparent with the data of Mayer (1903), although less well documented.

The Hyperiidea are an entirely pelagic marine group and mostly oceanic, although some species occur in coastal waters. Many species are known to be associated with medusae or salps (Bowman and Gruner, 1973; see review by Laval, 1980). Bowman (1973) summarized the distribution of species in the genus *Hyperia* and related genera. A few of the species, such as *Hyperia leptura,* appear to be known from a restricted locality, but the majority appear to be widely distributed. For example *Hyperia spinigera* (= *H. antarctica;* see Thurston, 1977) is known throughout the Atlantic, North Pacific, and Antarctic waters. *Hyperioides longipes* and *Lestrigonus shizogeneios* are known from warm waters around the world. As suggested by Bowman (1973), there is probably a great deal of similarity between the distributions of euphausiids and hyperiids.

Q. Isopoda

There are more than 4000 species of isopods distributed into 10 suborders. Isopods occur in virtually all habitats and include true terrestrial species. There is direct development within a brood pouch. The oldest group of isopods, the Phreatoicidea, extend back to the Middle Pennsylvanian (Schram, 1970), when they occurred in marine habitats. Today they have a freshwater Gondwana distribution (Schram, 1977) which includes South Africa, India, Australia, Tasmania, and New Zealand.

Kussakin (1973) summarized some distribution patterns of isopods and concluded that "they do not exhibit considerable reduction in species from tropical latitudes to the Poles and from the infralittoral to the ultraabyssal zone." The following summary is based on Kussakin unless otherwise noted. The Flabellifera is the most species rich of the suborders and is predominately tropical in distribution, with 770 of about 1050 species occurring in warm waters, and only 258 species occurring in cold and cold-temperate regions. Within the Flabellifera there is some variation among the families: the Cirolanidae and Sphaeromatidae are primarily (about 80% of the species) tropical, whereas 65% of the Serolidae species are found in the cold waters of the Southern Hemisphere. The distributional limits of some of the families tend to be asymmetrical. For example, the Cirolanidae is absent from the Arctic but is present in the Antarctic.

There are approximately 260 species of anthuridean isopods occurring in marine, estuarine, freshwater, and hypogean habitats (Brian Kensley, personal communication). They are known from the intertidal to the deep sea. The greatest number of species occurs in tropical shallow-water marine habitats (Kensley, 1980a,b).

The gnathiid isopods were reviewed by Monod (1926), and the data suggest that the greatest diversity of species occurs in the north and south temperate regions. In addition, there are about three times as many species in the Antarctic as in the Arctic region.

Most species (239 of 330) of Valvifera occur in cold and cold-temperate regions of the world. This is demonstrated in detail by the idoteid isopods of the northeast Pacific (Brusca and Wallerstein, 1979): there are 11 species of 60°N latitude, 18 species at 40°N latitude, and only 2 species from 10°N to 0° latitude.

The greatest number of Asellota species (702 out of 840) occur in cold-water regions. In addition, the most specialized taxa are also found in cold-water regions. The dominance of different groups results in taxonomic variation on a geographic basis. For example, the Flabellifera account for most tropical species (90% in the Philippines), while Asellota are uncommon in the tropics (2% in the Philippines). These values are reversed in the Arctic where Flabellifera account for 2.8% and Asellota for 78.9% of the isopod fauna.

The latitudinal pattern for the shelf isopods of the western North Atlantic (Schultz, 1969; Menzies and Frankenberg, 1966; Menzies and Glynn, 1968) confirms the observation of Kussakin (1973) that isopod species richness does not increase toward the tropics (Fig. 18A). In this region the greatest number of species occurs between 30° and 17°N latitude. As already noted, idoteid isopods of the northeastern Pacific have the greatest number of species at 40°N latitude. Along the coast of Chile (Fig. 18B; Menzies, 1962) the greatest number of species is found about 40°S latitude.

Although it is true that isopods occur at all depths, the conclusion of Kussakin (1973) that there is not a considerable reduction in species number with increasing depth is not correct (Menzies *et al.*, 1973). With the exception of certain families (e.g., Desmosomatidae), the number of isopod species decreases with depth. This is shown clearly by Kussakin's figures, and he was probably referring to the fact that, in contrast to many taxa, isopods are relatively abundant in the deep sea. For example, Hessler *et al.* (1979) report that only polychaetes are more diverse than isopods in the deep sea and that up to 100 isopod species may be collected in a single sample. The taxonomic composition of the isopod fauna changes dramatically with depth. Asellotes account for about 90% of the species in the abyssal regions but only about 8% in the intertidal-shelf region of the Carolinas (Menzies *et al.*, 1973).

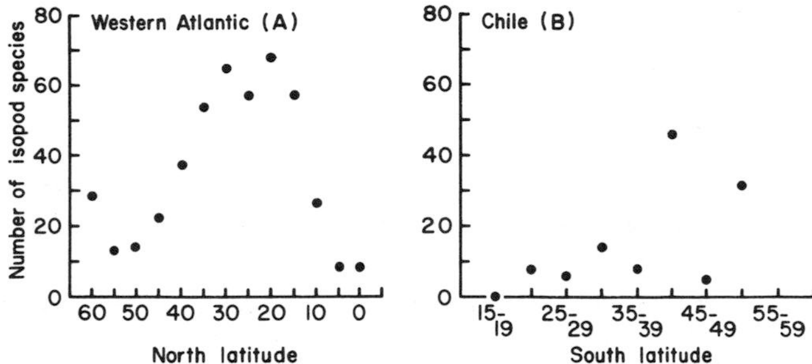

Fig. 18. Relationships between latitude and species number for isopods of the Western North Atlantic (A) and Chile (B). Data for the Western Atlantic from Menzies and Frankenberg (1966), Menzies and Glynn (1968), and Schultz (1969) and for Chile from Menzies (1962).

R. Tanaidacea

Tanaids are a relatively poorly known group of approximately 500 marine benthic species, with a few known from brackish water. Development is direct within a brood pouch. The fossil record extends back to the Lower Carboniferous. The order has recently been revised into three suborders and 17 families (Sieg, 1980a).

The literature on most of the families is scattered and few regions have been well studied. Richardson (1905) included tanaids in her study of North American isopods. Lang (e.g., 1968) contributed many papers to tanaid systematics, but the fauna of geographic regions remains poorly known. For example, Menzies (1953) reported 17 species from the subtropical Gulf of California, while Shiino (1978) reported a similar number, 19, from regions around Kerguelen in the subantarctic. Sieg (1977, 1980b) presents data on the Pseudotanaidae and the Tanaidae and recently (1981) summarized the known Tanaidacea of the world. Unfortunately, the paper is not available at this writing.

Gardiner (1975) reviewed the deep-sea family Neotanaidae. There are 28 species in this family which occur from 223 m (Antarctica) to 8300 m (Kermedec Trench). Sixteen species occur in the Atlantic, 17 in the Pacific, and 4 in both oceans. Four species are known from the Indian Ocean, one of which occurs also in the Atlantic. Within each ocean many species have very wide geographic ranges. This is in contrast to desmosomatid isopods (Hessler, 1970) and cumaceans (Jones and Sanders, 1972), two other peracarid crustacean groups which have many deep-sea species with narrow geographic ranges. Neotanaid species also appear to have broad depth ranges. One species, *Neotanais armiger,* has a depth range of 5500 m, if all specimens actually belong to a single species. The number of species at various depths is: 10 species from 200 to 2000 m, 24 species from 2000 to 6000 m, and 4 species below 6000 m.

S. Euphausiacea

Euphausiids, or krill, are marine planktonic eucarid crustaceans which are perhaps best known as prey of the baleen whales. Many of the 85 known species occur in great concentrations or swarms, especially at high latitudes. There is no fossil record. Eggs are released and development occurs in the plankton. Mauchline and Fisher (1969) and Mauchline (1980) reviewed the euphausiids, including their distributions, and the following data are derived from their excellent monograph. Additional detailed data on euphausiid distribution can be found in Brinton (1975, 1979).

The numbers of euphausiid species in various regions are summarized in Table VIII and Fig. 19. There are several general patterns that are interesting. The first is that many species tend to be widely distributed. For example, probably no less than 51 species occur in all oceans. Among the species that occur in all oceans are a number which have circumpolar distributions in the Antarctic or that are antitropical in distribution. In contrast, only 7 species are confined to the Atlantic, 17 species to the Pacific, and 1 to the Indian Ocean. Second, while the majority of euphausiids are widely distributed, some have comparatively restricted distributions. *Meganyctiphanes norvegica* occurs only in the North Atlantic north of the 15°C isotherm at 2000 m. Similarly, *Tessarabrachium oculatum* and some species of *Thysanoëssa* occur only in the North Pacific. Third, many euphausiids have spawning areas which are much more restricted than the total geographical distribution of adults (Einarsson, 1945).

A total of 62 species of krill has been reported from the Atlantic Ocean with 14 of these occurring north of 40°N. However, it appears that only 6 species (*Thysanopoda acutifrons, Nyctiphanes couchii, Meganyctiphanes norvegica, Thysanoëssa inermis, T. longicaudata, T. raschii*) actually breed

TABLE VIII

The Total Numbers of Euphausiids in Pairs of Oceans, the Percentages of the Total Shared by Each Member of the Pair, and the Percentages Endemic

	Total		Percentage		
Pairs of Oceans	Species	Shared	Atlantic	Pacific	Indian
Atlantic/Pacific	83	65	9.5	25.5	—
Atlantic/Indian	66	70	24.0	—	6.0
Pacific/Indian	75	67	—	33.0	0

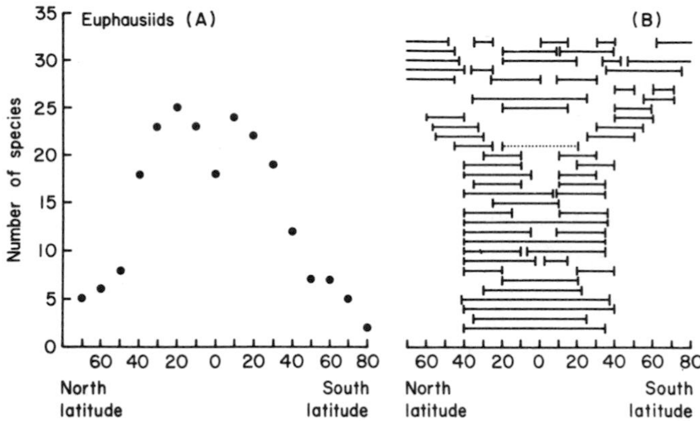

Fig. 19. Relationship between latitude and species number for the epipelagic euphausiids of the Pacific Ocean (A), and the range of each species (B). (Data from Mauchline and Fisher, 1969.)

in this region. There are 14 species known from the Mediterranean and none of these is endemic. In the region of 40°N to 40°S the eastern and western coastal regions have more species (38 and 37) than the central region (37). South of 40°S, 15 species are known to occur.

The Pacific Ocean, with 75 species, is probably the best known of the oceans with regard to euphausiids. The number of epipelagic species is plotted against latitude in Fig. 19A. If a species occurs "to 40°N," it is included in the count at 40°N. The number of species increases, north and south, toward the equator until about 10°-20° latitude where the maximum number of species occurs: there is then a decrease of species at the equator. The latitudinal range for the species is shown in Fig. 19B, arranged according to their relative depth distributions for epipelagic species only. Additional data on Pacific euphausiids is in Table VIII.

Fifty species have been reported from the Indian Ocean, but the deeper waters have not been adequately explored.

Twelve species have circumpolar distributions in the Antarctic Ocean: *Euphausia crystallorophias, E. superba, E. frigida, E. tricantha, E. vallentini, E. similis, E. longirostris, E. lucens, E. similis armata, E. spinifera, Thysanoëssa macrura,* and *T. vicina. Euphausia crystallorophias* is a neritic species living under the ice, while *E. superba* occurs under the ice and extends northward to the Antarctic convergence.

Table VIII lists the euphausiid fauna of pairs of oceans (excluding two species of questionable distribution) where it can be seen that the Pacific Ocean has the largest number of endemics, followed by the Atlantic. The Indian Ocean has none. All species known from the Indian Ocean (except one, *Stylocheiron indicum*, recently described) are also known from the Pacific.

The distribution of euphausiid species appears to change to a certain extent with the season (Mauchline and Fisher, 1969), probably because water masses shift during seasonal changes. Brinton (1962) has reviewed many cases of such changes in distribution.

The total number of euphausiid species and the volume of major bodies of water are highly correlated. The cube root of the volume and species numbers are as follows: Pacific, 75 species, 8.90×10^2 km; Atlantic, 62 species, 6.86×10^2 km; Indian, 50 species, 6.62×10^2 km; Mediterranean, 14 species, 1.61×10^2 km; and Red Sea, 11 species, 0.59×10^2 km. Taking the logs of the data linearizes the relationship to log species number = 0.747 log $\sqrt[3]{\text{volume}}$ + 1.23, where $r = 0.97$, $p < 0.01$.

T. Amphionidacea

The amphionids consist of a single circumtropical planktonic species, *Amphionides reynaudii*. The species is found in all oceans between 36°N and 36°S (Heegard, 1969). Larval stages tend to be found at the surface to about 30 m depth. Adults occur in deeper water from about 2000 to 5000 m, although specimens have been taken at 6000 m and as shallow as 60 m (Williamson, 1973; Heegard, 1969).

U. Decapoda

Decapods constitute the largest and most diverse order of malacostracans. There are probably 10,000 Recent species in marine, freshwater, and terrestrial habitats. The earliest record of a decapod is *Palaeopalaemon newberryi,* a pleocyemate with both astacidean and palinuran characteristics known from the Late Devonian (Schram *et al.,* 1978). Decapods are among the better known of the crustaceans and various authors have summarized their distribution. Griffin and Yaldwyn (1968) summarized the distribution of marine decapods of Australia while Bishop (1967) dealt with the freshwater species of Australia. Abele (1976b), Williams (1965), Coêlho (1967), and Boschi (1966, 1979) have dealt with distribution of decapods in the Western Atlantic, Monod (1956) has summarized some data on the West African fauna, and Schmitt (1921) and Rathbun (1904, 1910) have dealt with Eastern Pacific decapods. There are numerous summaries of individual taxa such as that of the Atyidae by Bouvier (1925). Special mention must be made of the remarkably modern study of Ortmann (1902). He incorporated the fossil record, phylogenetic relationships, paleogeography, and present distributions

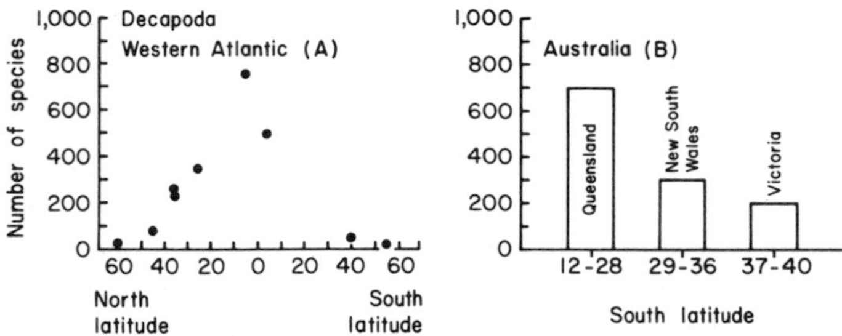

Fig. 20. Relationship between latitude and species number for decapods of the Western Atlantic (A) and of Australia (B). Western Atlantic data from Abele (1976b), Williams (1965), Coêlho (1967), and Boschi (1966); Australian data from Griffin and Yaldwyn (1968).

into a study of the freshwater decapods of the world. There are other studies on decapods that are cited by Briggs (1974).

There is a strong latitudinal gradient in species number of marine decapods, with the greatest number occurring just north of the equator. Fig. 20A shows this for the decapods of the Western Atlantic, and a similar relationship holds for the Eastern Pacific and Australia (Fig. 20B) (Griffin and Yaldwyn, 1968). Within the tropics, the number of marine species increases in the order Eastern Atlantic, Eastern Pacific, Western Atlantic (these two regions may be equal or reversed for some groups), and Indo-West Pacific. The large number of Indo-West Pacific species is due, in part, to the presence there of families and genera that are absent in the other regions. However, the greatest increase is the result of the larger number of congeneric species in the Indo-West Pacific relative to the other regions (Table I). For example, there are 19 to 25 species of *Alpheus* in each of the other regions and 128 species in the Indo-West Pacific. Similarly, the genus *Sesarma* is represented by more than 80 species in the Indo-West Pacific and from 6 to 15 species everywhere else.

There are approximately 1200 Recent decapod genera. The majority (56%) of these occur only in the tropics and they are distributed as follows: 21% are endemic in the Indo-West Pacific region, 10% are circumtropical, 4% are endemic in the tropical Western Atlantic, 3% in the tropical Eastern Pacific, 2% in tropical Australia, and 1% in the tropical Eastern Atlantic; 6% occur in the tropical Eastern Pacific and Atlantic Ocean, and about 2.5% occur in the Eastern Pacific and Indo-West Pacific; 2.5% occur in the Western Atlantic and Indo-West Pacific, 2% in the Eastern Atlantic and Indo-West Pacific, and the remaining 2% are distributed in various combinations of the tropical regions.

The remaining genera are distributed in various regions of the world and the following are approximate figures. Twenty percent occur in both temperate and tropical regions and were not included in the above figures for the tropics; another 10% occur only in deep water off the continental shelf, and 10% occur in freshwater. The remaining genera are endemic in various regions of the world.

One of the major freshwater groups, the astacidean crayfish, has a fossil record extending back to the Late Permian (Schram, Chapter 4 of this volume). This group also has direct development with some parental care (Little, 1975), which limits dispersal. The distribution of this group appears to be related to continental movements. For example, the Parastacidae contains representatives in Madagascar, Australia, Tasmania, New Guinea, New Zealand, and southern South America (Hobbs, 1974). The family Astacidae is represented in middle, western, and eastern Europe and in the Pacific drainages of North America. The Cambaridae is represented in North and Middle America, with the subfamily Cambaridinae represented in Eastern Asia. Thus, the superfamily Parastacoidea is primarily Gondwanian in distribution, whereas the Astacoidea is Laurentian in distribution.

It may be noted that while crayfish are primarily temperate in distribution, the freshwater crabs (Potamoidea) are tropical, so that the distributions of the two groups overlap very little.

An interesting distribution is shown by the caridean shrimp fauna of anchialine pools (pools with no surface connection with the sea, containing salt or brackish water which fluctuates with the tides; Holthuis, 1973). Eleven species in nine genera and five different families have been reported (Holthuis, 1973; Chace, 1975), and these species are often known only from widely separated localities. For example, *Calliasmata pholidota* is known from the anchialine pools on the Sinai Peninsula, in the Ellice Islands, and the Hawaiian Islands. All but one of the shrimps appears to be restricted to this habitat.

VIII. SUMMARY AND CONCLUSIONS

The absence of information in many areas hinders biogeographic analyses, and until more information is available, most conclusions will be tentative. Bearing this lack of information in mind, the distribution patterns of crustaceans appear to be best explained by a vicariance hypothesis under which previously wide-spread faunas have been modified by continental drift, opening and closing of seaways, speciation, and extinction. This is not a testable conclusion, but it appears to be consistent with the available facts.

(1) The number of crustaceans within the tropics generally increases in the order: Eastern Atlantic, Eastern Pacific, Western Atlantic, and Indo-West

Pacific. Peracarids have their greatest diversity in the temperate region and other crustaceans in the tropics. Large areas and spatially complex habitats contain more species than smaller or less complex habitats. Shallow water in general contains more species than deep-water regions.

(2) Migrations and dispersal of crustaceans appears related more to the size of the species pool and to ecological conditions than to competition. Marine and freshwater crustaceans have the fewest species between 3 and 5 ppt salinity.

(3) Size of crustaceans decreases with decreasing latitude and eye development decreases with increasing depth.

(4) Cephalocarids are widely distributed and probably achieved their present distribution in the distant past, although no fossils are known. Both *Sandersiella* and *Lightiella* are widely distributed; the former has representatives known from Japan, Peru, and southwest Africa and the latter from the Eastern Pacific, Gulf of Mexico, Western Atlantic, and New Caledonia. It seems unlikely that long distance dispersal has been important.

(5) Branchiopod genera and species tend to be very widely distributed in temporary and permanent freshwater ponds. Their wide distribution is usually attributed to passive dispersal of resistant eggs. However many genera (and a few species) have fossil records extending well back into the Mesozoic, and the Recent distribution of genera may be more the result of continental movements than long distance dispersal. For example, the ephippia of some cladocerans are attached to fixed objects and appear to be adapted more for remaining there than for dispersal (Fryer, 1972). Similarly, Hebert (1975) has shown that gene flow among populations of *Daphnia magna* is extremely limited. He found significant differences in allelic frequencies between populations isolated by only a few meters. I do not argue that passive dispersal does not occur; however, it has not influenced the world-wide distribution of the branchiopods.

(6) Marine ostracodes have their greatest diversity in the tropics and, within the tropics, in the Indo-West Pacific region. The wide distribution of pelagic species was probably achieved through seaways that are now closed. The distribution of freshwater ostracodes is consistent with explanations based on continental drift.

(7) Mystacocarids occur in and are adjacent to the Atlantic Ocean. They are thought to be a relatively young group which originated in the Atlantic through progenesis.

(8) Very little is known concerning the distribution of branchiurans. Most species of *Argulus* are found in temperate regions, and the occurrence of species of *Dolops* in Tasmania, South America and Africa suggests prior continental connections.

(9) Copepod species, both pelagic and benthic, tend to be widely distributed and have their greatest diversity in the tropics.

(10) The Cirrepedia have their greatest diversity in the tropics, particularly the Indo-West Pacific region.

(11) Phyllocarids were more taxonomically diverse during the Paleozoic. The single extant family, the Nebaliidae, contains a number of widely-distributed genera whose distributions were probably determined by tectonic events and by extinctions.

(12) Hoplocarids were also more diverse during the Paleozoic and are today represented by the stomatopods, which have their greatest diversity within the tropics, particularly the Indo-West Pacific region.

(13) Two orders of syncarids are extant, the Anaspidacea, which has a Gondawanan distribution, and the Bathynellacea, which occurs on all major continents. Anaspidaceans and the extinct palaeocaridaceans were more widely distributed in the past. All syncarids are primarily freshwater with very limited dispersal, suggesting that continental movements were important in their current distribution.

(14) Thermosbaenaceans occur in brackish to hypersaline water around the Mediterranean and in the West Indies, and in freshwater in Texas. This pattern suggests a distribution determined by sea level changes during the Miocene.

(15) Peracarid crustaceans all brood their young and include the Mysidacea, Amphipoda, Isopoda, Tanaidacea, and Cumacea. In general, the groups have their greatest diversity in temperate regions.

(16) Eucarids include the pelagic euphausiids and amphionidaceans and the primarily benthic decapods. The euphausiid species tend to be widely distributed in all oceans, with the greatest number occurring in low latitudes. There is a single circumglobal species of amphionidacean. Decapods have their greatest diversity within the tropics, particularly the Indo-West Pacific region.

ACKNOWLEDGMENTS

I thank the following individuals for their comments on the manuscript: J. L. Barnard, T. Bowman, F. A. Chace, Jr., Q. Cohen, R. Cressey, B. Felgenhauer, D. Frey, S. Gilchrist, K. Heck, R. R. Hessler, B. Kensley, R. Manning, F. Schram, D. Simberloff, H. Spivey, D. Strong, and A. Underwood. I thank also Ms. Arlee Montalvo for the illustrations.

REFERENCES

Abele, L. G. (1972). Introduction of two freshwater decapod crustaceans (Hymenosomatidae and Atyidae) into Central and North America. *Crustaceana* **23,** 209-218.

Abele, L. G. (1973). Taxonomy, distribution and ecology of the genus *Sesarma* (Crustacea, Decapoda, Grapsidae) in eastern North America, with special reference to Florida. *Am. Midl. Nat.* **90,** 375-386.

Abele, L. G. (1974). Species diversity of decapod crustaceans in marine habitats. *Ecology* **55,** 156-161.

Abele, L. G. (1976a). Comparative species richness in fluctuating and constant environments: Coral-associated decapod crustaceans. *Science* **192,** 461-463.

Abele, L. G. (1976b). Comparative species composition and relative abundance of decapod crustaceans in marine habitats of Panama. *Mar. Biol.* **38,** 263-278.

Abele, L. G. (1981). Biogeography colonization and experimental community structure of coral-associated decapod crustaceans. *In* "Community Ecology" (D. Strong, L. Abele, and D. Simberloff, eds.) Princeton Univ. Press (in press).

Abele, L. G., and Blum, N. (1977). Ecological aspects of the freshwater decapod crustaceans of the Perlas Archipelago, Panama. *Biotropica* **9,** 239-252.

Abele, L. G., and Patton, W. R. (1976). The size of coral heads and the community ecology of associated decapod crustaceans. *J. Biogeogr.* **3,** 35-47.

Abele, L. G., Heck, K. L., Jr., Simberloff, D., and Vermeij, G. J. (1982). Biogeography of crab claw size: Assumptions and a null hypothesis. *Syst. Zool.* **24,** 406-424.

Angel, M. V. (1972). Planktonic oceanic ostracods—historical, present and future. *Proc. R. Soc. Edinburgh B* **73,** 213-228.

Ball, I. R. (1975). Nature and formulation of biogeographic hypotheses. *Syst. Zool.* **24,** 407-430.

Barker, D. (1959). The distribution and systematic position of the Thermosbaenacea. *Hydrobiologica* **13,** 209-235.

Barnard, J. L. (1954). Marine Amphipoda of Oregon. *Oreg. State Monogr., Stud. Zool.* **8,** 1-103.

Barnard, J. L. (1962). South Atlantic abyssal amphipods collected by R. V. Vema. *Abyssal Crustacea, Vema Res. Ser.* **1,** 1-78.

Barnard, J. L. (1969a). Gammaridean Amphipoda of the rocky intertidal of California: Monterey Bay to La Jolla. *Bull.—U.S. Natl. Mus.* **258,** 1-230.

Barnard, J. L. (1969b). The families and genera of marine gammaridean Amphipoda. *Bull.—U.S. Natl. Mus.* **271,** 1-535.

Barnard, J. L. (1969c). A biological survey of Bahia de Los Angeles Gulf of California, Mexico. IV. Benthic Amphipoda (Crustacea). *Trans. San Diego Soc. Nat. Hist.* **15,** 175-228.

Barnard, J. L. (1970). Sublittoral Gammaridea (Amphipoda) of the Hawaiian Islands. *Smithson. Contrib. Zool.* **34,** 1-286.

Barnard, J. L. (1976). Amphipoda (Crustacea) from the Indo-West Pacific tropics: A review. *Micronesica* **12,** 169-181.

Barnard, J. L. (1979). Littoral gammaridean Amphipoda from the Gulf of California and the Galapagos Islands. *Smithson. Contrib. Zool.* **271,** 1-149.

Belyaev, G. M. (1972). "Hadal Bottom Fauna of the World Ocean." Inst. Oceanogr., Acad. Sci., USSR (Isr. Program Sci. Transl., Jerusalem).

Benson, R. H. (1964). Recent marine podocopid and platycopid ostracodes of the Pacific. *Pubbl. Stn. Zool. Napoli* **33,** Suppl., 387-420.

Benson, R. H. (1966). Recent marine podocopid ostracodes. *Oceanogr. Mar. Biol.* **4,** 213-232.

Bishop, J. A. (1967). The zoogeography of the Australian freshwater decapod Crustacea. *In* "Australian Inland Waters and Their Fauna" (A. H. Weatherly, ed.), pp. 107-122. Australian Nat. Univ. Press, Canberra.

Boschi, E. E. (1966). Preliminary note on the geographic distribution of the decapod crustaceans of the marine waters of Argentina (South-West Atlantic Ocean). *Symp. Crustacea, India, pt. 1,* pp. 449-456.

Boschi, E. E. (1979). Geographic distribution of Argentinian marine decapod crustaceans. *Bull. Biol. Soc. Washington* **3,** 134-143.

Bouvier, E. L. (1925). Recherches sur la morphologie, les variations et la distribution systématique des crevettes d'eau douce de la famille des atyidés. *Encycl. Entomol.* **4,** 1-370.

Bowman, T. (1971). The distribution of calanoid copepods of the southeastern United States between Cape Hatteras and southern Florida. *Smithson. Contrib. Zool.* **96,** 1-58.

Bowman, T. E. (1973). Pelagic amphipods of the genus *Hyperia* and closely related genera (Hyperiidea: Hyperiidae). *Smithson. Contrib. Zool.* **136,** 1-76.

Bowman, T. E., and H. E. Gruner (1973). The families and genera of Hyperiidea (Crustacea: Amphipoda). *Smithson. Contrib. Zool.* **146,** 1-64.

Brattegard, T. (1969). Marine biological investigations in the Bahamas. 10. Mysidacea from shallow water in the Bahamas and southern Florida. Part 1. *Sarsia* **39,** 17-106.

Brattegard, T. (1970). Marine investigations in the Bahamas. 13. Leptostraca from shallow water in the Bahamas and southern Florida. *Sarsia* **44,** 1-7.

Brattegard, T. (1973). Mysidacea from shallow water on the Caribbean coast of Colombia. *Sarsia* **54,** 1-66.

Brattegard, T. (1980). *Platymysis facilis* gen. et sp. nov. (Crustacea: Mysidacea: Heteromysini) from the Saba Bank, Caribbean Sea. *Sarsia* **65,** 49-52.

Briggs, J. C. (1966). Zoogeography and evolution. *Evolution* **20,** 282-289.

Briggs, J. C. (1974). "Marine Zoogeography." McGraw-Hill, New York.

Brinton, E. (1962). The distribution of Pacific euphausiids. *Bull. Scripps Inst. Oceanogr.* **8,** 51-170.

Brinton, E. (1975). Euphausiids of southeast Asian waters. *Naga Report,* **4**(5), 1-287.

Brinton, E. (1979). Parameters relating to the distributions of planktonic organisms, especially euphausiids in the eastern tropical Pacific. *Progress in Oceanography* **8,** 125-189.

Brooks, J. L. (1950). Speciation in ancient lakes. *Q. Rev. Biol.* **25,** 30-60, 131-176.

Brusca, R. C., and Wallerstein, B. R. (1979). Zoogeographic patterns of idoteid isopods in the northeast Pacific, with a review of shallow water zoogeography of the area. *Bull. Biol. Soc. Wash.* **3,** 69-105.

Calman, W. T. (1906). Zoological results of the third Tanganyika expedition, conducted by Dr. W. A. Cunnington, 1904-1905. Report on the macrurous Crustaca. *Proc. Zool. Soc. London* **1,** 187-206, pls. XI-XIV.

Calman, W. T. (1917). Crustacea, 4—Stomatopoda, Cumacea, Phyllocarida and Cladocera. *Br. Antarct. "Terra Nova" Exped. Zool.* **3,** 137-162.

Calman, W. T. (1927). Report on Phyllocarida, Cumacea and Stomatopoda. *Trans. Zool. Soc. London* **22,** 399-401.

Cannon, H. G. (1931). Nebaliacea. *Discovery Rep.* **3,** 199-222.

Cannon, H. G. (1960). Leptostraca. *Bronn's Klassen* **5**(1), 4(1), 1-81.

Carlisle, D. B. (1968). *Triops* eggs killed only by boiling. *Science* **161,** 279.

Carlton, J. T. (1975). Introduced intertidal invertebrates. *In* "Light's Manual, Intertidal Invertebrates of the Central California Coast" (R. I. Smith and J. T. Carlton, eds.), pp. 17-25. Univ. of California Press, Berkeley.

Castro, P. (1981). Notes on symbiotic decapod crustaceans from Gorgona Island, Colombia, with a preliminary revision of the eastern Pacific species of *Trapezia* (Brachyura, Xanthidae), symbionts of scieractinian corals. *An. Inst. Invest. Mar. Punta Betin* **12.**

Chace, F. A., Jr. (1972). The shrimps of the Smithsonian-Bredin Caribbean expeditions with a summary of the West Indian shallow-water species (Crustacea: Decapoda: Natantia). *Smithson. Contrib. Zool.* **98,** 1-179.

Chace, F. A., Jr. (1975). Cave shrimps (Decapoda: Caridea) from the Dominican Republic. *Proc. Biol. Soc. Washington* **88,** 29-44.

Chace, F. A., Jr., and Hobbs, H. H., Jr. (1969). The freshwater and terrestrial decapod crustaceans of the West Indies with special reference to Dominica. *Bull.—U.S. Natl. Mus.* **292,** 1-258.

Chapman, M. A. (1961). The terrestrial ostracod of New Zealand, *Mesocypris audax* sp. nov. *Crustaceans* **2,** 255-261.

Chilton, C. (1911). Note on the dispersal of marine crustacea by means of ships. *Trans. Proc. N. Z. Inst.* **43,** 131-133.

Clark, A. E. (1932). *Nebaliella caboti* n. sp. with observations on other Nebaliacea. *Trans. R. Soc. Can., Sect. 3* **26,** 217-235.

Clarkson, E.N.K. (1967). Environmental significance of eye reduction in Trilobites and Recent arthropods. *Mar. Geol.* **5**(5/6), 367-376.

Coêlho, P. A. (1967). A distribução dos crustáceos decápodos reptantes do norte do Brasil. *Trab. Oceanogr. Univ. Fed. Pe., Recife* **9/11,** 223-238.

Cohen, A. (1982). Ostracoda. *In* "Synopsis and classification of Living Organisms." McGraw-Hill (in press).

Connor, E. F., and McCoy, E. D. (1979). The statistics and biology of the species-area relationship. *Am. Nat.* **113,** 791-833.

Coull, B. C. (1972). Species diversity and faunal affinities of meiobenthic Copepoda in the deep sea. *Mar. Biol.* **14,** 48-51.

Coull, B. C., and Herman, S. S. (1970). Zoogeography and parallel-level bottom communities of the meiobenthic Harpacticoida (Crustacea Copepoda) of Bermuda. *Oecologia* **5,** 392-399.

Crawford, A. R. (1974). A greater Gondwanaland. *Science* **184,** 1179-1181.

Cressey, R. F. (1971). Two new argulids (Crustacea: Branchiura) from the Eastern United States. *Proc. Biol. Soc. Washington* **84,** 253-258.

Cressey, R. F. (1972). "The Genus *Argulus* (Crustacea: Branchiura) of the United States. Biota of Freshwater Ecosystems," Id. Manual No. 2, pp. 1-11. US Govt. Printing Office, Washington, D.C.

Cressey, R. F., and Collette, B. B. (1970). Copepods and needlefishes: a study in host-parasite relationships. *Fish. Bull.* **68**(3), 347-432.

Cressey, R. F., and Patterson, C. (1973). Fossil parasitic copepods from a Lower Cretaceous fish. *Science* **180,** 1283-1285.

Cressey, R., and Cressey, H. B. (1980). Parasitic copepods of mackeral- and tuna-like fishes (Scombridae) of the world. *Smithsonian Contr. Zool.* **311,** 1-186.

Crossland, S. O. (1965). The pest status and control of the tadpole shrimp, *Triops granarius,* and of the snail, *Lanistes ovum,* in Swaziland rice fields. *J. Appl. Ecol.* **2,** 115-120.

Dahl, E. (1963). Some aspects of the ecology and zonation of the fauna of sandy beaches. *Oikos* **4,** 1-27.

Dana, J. D. (1852). "Crustacea. United States Exploring Expedition During the Years 1838, 1839, 1840, 1841, 1842 Under the Command of Charles Wilkes, U. S. N.," Vol. 13.

De Deckker, P. (1977). The distribution of the "giant" ostracods (Family Cyprididae Baird, 1845) endemic to Australia. *In* "Aspects of Ecology and Zoogeography of Recent and Fossil Ostracoda" (J. Löffler and D. Danielopol, eds.), pp. 285-294. Junk, The Hague.

Delomore, L. D., and Donald, D. (1969). Torpidity of freshwater ostracodes. *Can. J. Zool.* **47,** 997-999.

Dexter, D. (1972). Comparison of the community structure in a Pacific and Atlantic Panamanian sandy beach. *Bull. Mar. Sci.* **22,** 449-462.

Edmondson, C. H. (1962). Xanthidae of Hawaii. *Occas. Pap. Bishop Mus.* **22,** 215-309.

Einarsson, H. (1945). Euphausiacea. I. North Atlantic species. *Dana Rep.* **27,** 1-185.

Ekman, S. (1953). "Zoogeography of the Sea." Sidgwick & Jackson, London.

Fisher, A. G. (1960). Latitudinal variation in organic diversity. *Evolution* **14,** 64-81.

Fleminger, A., and Hulsemann, K. (1973). Relationship of Indian Ocean epiplanktonic calanoids to the world oceans. *Ecol. Stud.* **3,** 339-348.

Foster, B. A. (1978). The marine fauna of New Zealand: Barnacles (Cirripedia: Thoracica). *N. Z. Oceanogr. Inst. Mem.* **69,** 1-143.

Fox, H. M. (1949). On *Apus,* its rediscovery in Britain. *Proc. Zool. Soc. London, Ser. B* **119,** 693-702.

Frey, D. G. (1980). The non-swimming chydorid Cladocera of wet forests, with descripttions of a new genus and two new species. *Int. Rev. Gesamten Hydrobiol.* **65,** 613-641.

Friauf, J. J., and Bennett, L. (1974). *Derocheilocaris hessleri* a new mystacocarid (Crustacea) from the Gulf of Mexico. *Vie Milieu* **24,** *3, ser. A,* 487-496.

Fryer, G. (1953). Notes on certain freshwater crustaceans. *Naturalist* No. 846, 101-109.

Fryer, G. (1956). A report on the parasitic Copepoda and Branchiura of the fishes of Lake Nyasa. *Proc. Zool. Soc. London* **127,** 293-344.

Fryer, G. (1965). Studies on the functional morphology and feeding mechanism of *Monodella argentarii* Stella (Crustacea: Thermosbaenacea). *Trans. R. Soc. Edinburgh* **66**(4), 49-90.

Fryer, G. (1968). The parasitic Crustacea of African freshwater fishes; their biology and distribution. *J. Zool. (London)* **156,** 45-95.

Fryer, G. (1969). A new freshwater species of the genus *Dolops* (Crustacea: Branchiura) prasitic on a galaxiid fish of Tasmania—with comments on disjunct distribution patterns in the southern hemisphere. *Aust. J. Zool.* **17,** 49-64.

Fryer, G. (1972). Observations on the ephippia of certain macrothricid cladocerans. *J. Linn. Soc. London, Zool.* **51,** 79-96.

Fryer, G. (1974). Evolution and adaptive radiation in the Macrothricidae (Crustacea: Cladocera): A study in comparative functional morphology and ecology. *Philos. Trans. R. Soc. London, Ser. B* **269,** 137-274.

Fulton, S. W., and Grant, F. E. (1900). Note on the occurrence of the European crab, *Carcinus maenas,* Leach, in Port Phillip. *Victorian Nat.* **17,** 145-146.

Gainey, L. F., Jr., and Greenberg, M. J. (1977). Physiological basis of the species abundance-salinity relationship in molluscs: A speculation. *Mar. Biol.* **40,** 41-49.

Gardiner, L. F. (1975). The systematics, postmarsupial development and ecology of the deep-sea family *Neotanaidae* (Crustacea: Tanaidacea). *Smithson. Contrib. Zool.* **170,** i-iv, 1-265.

Garth, J. S. (1957). The Crustacea Decapoda Brachyura of Chile. *In* Reports of the Lund University Chile Expedition 1948-49. 29. *Lunds Univ. Arsskr., Avd. 2* [N.S.] **53**(7), 3-131.

Garth, J. S. (1974). On the occurrence in the Eastern Tropical Pacific of Indo-West Pacific decpod crustaceans commensal with reef-building corals. *Proc. Int. Coral Reef Symp., 2nd, 1974* Vol. I, pp. 397-404.

Giesbrecht, W. (1892). Systematik und Faunistik der pelagischen Copepoden des Golfes von Neapel und der angrenzen den Meeresabschnitte. *Fauna Flora Golfes Neapel* **19,** 1-831.

Gordon, I. (1957). *Spelaeogriphus,* a new cavernicolous crustacean from South Africa. *Bull. Br. Mus. (Nat. Hist.), Zool.* **5,** 31-47.

Gould, S. J. (1977). "Ontogeny and Phylogeny." Harvard Univ. Press, Cambridge, Massachusetts.

Griffin, D.J.G., and Yaldwyn, J. C. (1968). The constitution, distribution and relationships of the Australian decapod Crustacea. *Proc. Linn. Soc. N. S. W.* **93,** 164-183.

Grigarick, A. A., Lange, W. H., and Finfrock, D. C. (1961). Control of the tadpole shrimp, *Triops longicaudatus,* in California rice fields. *J. Econ. Entomol.* **54,** 36-40.

Hall, J. R., and Hessler, R. R. (1971). Aspects in the population dynamics of *Derocheilocaris typica* (Mystacocarida, Crustacea). *Vie Milieu, Ser. A* **22,** 305-326.

Hansen, H. J. (1920). Crustacea Malacostraca. IV. *Dan. Ingolf-Exped.* **3**(6), 1-86.

Harding, J. P. (1953). The first known example of a terrestrial ostracod, *Mesocypris terrestris* sp. nov. *Ann. Natal Mus.* **12,** 359-365.

Hart, D., and Hart, C. (1974). The ostracod family Entocytheridae. *Acad. Nat. Sci. Philadelphia Monogr.* **18,** ix, 1-239.

Hebert, P.D.N. (1975). Enzyme variability in natural populations of *Daphnia magna* I. Population structure in East Anglia. *Evolution* **28,** 546-556.

Heck, K. L., Jr. (1979). Some determinants of the composition and abundance of motile macroinvertebrate species in tropical and temperate seagrass meadows (*Thalassia testudinum*). *J. Biogeogr.* **6,** 183-197.

Heck, K. L., Jr., and McCoy, E. D. (1978). Long distance dispersal and the reef-buidling corals of the Eastern Pacific. *Mar. Biol.* **48,** 349-356.

Heck, K. L., Jr., and Wetstone, G. S. (1977). Habitat complexity and invertebrate species richness and abundance in tropical seagrass meadows. *J. Biogeogr.* **4,** 135-142.

Heegard, P. (1969). Larvae of decapod Crustacea. The Amphionidae. *Dana Rep.* **77,** 1-82.

Henry, D. P., and McLaughlin, P. A. (1975). The barnacles of the *Balanus amphitrite* complex (Cirripedia Thoracica). *Zool. Verh.* **141,** 1-254.

Hessler, A. Y., Hessler, R. R., and Sanders, H. L. (1970). Reproductive system of *Hutchinsoniella macrantha. Science* **168,** 1464.

Hessler, R. R. (1964). The Cephalocarida. Skeletomusculature. *Mem. Conn. Acad. Arts Sci.* **16,** 1-97.

Hessler, R. R. (1970). The Desmosomatidae (Isopoda, Asellota) of the Gay Head-Bermuda transect. *Bull. Scripps Inst. Oceanogr.* **15,** 1-185.

Hessler, R. R. (1971). Biology of the Mystocarida: A prospectus. *Smithson. Contrib. Zool.* **76,** 87-90.

Hessler, R. R. (1972). New species of Mystacocarida from Africa. *Crustaceana* **22,** 259-273.

Hessler, R. R., and Sanders, H. L. (1965). Bathyal Leptostraca from the continental slope of the northeastern United States. *Crustaceana* **9,** 71-74.

Hessler, R. R., and Sanders, H. L. (1973). Two new species of *Sandersiella* (Cephalocarida), including one from the deep sea. *Crustaceana* **24,** 181-196.

Hessler, R. R., Wilson, G. D., and Thistle, D. (1979). The deep-sea isopods: A biogeographic and phylogenetic overview. *Sarsia* **64,** 67-75.

Hobbs, H. H., Jr. (1974). Synopsis of the families and genera of crayfishes (Crustacea: Decapoda). *Smithson. Contrib. Zool.* **164,** 1-32.

Holsinger, J. R., and Longley, G. (1980). The subterranean amphipod crustacean fauna of an Artesian Well in Texas. *Smithson. Contrib. Zool.* **308,** 1-62.

Holthuis, L. B. (1958). Crustacea Decapoda from the northern Red Sea (Gulf of Aqaba and Sinai Peninsula). I. Macrura. *Sea Fish. Res. Stn., Haifa, Bull.* **17,** 1-40.

Holthuis, L. B. (1973). Caridean shrimps found in land-locked saltwater pools at four Indo-West Pacific localities (Sinai Peninsula, Funafuti Atoll, Maui and Hawaii Islands), with the description of one new genus and four new species. *Zool. Verh.* **128,** 1-48, pls. 1-7.

Holthuis, L. B., and Provenzano, A. J., Jr. (1970). New distribution records for species of Macrabrachium with notes on the distribution of the genus in Florida. *Crustaceana* **19,** 211-213.

Humes, A. G., and Stock, A. G. (1973). A revision of the family Lichomolgidae Kossman, 1877, cyclopoid copepods mainly associated with marine invetebrates. *Smithson. Contrib. Zool.* **127,** 1-368.

Hurley, D. E. (1959). Notes on the ecology and environmental adaptations of the terrestrial Amphipoda. *Pac. Sci.* **13,** 107-129.

Hurley, D. E. (1968). Transition from water to land in amphipod crustaceans. *Am. Zool.* **8,** 327-353.

Jazdzewski, K. (1980). Range extensions of some gammaridean species in European inland waters caused by human activity. *Crustaceana, Suppl.* **6,** 85-107.

John, D. M., and Lawson, G. W. (1974). Observations of the marine algal ecology of Gabon. *Bot. Mar.* **17,** 249-254.

Jones, N. S. (1963). The marine fauna of New Zealand: Crustaceans of the order Cumacea. *Bull.—N. Z. Dep. Sci. Ind. Res.* **152,** 9-80.

Jones, N. S. (1969). The systematics and distribution of Cumacea from depths exceeding 200 meters. *Galathea Rep.* **10,** 99-180.

Jones, N. S., and Sanders, H. L. (1972). Distribution of Cumacea in the deep Atlantic. *Deep-Sea Res.* **19,** 737-745.

Kaestner, A. (1970). "Invertebrate Zoology," Vol. III. Wiley (Interscience), New York.

Kensley, B. (1980a). Records of anthurids from Florida, Central America, and South America (Crustacea:Isopoda:Anthuridae). *Proc. Biol. Soc. Washington* **93,** 725-742.

Kensley, B. (1980b). Anthuridean isopod crustaceans from the International Indian Ocean Expedition, 1960-1965, in the Smithsonian Collections. *Smithsonian Contr. Zool.* **304,** 1-37.

Kent, B. W. (1979). Interoceanic and latitudinal patterns in spiny lobster mandible size. *Crustaceana, Suppl.* **5,** 142-146.

Klie, W. (1939). Ostracoden aus dem Kenia-Gebiet, vornehmlich von dessen Hochgebirgen. *Int. Rev. Gesamten Hydrobiol. Hydrogr.* **39,** 99-161.

Knott, B., and Lake, P. S. (1980). *Eucrenonaspides oinotheke* gen. et sp. n. (Psammaspididae) from Tasmania, and a new taxonomic scheme for Anaspidacea (Crustacea, Syncarida). *Zool. Scr.* **9,** 25-33.

Knox, G. A., and Fenwick, D. G. (1977). *Chiltoniella elongata* n. gen. et sp. (Crustacea: Cephalocarida) from New Zealand. *J. R. Soc. N. Z.* **7,** 425-432.

Kohzov, M. (1963). Lake Baikal and its life. *Monogr. Biol.* **11,** 1-352.

Kornfield, I. L., and Nevo, E. (1976). Likely pre-Suez occurrence of a Red Sea fish *Aphanius dispar* in the Mediterranean. *Nature (London)* **264,** 289-290.

Kornicker, L. S. (1975). Antarctic Ostracoda (Myodocopina) [in two parts]. *Smithson. Contrib. Zool.* **163,** 1-720.

Kornicker, L. S. (1977). Diversity of benthic myodocopid ostracodes. *In* "Aspects of Ecology and Zoogeography of Recent and Fossil Ostracoda" (H. Löffler and D. Danielopol, eds), pp. 159-173. Junk, The Hague.

Kornicker, L. S., and Sohn, I. G. (1971). Viability of ostracode eggs egested by fish and effect of digestive fluids on ostracode shells: Ecological and paleoecologic implications. *In* "The Paleoecology of Ostracods" (H. Oertli, ed.), pp. 125-135.

Krömmelbein, K. (1966). On "Gondwana Wealden" ostracoda from NE Brazil and West Africa. *Proc. West. Afr. Micropaleontol. Colloq., 2nd, 1965* pp. 113-119.

Kussakin, O. G. (1973). Peculiarities of the geographical and vertical distribution of marine isopods and the problem of deep-sea fauna origin. *Mar. Biol.* **23,** 19-34.

Lachner, E. A., Robins, C. R., and Courtenay, W. R., Jr. (1970). Exotic fishes and other aquatic organisms introduced into North America. *Smithson. Contrib. Zool.* **59,** 1-29.

Lang, K. (1968). Deep-sea Tanaidacea. *Galathea Rep.* **9,** 23-209.

Laval, P. (1980). Hyperiid amphipods as crustacean parasitoids associated with gelatinous zooplankton. *Oceanogr. Mar. Biol.* **18,** 11-56.

Linder, F. (1941). Morphology and taxonomy of the Branchiopod Anostraca. *Zool. Bidr. Uppsala* **20,** 101-302.

Linder, F. (1952). The morphology and taxonomy of the Branchiopod Notostraca, with special reference to the North American species. *Proc. U. S. Natl. Mus.* **102,** 1-69.

Little, E. E. (1975). Chemical communication in maternal behaviour of crayfish. *Nature (London)* **255,** 400-401.

Löffler, H. (1964). Vogelzung und Crustaceenverbreitung. *Zool. Anz., Suppl.* **27,** 311-316.
Longhurst, A. R. (1954). Reproduction in Notostraca. *Nature (London)* **173,** 781-782.
Longhurst, A. R. (1955a). Evolution in the Notostraca. *Evolution (Lawrence, Kans.)* **9,** 84-86.
Longhurst, A. R. (1955b). A review of the Notostraca. *Bull. Br. Mus. (Nat. Hist.) Zool.* **3,** 1-57.
Lynch, J. E. (1966). *Lepidurus lemmoni* Holmes: a redescription with notes on variation and distribution. *Trans. Am. Microsc. Soc.* **85,** 181-192.
Lynch, J. E. (1972). *Lepidurus couesii* Packard (Notostraca) redescribed with a discussion of specific characters in the genus. *Crustaceana* **23,** 43-49.
McCain, J. C. (1968). The Caprellidae (Crustacea: Amphipoda) of the western north Atlantic, *Bull.—U. S. Natl. Mus.* **278,** 1-147.
McGowan, J. A. (1974). The nature of oceanic ecosystems. *In* "The Biology of the Oceanic Pacific" (C. B. Miller, ed.), pp. 9-28. Oregon State Univ. Press, Corvallis.
McKenzie, K. G. (1967). The distribution of Caenozoic marine Ostracoda from the Gulf of Mexico to Australia. *Syst. Assoc. Publ.* **7,** 219-238.
McKenzie, K. G. (1971a). Entomostraca of Aldabra, with special reference to the genus *Heterocypris* (Crustacea, Ostracoda). *Philos. Trans. R. Soc. London, Ser. B* **260,** 257-297
McKenzie, K. G. (1971b). Palaeozoogeography of freshwater Ostracoda. *Bull. Cent. Rech. Pau* **5,** 207-237.
McKenzie, K. G., and Hussainy, S. U. (1968). Relevance of a freshwater cytherid (Crustacea Ostracoda) to the continential drift hypothesis. *Nature (London)* **220,** 806-808.
McLaughlin, P. A. (1976). A new species of *Lightiella* (Crustacea: Cephalocarida) from the west coast of Florida. *Bull. Mar. Sci.* **26,** 593-599.
Maguire, B., Jr. (1964). Crustacea: A primitive Mediterranean group also occurs in North America. *Science* **146,** 931-932.
Maguire, B., Jr. (1965). *Monodella texana* n. sp., an extension of the range of the crustacean order Thermosbaenacea to the Western Hemisphere. *Crustaceana* **9,** 149-154.
Manning, R. B. (1977). A monograph of the West African stomatopod Crustacea. *Atlantide Rep.* **12,** 25-181.
Manning, R. B. (1980). The superfamilies, families and genera of Recent stomatopod Crustacea, with diagnoses of six new families. *Proc. Biol. Soc. Wash.* **93,** 362-372.
Manning, R. B., and Struhsaker, P. (1976). Occurrence of the Caribbean stomatopod, *Bathysquilla microps,* off Hawaii, with additional records for *B. microps* and *B. crassispinosa. Proc. Biol. Soc. Wash.* **89,** 439-450.
Marchand, L. J. (1946). The saber crab, *Platychirograpsus typicus* Rathbun, in Florida: A case of accidental dispersal. *Q. J. Fla. Acad. Sci.* **9,** 93-100.
Mauchline, J. (1980). The biology of mysids and euphausiids. *Adv. Mar. Biol.* **18,** 1-681.
Mauchline, J., and Fisher, L. R. (1969). The biology of euphasiids. *Adv. Marine Biol.* **7,** 1-454.
Mauchline, J., and Murano, M. (1977). World list of the Mysidacea, Crustacea. *J. Tokyo Univ. Fish.* **64,** 39-88.
Mayer, P. (1903). Die Caprellidae der *Siboga* Expedition. *Siboga Exped.* **34,** 1-160.
Menzies, R. J. (1953). The apseudid Chelifera of the eastern tropical and north temperate Pacific Ocean. *Bull. Mus. Comp. Zool.* **107,** 443-496.
Menzies, R. J. (1962). The zoogeography, ecology and systematics of the Chilean marine isopods. *Fysiogr. Saellsk. I Lund* [N.S.] **72,** 1-162.
Menzies, R. J., and Frankenberg, D. (1966). "Handbook on the Common Marine Isopod Crustacea of Georgia." Univ. of Georgia Press, Athens.
Menzies, R. J., and George, R. Y. (1967). A re-evaluation of the concept of hadal or ultra-abyssal fauna. *Deep-Sea Res.* **14,** 703-723.
Menzies, R. J., and Glynn, P. W. (1968). The common marine isopod Crustacea of Puerto Rico. *Stud. Fauna Curacao Other Caribb. Is.* **27,** 1-133.

Menzies, R. J., and Mohr, J. L. (1952). The occurrence of the wood-boring crustacean *Limmoria* and of Nebaliacea in Morro Bay, Calif. *Wasmann J. Biol.* **10,** 81-86.

Menzies, R. J., George, R. Y., and Rowe, G. (1968). Vision index for isopod Crustacea related to latitude and depth. *Nature (London)* **217,** 93-95.

Menzies, R. J., George, R. Y., and Rowe, G. T. (1973). "Abyssal Environment and Ecology of the World Oceans." Wiley (Interscience), New York.

Mileikovsky, S. A. (1971). Types of larval development in marine bottom invertebrates, their distribution and ecological significance: A reevaluation. *Mar. Biol.* **10,** 193-213.

Monod, T. (1924). Sur un type nouveau de Malacostracé: *Thermosbaena mirabilis* nov. gen. nov. sp. *Bull. Soc. Zool. Fr.* **49,** 58-68.

Monod, T. (1926). Les Gnathiidae. *Mem. Soc. Sci. Nat. Maroc* **13,** 1-667.

Monod, T. (1928). Les Argulides du musée du Congo. Inventaire systématique comprenent la description d'*Argulus schoutedeni* nov. sp., et liste générale critique des branchiures africains, tant marins que dulcaquicoles. *Rev. Zool. Bot. Afr.* **16,** 242-274.

Monod, T. (1956). Hippidea et Brachyura ouest-africains. *Mem. Inst. Fr. Afr. Noire* **45,** 1-674.

Naylor, E. (1957). Introduction of a grapsoid crab, *Brachynotus sexdentatus* (Risso), into British waters. *Nature (London)* **180,** 616-617.

Naylor, E. (1960). A North American Xanthoid crab new to Britain. *Nature (London)* **187,** 256-257.

Nelson, W. G. (1980). A comparative study of amphipods in seagrasses from Florida to Nova Scotia. *Bull. Mar. Sci.* **30,** 80-89.

Newman, W. A. (1963). On the introduction of an edible oriental shrimp (Caridea, Palaemonidae) to San Francisco Bay. *Crustaceana* **5,** 119-132.

Newman, W. A. (1979). A new scalpellid (Cirripedia); a Mesozoic relic living near an abyssal hydrothermal spring. *Trans. San Diego Soc. Nat. Hist.* **19,** 153-167.

Newman, W. A., and Ross, A. (1971). Antarctic Cirripedia. *Antarct. Res. Ser.* **14,** 1-257.

Nilsson-Cantell, C. A. (1938). Cirripedes from the Indian Ocean in the collection of the Indian Museum, Calcutta. *Mem. Indian Mus.* **13,** 1-81.

Nilsson-Cantell, C. A. (1978). Cirripedia Thoracica and Acrothoracica. *Mar. Invertebr. Scand.* **5,** 1-133.

Noodt, W. (1965). Naturliches System und Biogeographie der Syncarida (Crustaca, Malacostraca). *Gewaesser Abwasser* **37/38,** 77-186.

Ortmann, A. E. (1902). The geographical distribution of freshwater decapods and its bearing upon ancient geography. *Proc. Am. Philos. Soc.* **41,** 267-400.

Pianka, E. R. (1966). Latitudinal gradients in species diversity: A review of concepts. *Am. Nat.* **100,** 33-46.

Por, F. D. (1971). One hundred years of Suez Canal—a century of Lessepsian migration: Retrospect and viewpoints. *Syst. Zool.* **20,** 138-159.

Por, F. D. (1973). The nature of the Lessepsian migration through the Suez Canal. *Rapp. P.-V. Reun.—Comm. Int. Explor. Sci. Mer Mediterr.* **21**(9), 697-682.

Poulsen, E. M. (1962). Ostracoda-Myodocopa. Part I. Cypridiniformes-Cypridinidae. *Dana Rep.* **57,** 1-414.

Poulsen, E. M. (1965). Ostracoda-Myodocopa. Part II. Rutidermatidae, Sarsiellidae and Asteropidae. *Dana Rep.* **65,** 1-484.

Poulsen, E. M. (1969). Ostracoda-Myodocopa. Part IIIA. Halocypriformes-Thaumatocypridae and Halocypridae. *Dana Rep.* **75,** 1-100.

Poulsen, E. M. (1977). Zoogeograhical remarks on marine pelagic Ostracoda. *Dana Rep.* **87,** 1-34.

Proctor, V. W. (1964). Viability of crustacean eggs recovered from ducks. *Ecology* **45,** 656-658.

Proctor, V. W., and Malone, C. R. (1965). Further evidence of the passive dispersal of small aquatic organisms via the internal tract of birds. *Ecology* **46,** 728-729.

Puri, H. S. (1967). Ecologic distribution of Recent Ostracoda. *Proc. Symp. Crustacea (India)* **1,** 457–495.

Rathbun, M. J. (1904). "Harriman Alaska Series," Vol. X, pp. 1-210. Smithson. Inst. Press, Washington, D.C.

Rathbun, M. J. (1910). The stalk-eyed Crustacea of Peru and the adjacent coast. *Proc. U. S. Natl. Mus.* **38,** 531-620.

Reaka, M. L. (1980). Geographic range, life history patterns, and body size in a guild of coral-dwelling mantis shrimps. *Evolution* **34,** 1019-1030.

Reaka, M. L., and Manning, R. B. (1980). The distributional ecology and zoogeographic relationships of stomatopod Crustacea from Pacific Costa Rica. *Smithson. Contrib. Mar. Sci.* **7,** 1-29.

Remane, A. (1934). Die Brackwasserfauna. *Verh. Dtsch. Zool. Ges.* **7,** 34-74.

Renaud-Mornant, J. (1976). Un nouveau genre de Crustacé mystacocaride de la zone neotropicale: *Ctenocheilocaris claudiae* n. g., n. sp. *C. R. Hebd. Seances Acad. Sci., Ser. D* **282,** 863-866.

Renaud-Mornant, J., and Delamare-Deboutteville, C. (1977). L'originalité de la sous-classe des mystacocarides (Crustacea) et le probléme de leur repartition. *Ann. Speleol.* **31,** 75-83.

Richardson, H. (1905). Monograph on the Isopods of North America. *Bull.—U. S. Natl. Mus.* **54,** 1-727.

Ricklefs, R. E. (1979). "Ecology." Chiron Press, New York.

Ringuelet, R. (1943). Revisión de los Argúlidos Argentinos (Crustacea: Branchiura) con el catálogo de las especies neotropicales. *Reuta Mus. La Planta (Seccion Zoologia)* **3,** 43-99.

Rolfe, W.D.I. (1969). Phyllocarida. *In* "Treatise on Invertebrate Paleontology" (R. C. Moore, ed.), Vol. I, pp. R296-R331. Geol. Soc. Am., Boulder, Colorado, and the Univ. of Kansas Press, Lawrence.

Rosen, D. E. (1978). Vicariant patterns and historical explanation in biogeography. *Syst. Zool.* **27,** 159-188.

Rouch, R. (1965). Contribution à la connaissance du genre *Monodella* (Thermosbaenacés). *Ann. Speleol.* **19**(4), 717-727.

Rzoska, J. (1961). Observations on tropical rainpools and general remarks on temporary waters. *Hydrobiologia* **17,** 265-286.

Sandberg, P. A. (1964). The ostracod genus *Cyprideis* in the Americas. *Stockholm Contrib. Geol.* **12,** 10178.

Sandberg, P. A., and Plusquellec, P. L. (1974). Notes on the anatomy and passive dispersal of *Cyprideis* (Cytheracea, Ostracoda). *Geosci. Man* **6,** 1-26.

Sanders, H. L. (1968). Marine benthic diversity: A comparative study. *Am. Nat.* **102,** 243-282.

Sars, G. O. (1895). On some South-African Entomostraca raised from dried mud. *Christiania Selsk. Skr., Math. Nat. Kl.* **1,** 28-56.

Schminke, H. K. (1974a). *Psammaspides williamsi* gen. n., sp. n., ein vertreter einer neuen Familie mesopsammaler Anaspidacea (Crustacea, Syncarida). *Zool. Scr.* **3,** 177-183.

Schminke, H. K. (1974b). Mesozoic intercontinental relationships as evidenced by bathynellid Crustacea (Syncarida: Malacostraca). *Syst. Zool.* **23,** 17-164.

Schminke, H. K. (1980). Zur systematik der Stygocarididae (Crustacea, Syncarida) und Beschreibung zweier neuer arten (*Stygocavella pleotelson* gen. n., sp. n. und *stygocaris giselae* sp. n). *Beaufortia* **30,** 139-154.

Schminke, H. K., and Noodt, W. (1968). Discovery of Bathynellacea, Stygocaridacea, and other interstial Crustacea in New Zealand. *Naturwissenschaften* **55,** 184-185.

Schmitt, W. L. (1921). The marine decapod Crustacea of California, with special reference to the decapod Crustacea collected by the United States Bureau of Fisheries Steamer "Albatross" in connection with the biological survey of San Francisco Bay during the years 1912-1913. *Univ. Calif., Berkeley, Publ. Zool.* **23,** 1-470.

Schram, F. R. (1970). Isopod from the Pennsylvanian of Illinois. *Science* **169,** 854-855.

Schram, F. (1974a). Late Palaeozoic Peracarida of North America. *Fieldiana, Geol.* **33,** 95-124.

Schram, F. R. (1974b). Convergence between Late Palaeozoic and modern caridoid Malacostraca. *Syst. Zool.* **23,** 323-332.

Schram, F. R. (1977). Palaeozoogeography of Late Palaeozoic and Triassic Malacostraca. *Syst. Zool.* **26,** 367-379.

Schram, F. R., Feldman, R. M., and Copeland, M. J. (1978). The Late Devonian Palaeopalaemonidae and the earliest decapod crustaceans. *J. Paleontol.* **52,** 1375-1387.

Schultz, G. A. (1969). "How to Know the Marine Isopod Crustaceans." The Pictured-Key Nature Series, W. C. Brown, Dubuque, Iowa.

Sewell, R.B.S. (1940). Copepoda, Harpacticoida. *John Murray Exped. Sci. Rep.* **6,** No. 2, 117-382.

Sewell, R.B.S. (1948). The free-swimming planktonic Copepoda. Geographic distribution. *John Murray Exped. Sci. Rep.* **8,** No. 3, 317-592.

Shiino, S. M. (1978). Tanaidacea collected by French scientists on board the survey ship "Marion-Dufresne" in the regions around the Kerguelen Islands and other subantarctic islands in 1972, '74, '75, '76. *Sci. Rep. Shima Marineland* **5,** 1-122.

Sieg, J. (1977). Taxonomische monographie der Pseudotanaidae. *Mitt. Zool. Mus. Berlin* **53,** 3-109.

Sieg, J. (1980a). Sind die Dikonophora eine polyphyletische Gruppe. *Zool. Anz., Jena,* **205**(5/6), 401-416.

Sieg, J. (1980b). Taxonomische Monographie der Tanaidae Dana, 1849. *Abh. Ges. Senckenberg. Naturforsch. Ges.* **537,** 1-276.

Sieg, J. (1981). Tanaidacea. *In* "Catalogus Crustaceorum" (H. E. Gruner and L. B. Holthuis, eds.) (in press).

Simberloff, D. S., Heck, K. L., Jr., McCoy, E. D., and Connor, E. F. (1981). There have been no statistical tests of cladistic biogeographic hypotheses! *In* "Vicariance Biogeography: A Critique" (G. Nelson and D. Rosen, eds.) Columbia Univ. Press, New York.

Smirnov, N. N. (1970). Cladocera (Crustacea) iz permskikh otlozheniy Vostochnogo Kazakhsta na. *Paleontol. Zh.* **3,** 95-100.

Spivey, H. R. (1981). Origins, distribution and zoogeographic affinities of the Cirripedia (Crustacea) of the Gulf of Mexico. *J. Biogeogr.* **8,** 153-176.

Stehli, F. G., and Wells, J. W. (1971). Diversity and age patterns in hermatypic corals. *Syst. Zool.* **20,** 115-126.

Steuer, A. (1933). Zur planmässigen Erforschung des geographischen Verbreitung des Haliplanktons, besonders der Copepoden. *Zoogeografica* **1,** 269-302.

Stock, J. H. (1976). A new genus and two new species of the crustacean order Thermosbaenacea from the West Indies. *Bijdr. Dierkd.* **46**(1), 47-70.

Stoner, A. W. (1981). Occurrence of the cephalocarid crustacean *Lightiella floridana* in the northern Gulf of Mexico with notes on its habitat. *Northeast Gulf Sci.* **4,** 105-107.

Strong, D. R., Jr. (1974). Nonasymptotic species richness models and the insects of British trees. *Proc. Natl. Acad. Sci. U.S.A.* **71,** 2766-2769.

Stubbings, H. G. (1967). The cirriped fauna of tropical West Africa. *Bull. Br. Mus. (Nat. Hist.), Zool.* **15,** 229-319.

Tasch, P. (1969). Branchiopoda. *In* "Treatise on Invertebrate Paleontology" (R. C. Moore, ed.), Vol. I, pp. R128-R191. Geol. Soc. Am., Boulder, Colorado, and the Univ. of Kansas Press, Lawrence.

Tasch, P. (1979). Crustacean branchiopod distribution and speciation in Mesozoic lakes of the southern continents. *Antarct. Res. Ser.* **30,** 65-74.

Tasch, P., and Zimmerman, J. R. (1961). Fossil and living conchostracan distribution in Kansas-Oklahoma across a 200 million year time gap. *Science* **133,** 584-586.

Tattersall, W. M. (1905). On *Nebalia typhlops* G. O. Sars. *Annu. Rep. Fish. Ireland 1902-1903,* p. 210.

Tasch, P., and Zimmerman, J. R. (1961). Fossil and living conchostracan distribution in Kansas-Oklahoma across a 200 million year time gap. *Science* **133,** 584-586.

Tattersall, W. M. (1905). On *Nebalia typhlops* G. O. Sars. *Annu. Rep. Fish. Ireland, 1902-1903,* p. 210.

Tattersall, W. M. (1951). A review of the Mysidacea of the United States National Museum. *Bull.—U. S. Natl. Mus.* **201,** 1-292.

Thiele, J. (1904). Die Lepostraca. *Wiss. Ergeb. Dtsch. Tiefsee-Exped. Valdavia* **8,** 1-26.

Thorson, G. (1950). Reproductive and larval ecology of marine bottom invertebrates. *Biol. Rev. Cambridge Philos. Soc.* **25,** 1-45.

Thurston, M. H. (1977). Depth distributions of *Hyperia spinigera* Bovallius, 1889 (Crustacea: Amphipoda) and medusae in the North Atlantic Ocean, with notes on the associations between *Hyperia* and coelenterates. *In* "A Voyage of Discovery: George Deacon 70th Anniversary Volume" (M. Angel, ed.), pp. 499-536. Pergamon, Oxford.

Tomlinson, J. T. (1969). The burrowing barnacles (Cirripedia: order Acrothoracica). *Bull.—U. S. Natl. Mus.* **296,** 1-162.

Tomlinson, J. T. (1973). Distribution and structure of some burrowing barnacles, with four new species (Cirripedia: Acrothoracica). *Wasmann J. Biol.* **31,** 263-288.

Topp, R. W. (1969). Interoceanic sea-level canal: Effects on the fish faunas. *Science* **165,** 1324-1327.

Turner, J. T. (1981). Longitudinal patterns of calanoid and cyclopoid copepod diversity in estuarine waters of eastern North America. *J. Biogeogr.* **8**(5), 369-382.

van Morkhoven, F.P.C.M. (1972). Bathymetry of Recent Marine Ostracoda in the Northwest Gulf of Mexico. *Trans.—Gulf Coast Assoc. Geol. Soc.* **12,** 241-252.

Vermeij, G. J. (1977). Patterns in crab claw size: The geography of crushing. *Syst. Zool.* **26,** 138-151.

Vermeij, G. J. (1978), "Biogeography and Adaptation." Harvard Univ. Press, Cambridge, Massachusetts.

Wakabara, Y. (1965). On *Nebalia* sp. from Brazil (Leptostraca). *Crustaceana* **9,** 245-248.

Wakabara, Y., and Mizoguchi, S. M. 1976). Record of *Sandersiella bathyalis* Hessler and Sanders, 1973 (Cephalocarida), from Brazil. *Crustaceana* **30,** 220-221.

Walton, M., and Hobbs, H. H., Jr. (1971). The distribution of certain entocytherid ostracods on their crayfish hosts. *Proc. Acad. Nat. Sci. Philadelphia* **123,** 87-103.

Watling, L. (1979). Zoogeographic affinities of northeastern North American Gammaridean Amphipoda. *Bull. Biol. Soc. Wash.* **3,** 256-282.

Wells, J.B.J. (1967). The littoral Copepoda (Crustacea) of Inhaca Island, Mozambique. *Trans. R. Soc. Edinburgh* **67,** 189-358.

Whiteside, M. C., and Harmsworth, R. V. (1967). Species diversity in chydorid (Cladocera) communities. *Ecology* **48,** 664-667.

Wigley, R. L., and Burns, B. R. (1971). Distribution and biology of mysids (Crustacea, Mysidacea) from the Atlantic coast of the United States in the NMFS Woods Hole collection. *Fish. Bull.* **69,** 717-746.

Williams, A. B. (1965). Marine decapod crustaceans of the Carolinas. *Fish. Bull.* **65,** 1-298.

Williams, A. B. (1972). A ten-year study of meroplankton in North Carolina estuaries: Mysid shrimps. *Chesapeake Sci.* **13,** 254-262.

Williams, W. D. (1968). Distribution of *Triops* and *Lepidurus* in Australia. *Crustaceana* **14,** 119-126.

Williamson, D. I. (1973). *Amphionides reynaudii* (H. Milne Edwards), representative of a proposed new order of eucaridian Malacostraca. *Crustaceana* **25,** 35-50.

Wilson, C. B. (1903). North American parasitic copepods of the family Argulidae, with a bibliography of the group and a systematic review of all known species. *Proc. U. S. Natl. Mus.* **25,** 635-742.

Wolff, T. (1954). Occurrence of two East American species of crabs in European waters. *Nature (London)* **174,** 188-189.

Wolff, T. (1956a). Isopoda from depths exceeding 6000 meters. *Galathea Rep.* **2,** 85-157.

Wolff, T. (1956b). Crustacea Tanaidacea from depths exceeding 6000 meters. *Galathea Rep.* **2,** 187-241.

Yager, J. (1981). A new class of Crustacea from a marine cave in the Bahamas. *J. Crustacean Biol.* **1**(3), 328-333.

Yager, J., and Turner, R. L. (1980). A new class of Crustacea from the Bahamas. *Am. Zool.* **20**(4), 815.

Yamaguti, S. (1963). "Parasitic Copepoda and Branchiura of Fishes." Wiley (Interscience), New York.

Zimmer, C. (1980). Cumaceans of the Atlantic Boreal coast region (Crustacea:Peracarida). (T. E. Bowman and L. Watling, eds.). *Smithsonian Contr. Zool.* **302,** v, 1-29.

Systematic Index

Subject Index

G

H

I

J

K

L

M